AF807765
WITHDRAWN

The Design of Drugs to Macromolecular Targets

The cover picture shows the active site structure of dihydrofolate reductase and differences in the space occupied when two inhibitors are separately bound. The display software is a modified form of FRODO (see Jones, 1978: Chapter 2) and the structure was determined by Kuyper *et al.* 1985: Chapters 2 and 4.

The Design of Drugs to Macromolecular Targets

Edited by
C. R. BEDDELL
The Wellcome Research Laboratories, Beckenham, UK

JOHN WILEY & SONS
Chichester · New York · Brisbane · Toronto · Singapore

RS
420
.D46
1992

Copyright © 1992 by John Wiley & Sons Ltd,
Baffins Lane, Chichester,
West Sussex PO19 1UD, England

All rights reserved.

No part of this book may be reproduced by any means, or transmitted, or translated into a machine language without the written permission of the publisher.

Other Wiley Editorial Offices

John Wiley & Sons, Inc., 605 Third Avenue,
New York, NY 10158-0012, USA

Jacaranda Wiley Ltd, G.P.O. 859, Brisbane,
Queensland 4001, Australia

John Wiley & Sons (Canada) Ltd, 22 Worcester Road,
Rexdale, Ontario M9W 1L1, Canada

John Wiley & Sons (SEA) Pte Ltd, 37 Jalan Pemimpin #05-04,
Block B, Union Industrial Building, Singapore 2057

Library of Congress Cataloging-in-Publication Data

The Design of drugs to macromolecular targets / edited by C. R. Beddell.
p. cm.
Includes bibliographical references and index.
ISBN 0 471 92080 0
1. Drugs—Design. 2. Macromolecules. 3. Molecular recognition.
I. Beddell, C. R.
[DNLM: 1. Drug Design. 2. Macromolecular Systems. QV 744 D4572]
RS420.D46 1992
615'.19—dc20
DNLM/DLC
for Library of Congress 92-5078
CIP

British Library Cataloguing in Publication Data

A catalogue record for this book is available from the British Library

ISBN 0 471 92080 0

Typeset in Times 10/12pt by APS, Salisbury, Wilts
Printed and bound in Great Britain by Biddles Ltd, Guildford, Surrey

Contents

List of Contributors

C. C. F. Blake
Laboratory of Molecular Biophysics, The Rex Richards Building, South Parks Road, Oxford OX1 3QU, UK

J. M. Burridge
IBM United Kingdom Ltd, 389 Chiswick High Road, London W4 4AL, UK

J. N. Champness
The Wellcome Research Laboratories, Langley Court, Beckenham, Kent BR3 3BS, UK

P. de la Paz
BBC, Open University Production Centre, Walton Hall, Milton Keynes MK7 6BH, UK

A. S. Gilbert
The Wellcome Research Laboratories, Langley Court, Beckenham, Kent BR3 3BS, UK

W. G. J. Hol
BIOSON Research Institute, Department of Chemistry, University of Groningen, Nijenborgh 16, 9747 AG Groningen, The Netherlands

R. M. Hyde
The Wellcome Research Laboratories, Langley Court, Beckenham, Kent BR3 3BS, UK

G. R. Marshall
Center for Molecular Design, Washington University, St Louis, MO 63110, USA

S. Neidle
Cancer Research Campaign Biomolecular Structure Unit, The Institute of Cancer Research, Sutton, Surrey SM2 5NG, UK

S. J. Oatley
(deceased)

B. Roth
Department of Chemistry, University of North Carolina, Chapel Hill, NC 27599 USA

D. K. Stammers
The Wellcome Research Laboratories, Langley Court, Beckenham, Kent BR3 3BS, UK

R. K. Wierenga
EMBL, Postfach 10.2209, Meyerhofstrasse 1, 6900 Heidelberg, Germany

R. Wootton
The Wellcome Research Laboratories, Langley Court, Beckenham, Kent BR3 3BS, UK

Preface

It is now widely recognized that the combined research and development time for a novel medicine is substantial (e.g. 10–12 years) and the ultimate success of new drug-design methods can therefore only be assessed on a long time scale. However, no research-based pharmaceutical company would wish to ignore new methods and the innovative climate which they help to create.

Over 30 years have elapsed since the three-dimensional structure of a protein was first established by X-ray crystallography and the generation of detailed molecular models of hydrated crystalline proteins is now a relatively commonplace event. In addition nuclear magnetic resonance spectroscopy has recently started to provide information for some smaller proteins in aqueous solution.

Since the effects of the active constituents of medicines are generally mediated through specific physicochemical interaction with macromolecules such as proteins and nucleic acids there has arisen hope that the development of new drugs might be assisted by tailoring chemical structures to fit into clefts in the surface of target macromolecules. This belief has been nurtured by the increased ease with which the macromolecules themselves can be identified and prepared, through the advent of genetic engineering methodologies.

Thus the scene is set for the direct design of drugs to selected macromolecules to take its place as one component of the overall complex of procedures by which new medicines are developed. The primary aim of this book is to indicate the current state of expertise and the ways in which this expertise might develop in the immediate future for molecular tailoring to be as successful as possible. In addition the relationship of this approach to longer established methods of drug design is reviewed, since any application would not occur in isolation, but in combination with other techniques.

The chapters stand independently, but they are ordered to provide a logical progression, from consideration of the role of macromolecules in drug action, through the physicochemical principles and computational techniques by which small chemicals might be designed, to experimental studies which illustrate the degree of success and understanding that has been achieved and a final overview of the way in which molecular tailoring complements other approaches to the selection of chemicals as medicines. No attempt has been made to ensure uniformity of viewpoints as in such a rapidly developing field it is inevitable that there will be differing perspectives. The possibilities revealed for drug design are exciting, and it is increasingly recognized that even outline models of macromolecules can sometimes assist the medicinal chemist to select new molecular targets for synthesis.

C.R.B.
1992

Introduction

The introduction of new medicines this century has involved a general shift in emphasis away from natural sources of active materials towards molecules synthesized chemically (and recently through biotechnology), though the natural domain remains a rich resource of diverse chemical entities which are still exploited. The process of discovery and selection of chemicals for medicinal purposes is lengthy, complex and expensive, including evaluation in many, often novel, biological and biochemical systems, and considerations of efficacy, stability, safety and administration within national regulatory frameworks. There is thus continual pressure to improve efficiency in the ways that relevant molecules are found. Expertise in chemical synthesis continues to expand, enabling the medicinal chemist to have more choice in what chemicals to make and it is therefore crucial that selection procedures develop to take account of the opportunities afforded.

This book is concerned with just one of many developing methods which now influence the selection process. It arises because the active components in many medicines initiate biological effects by interaction with various specific macromolecules in the body. Relatively recent advances in the chemical and physical sciences have enabled the architecture and spatial details of such macromolecules to be established, whilst genetic engineering techniques are now transforming the ease with which they can be obtained and studied. The book therefore encompasses various aspects of how target chemical structures for synthesis can be derived from consideration of spatial arrangements of atoms within macromolecules.

THE ROLE OF MACROMOLECULES IN DRUG ACTION

Essential to the overall approach which is the subject matter of this book is the concept of the role of macromolecules in drug action. The introductory chapter reviews the diverse ways in which macromolecules are involved and the variety of macromolecules, for example, proteins, free or membrane bound, ribonucleic and deoxyribonucleic acids and complexes of these. Although the molecular site of action of many drugs remains to be established, there is already much evidence to support the concept that macromolecules are the principal targets for drug action.

LIGAND FITTING METHODOLOGY

Despite shortcomings in current understanding, sufficient concepts and rules-of-thumb exist for the invention of chemical structures to interact at defined protein sites, although in view of the fact that binding site architecture (which may be flexible) can never be precisely determined with consequent uncertainties in posited interactions, success is likely to be increased by empirical incorporation of the results of successive cycles of design. Practical procedures have been developed which, provided they can be applied with skill and imagination, can generate novel ligands for macromolecules.

COMPOUNDS DESIGNED TO BIND TO HAEMOGLOBIN

Some of the earliest studies of the practicality of ligand design to macromolecules from considerations of the molecular architecture have been performed with the macromolecule haemoglobin. This molecule binds oxygen reversibly and thereby increases the oxygen conveying capacity of blood beyond that achieved by solution in plasma. The affinity of haemoglobin for oxygen is important for its successful functioning, and haemoglobin can exist in different conformational states with differing affinity for oxygen. Study of the architecture of these different conformations enabled ligands to be designed to respectively increase or decrease the affinity of haemoglobin for oxygen, and these studies have potential physiological and clinical relevance.

DRUG INTERACTIONS WITH TARGET ENZYMES OF KNOWN STRUCTURE

Many macromolecules are enzymes and a few have been studied in detail at the molecular level in order to elucidate the interaction with drugs. The enzyme dihydrofolate reductase has been the most studied. It is widespread in nature and is important to the functioning of many organisms, including man. A variety of drugs which act by inhibiting the enzyme of certain cancerous or infective entities have been developed and the interactions between some of these drugs and the enzyme from different sources have been visualized. This information has been used to design molecules of greater inhibitory potency and the enzyme is likely to provide a model system for testing the principles of design for many years to come.

MULTIPLE MODES OF BINDING OF THYROID HORMONES AND OTHER IODOTHYRONINES TO HUMAN PLASMA TRANSTHYRETIN

Some drugs are hormones, or chemically and functionally related to hormones, and details are now known of how one group of hormones binds to a plasma protein, transthyretin, which is involved in hormone transport processes. The detailed studies show both the subtlety of interactions of closely related chemical structures and the power of current experimental techniques to expose the details. The situation in which multiple modes of ligand binding exist complicates the derivation of contributions to binding from individual substituents, but nevertheless such complexities must be accommodated within any comprehensive analysis.

COMPUTED MODELLING OF DRUG-DNA INTERCALATIVE INTERACTIONS

Nucleic acids are the principal or exclusive site of action of many drugs and for a number of such drugs the way they interact with nucleic acids has been increasingly defined. Some drugs bind in a relatively external fashion, but some become partially inserted into the nucleic acid non-covalently, inducing conformational changes in the nucleic acid to enable them to be so accommodated. Despite the complexities arising from the need to define the relevant conformational states in the nucleic acid as well as in the ligand, theoretical prediction of binding modes is becoming increasingly successful; extension to precise predictions *in vivo* remains a major challenge.

CRYSTALLOGRAPHIC INVESTIGATIONS OF GLYCOLYTIC ENZYMES FROM *TRYPANOSOMA BRUCEI*: POTENTIAL STARTING POINTS FOR THE DESIGN OF NEW SLEEPING SICKNESS DRUGS

The enzymes of organisms are often grouped according to the functions to which they contribute. For example, in the trypanosome parasite enzymes of glycolysis, which lead to the degradation of sugar to smaller molecules and the conservation of some of the chemical energy from this in the form of ATP for subsequent liberation in other reactions, are grouped in a body called the glycosome. This chapter describes the approach to the concept of exploring in a systematic fashion the feasibility of targeting a metabolic pathway for drug generation by specific consideration of the individual enzymes and selection of the most promising candidate targets.

DRUG DISCOVERY AND INVENTION : SOME APPROACHES COMPARED

Targeting of compounds to macromolecules as ligands is only one part of the process of optimisation of chemical structure selection. In the absence of direct information about the macromolecular target, progress can be made by searching for correlations between the physicochemical properties of the ligands and their performance in relevant biological assays. The knowledge gained by such exercises can however give an alternative and complementary perspective which can enhance the efficiency of the search for optimal structure even where the macromolecule is known. These considerations are reviewed to show the broader context of the ongoing efforts to generate new and improved medicines.

1 The Role of Macromolecules in Drug Action

GARLAND R. MARSHALL
Center for Molecular Design, Washington University, St Louis, MO, 63110, USA

1.1 INTRODUCTION

Receptors are, at least functionally, those substances in biological systems which combine with other chemicals to produce a biological effect. Traditionally, the term receptor has been reserved for those macromolecules which are usually found on the cell membrane and which act to transduce the chemical messages within the body. The term is most closely associated with the endocrine system, i.e. hormone receptors, or the nervous system, i.e. neurotransmitter receptors. If one considers the whole spectrum of therapeutic intervention, then the concept of drug receptor is a natural extension; substrates, enzymes, nucleic acids, membranes and structural proteins as well have all been shown to function as receptors for some drug or another. We will consider several examples as a means of illustrating the various means by which a drug can exert its desired therapeutic effect. These examples can also indicate the rather sophisticated

The Design of Drugs to Macromolecular Targets. Edited by C. R. Beddell
© 1992 John Wiley & Sons Ltd

state of our knowledge regarding drug-receptor interactions in certain specific cases, and the speed with which three-dimensional information about therapeutic targets can be obtained. Macromolecules also play an important role in determining the fate of drugs within the body by their enzymatic role in metabolism and excretion as well as a passive role in binding and transport.

1.2 ROLE OF MACROMOLECULES AS RECEPTORS

1.2.1 SMALL MOLECULES AS RECEPTORS

In order to understand the process of molecular recognition and the mechanisms of enzymatic catalysis, chemists have attempted to construct simplified model systems in which both components can be synthetically manipulated. As the first component of the system, specificity for a class of structures has been explored through the use of host–guest complexes. This arena of chemistry has exploded since the publication of discovery of crown ethers and their affinity for metals (reviewed by Pedersen, 1988) at Dupont in 1967. Others have attempted to use these (Stoddart, 1987) and other recognition cavities (Tabushi, 1982; Rebek, 1987) as a scaffold for orienting appropriate functional groups to generate model enzymes. This area of investigation has generated an enormous literature with many model systems (Cram, 1988) which are of considerable interest to the theoretical chemist and modeller as a source of calibration data. As the synthetic chemist has increased the complexity of the host molecules in order to enhance discrimination, the structures of newer compounds are now virtually indistinguishable in complexity from those of natural products.

While most therapeutic targets, such as enzymes or nucleic acids, are macromolecules, molecular size is not a requirement for playing a significant biological role and one can certainly envisage therapeutic targets, such as crucial intermediates in metabolism, which are not macromolecules. It is of interest that nature has created a series of small molecular weight antibiotics, the vancomycin family (Mueller *et al.*, 1989; Waltho and Williams 1989) which function by binding to a N-terminal dipeptide essential for the biosynthesis of bacterial cell walls and which resemble compounds in use in host–guest studies by bioorganic chemists. Certainly, one of the best studied cases of molecular recognition by NMR (Mueller *et al.*, 1989; Waltho and Williams, 1989) is the family of antibiotics (vancomycin, ristocitin, aricidin and avoparcin) which inhibit formation of the cell wall in Gram-positive bacteria. The cell wall is composed of cross-linked peptidoglycans and the antibiotic functions by binding to substrate molecules terminating in D-Ala-D-Ala during the biosynthesis of the cell wall. NMR studies have clearly established both the conformation of several of the members of this family as well as the conformation of the complex with model peptides ending in D-Ala-D-Ala (Figures 1.1 and 1.2).

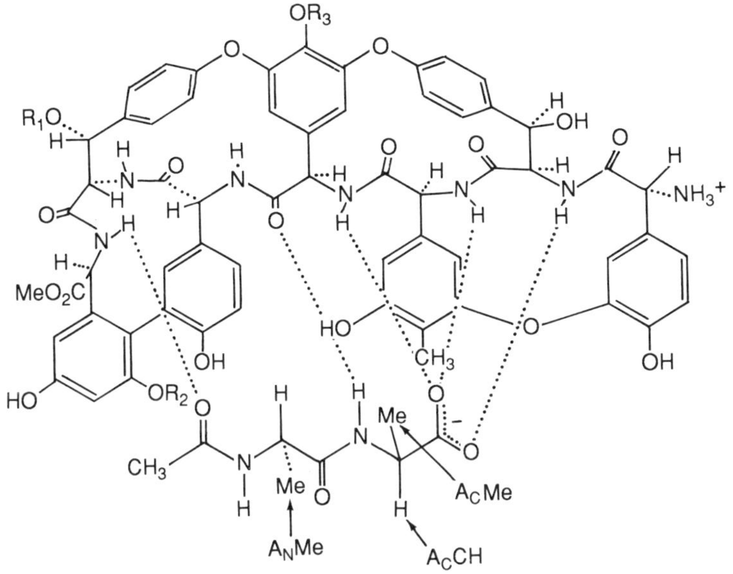

Figure 1.1 Schematic representation of complex formed between model peptide, *N*-acetyl-D-alanyl-D-alanine, and ristocetin, a member of the vancomycin antibiotic family; broken lines represent intermolecular hydrogen bonds. (Reprinted with permission from Waltho, J. P. and Williams, D. H., *J. Am. Chem. Soc.*, **111**, 2475–2480. Copyright (1989) American Chemical Society)

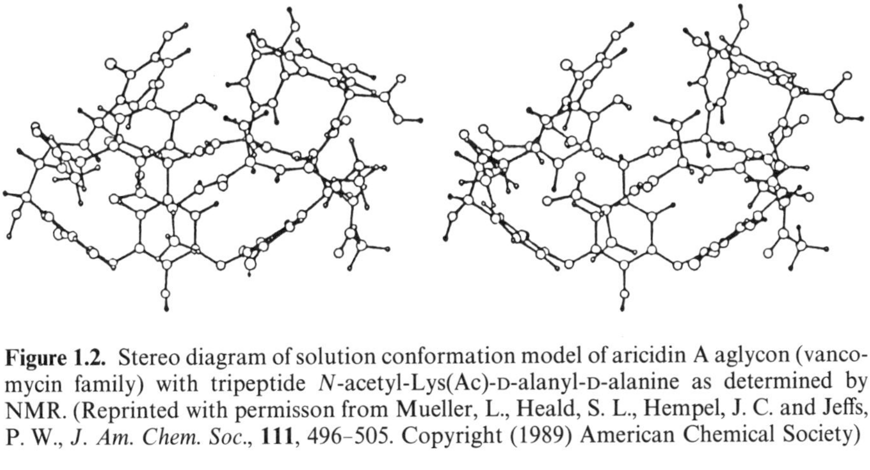

Figure 1.2. Stereo diagram of solution conformation model of aricidin A aglycon (vancomycin family) with tripeptide *N*-acetyl-Lys(Ac)-D-alanyl-D-alanine as determined by NMR. (Reprinted with permisson from Mueller, L., Heald, S. L., Hempel, J. C. and Jeffs, P. W., *J. Am. Chem. Soc.*, **111**, 496–505. Copyright (1989) American Chemical Society)

In this case, the drug has targeted a critical metabolite in a biochemical pathway which is essential for microbial viability. There are not many other examples of direct removal of a critical metabolite by complexation, unless one considers metal complexation which can inhibit enzymes by removal of active site metal catalysts. A variation on this theme is the use of chelating agents for detoxification of heavy metal poisoning (Aposhian and Aposhian, 1990).

1.2.2 NUCLEIC ACIDS AS DRUG RECEPTORS

Nucleic acids are the repository of genetic information, and the selective reading of the genetic code is the basis of differentiation and embryology. Only now are we beginning to unravel the details of the multiple mechanisms by which the diversity of protein products such as antibodies are produced, or by which the expression of genes is regulated to unfold the natural order and progression seen in the development from a single fertilized cell to an organism. Aberrations in that regulation lead to cancer, birth defects, and autoimmune disease. The detailed machinery of regulation, expression and translation are intimately involved with nucleic acids and enzymes involved in their modification and maintenance. As our knowledge increases about the molecular details of regulation and expression, differences resulting in genetic disease states are uncovered and a basis for therapeutic intervention becomes apparent. Gene therapy is a feasible approach to many diseases with only technical or logistical problems remaining to be overcome. The discovery of oncogenes responsible for cellular transformation seen in cancer and the molecular basis of their activation provide a great stimulus to the development of novel therapeutic approaches.

DNA itself has been shown to be the receptor for many drugs used in cancer chemotherapy (Hurley and Boyd, 1988; Hurley, 1989). These work through a variety of mechanisms including chemical modification and cross-linking of the DNA (cisplatin), or cleavage of the DNA (bleomycin). Many work, however, by reversible association, either by intercalation of a polyaromatic ring system into the double stranded helix (actinomycin D, ethidium, proflavine), or by binding to either the major or minor grooves of DNA. Numerous examples of such drugs exist and the three-dimensional structure of the complex has been determined by X-ray crystallography in the case of triostin (Wang *et al.*, 1984) as seen in Figure 1.3 and actinomycin D (Takusagawa *et al.*, 1982) or by NMR spectroscopy (Klevit *et al.*, 1986) for distamycin (Pelton and Wemmer, 1990) as seen in Figure 1.4, echinomycin and chromomycin by Gao and Patel (1989), and anthramycin (Boyd *et al.*, 1990). The triostin complex has also been examined by molecular mechanics (Singh *et al.*, 1986) to understand the stabilization of the A–T base pairs in Hoogsteen pairing.

Because of the potential therapeutic advantages of controlling the transcription and translation machinery of the cell, the molecular recognition (Brennan and Matthews, 1989a) of specific sequences of DNA has become a topic of keen

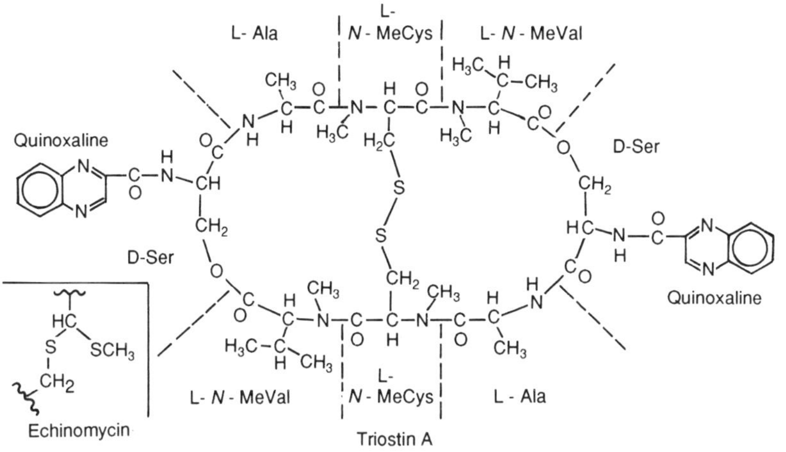

(a)

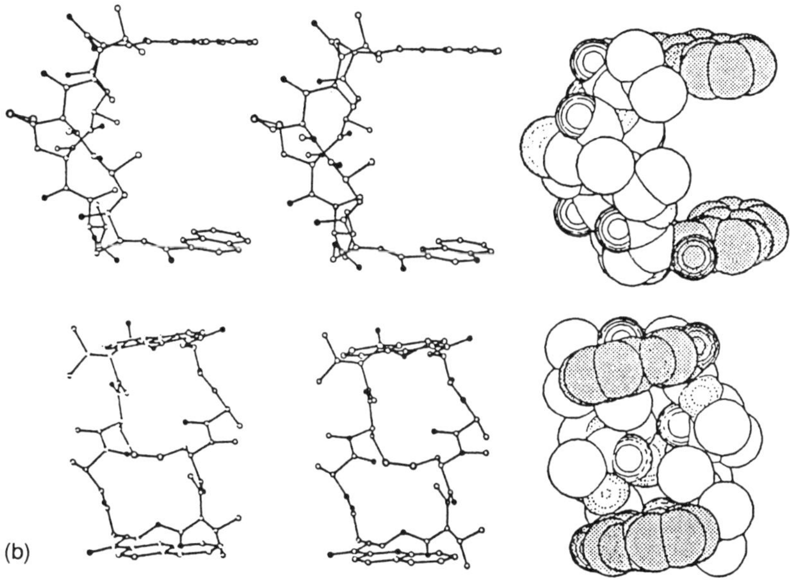

(b)

Figure 1.3. Interaction of antibiotic triostin A by intercalation with DNA as seen in crystal structure of complex. (a) Molecular structure of triostin A with inset showing modification found in echinomycin. (b) Stereo and space-filling views of triostin A as seen in complex, quinoxaline rings which intercalate are shaded. (c) Diagram of complex of triostin A with three base pairs of oligonucleotide, hydrogen bonds indicated by broken lines. (Reproduced with permission from Wang *et al.*, 1984. Copyright (1984) by the AAAS)

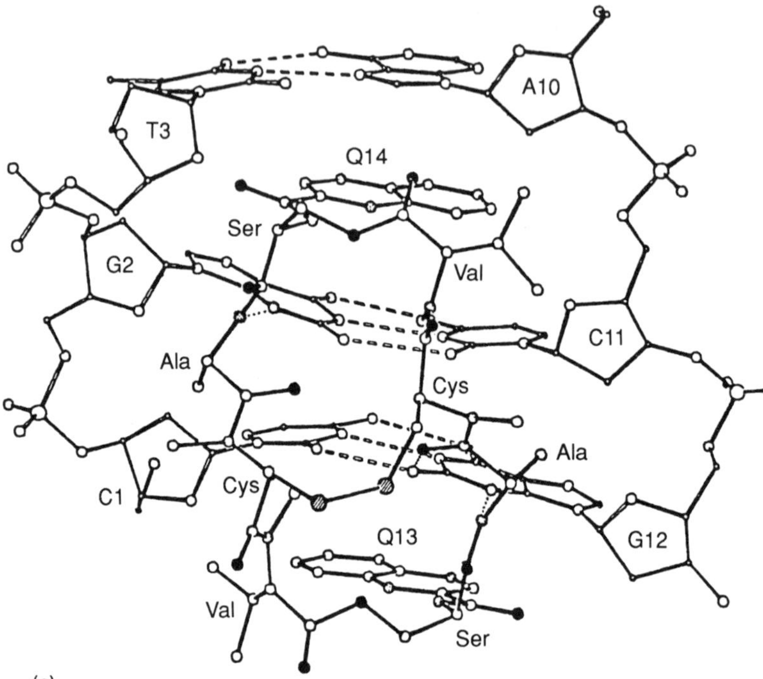

(c)

Figure 1.3 (*continued*)

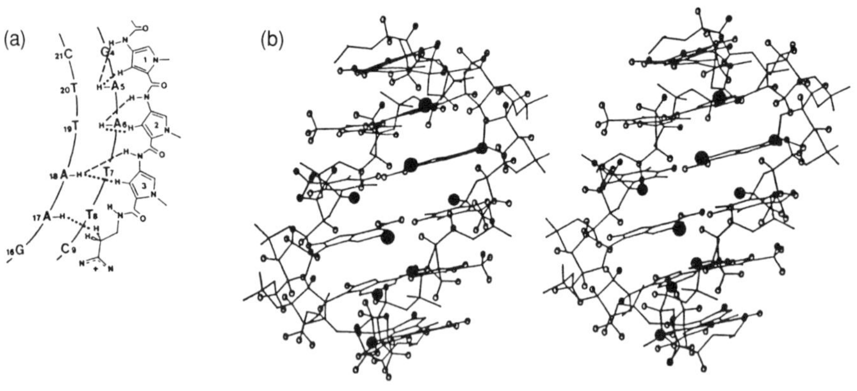

Figure 1.4. Schematic diagram (a) of distamycin binding to minor groove of oligonucleotide as determined by NMR. Stereoview of complex (b) formed. (Reprinted with permission from Klevit, R. E., Wemmer, D. E. and Reid, B., *Biochemistry* **25**, 3296–3303. Copyright (1986) American Chemical Society)

interest. Regulation of gene expression is the basis of differentiation and response to environmental change as well as transformation. Viral infection often results in incorporation of the viral genome into the host cell's DNA and the ability to differentiate between host and viral DNA is certainly a desirable therapeutic goal. Insight into potential therapeutic strategies may come from an understanding of the normal mechanisms involved in the regulation and expression of genes. Proteins involved in the regulation of transcription in eukaryotes show several distinct structural motifs; helix–turn–helix (Brennan and Matthews, 1989b), two kinds of zinc fingers (Berg, 1990) and the leucine zipper (Vinson *et al.*, 1989; Hu *et al.*, 1990). The helix–turn–helix motif (Brennan and Matthews, 1989b) was first described in 1981 for the cro repressor and the catabolic gene activator protein by the groups of Matthews and Steitz, respectively, see Figure 1.5. Crystal structures of members of this class complexed with nucleic acid fragments including the Trp repressor (Otwinowski *et al.*, 1988) have been reported (Brennan and Matthews, 1989b). The Trp repressor is a fascinating story where one can examine the details by which a small molecular weight ligand can determine the interaction between DNA and a protein regulator. The study of Marmorstein and Sigler (1989) on the

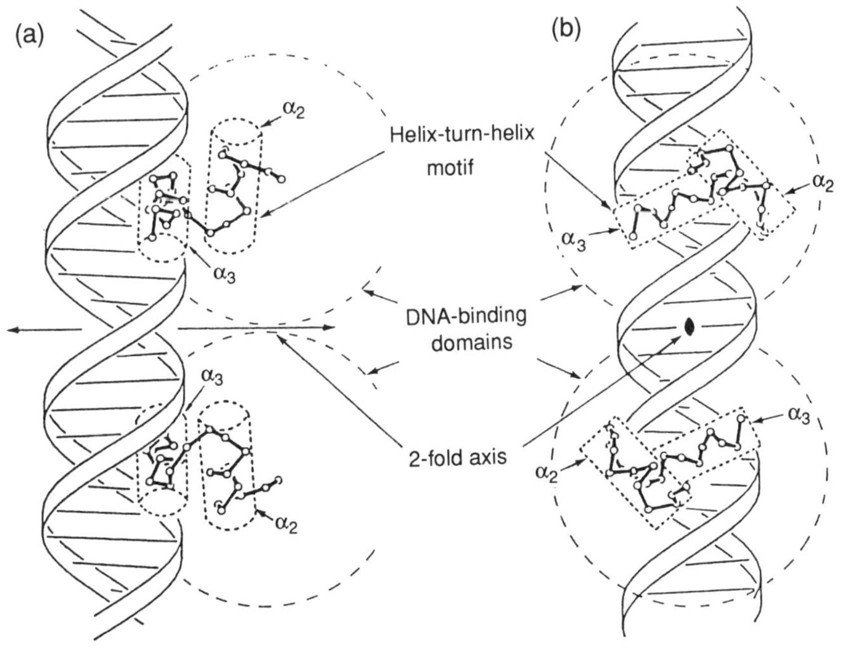

Figure 1.5. Diagram illustrating interaction between 'helix–turn–helix' motif and right-handed B-form DNA as occurs in many forms of DNA-regulatory proteins. (Reproduced from Brennan and Matthews, 1989a, by permission of the Elsevier Trends Journals)

structure–activity relationships of tryptophan analogs and their interpretation in terms of the crystal structure of the DNA–repressor complex should be required reading for medicinal chemists. A molecular dynamics simulation of the lac repressor headpiece and its operator in solution (de Vlieg *et al.*, 1989) gave an intrinsically stable structure without use of NOE constraints over the 150 picosecond simulation. Qian *et al.* (1989) have shown that a eukaryotic DNA binding protein of *Drosophila* shows a similar conformational motif, the helix–turn–helix, in solution to that seen by X-ray with prokaryotic proteins.

Several other modes of binding of regulatory proteins to DNA have recently been described or proposed. These include major groove helix binding with recognition of symmetry-related α-helices or β-sheets for the Met repressor (Rafferty *et al.*, 1989) similar to that seen with helix–turn–helix proteins. The mode of interaction with the regulator, *S*-adenosylmethionine, is entirely different from that seen with the Trp repressor. A model for DNA recognition by leucine zipper motifs, the scissors-grip model, has recently been proposed by Vinson *et al.* (1989). This model has received experimental support (O'Neil *et al.*, 1990; Talanian *et al.*, 1990) through the synthesis of two much simplified analogues which still show sequence-specific DNA binding. A new motif, the copper fist, has been proposed for the regulatory protein, ACE1, in yeast which controls the expression of metallothionein, which binds excess copper (Furst and Hammer, 1989). Cu ions activate transcription of the metallothionein gene in yeast by altering conformation and DNA binding activity of ACE1 transcription factor (Furst and Hammer, 1989). The 'copper fist' conformation in which cysteine residues are clustered by the copper ions produces a fist-like set of loops which interact with the DNA.

Other traditional hormones, such as oestrogen or thyroid hormone as well as the vertebrate morphogen, retinoic acid, bind to an intracellular receptor which functions to regulate transcription. The DNA-binding domains show strong homology which can be configured as two zinc-binding fingers with each zinc ion forming a tetrahedral coordination complex with four Cys residues (Evans, 1988). This places the steroid hormone receptor clearly in the category of a protein which modifies eukaryotic transcription as the zinc finger is one of four motifs identified with that function (Berg, 1990). Binding of the hormone to a separate domain leads to exposure of the DNA-binding domain by a hypothetical allosteric transition. By site-directed mutagenesis, those residues in the zinc finger region responsible for target gene specificity have been identified in the case of steroid/thyroid hormones (Mader *et al.*, 1989; Umesono and Evans, 1989) as shown in Figure 1.6. The three-dimensional structure of a single zinc finger has been determined by NMR (Lee *et al.*, 1989). It consists of a well-defined helix (mixed α-3_{10}) which packs against two β strands that are arranged in a hairpin structure. A high density of basic and polar amino acid residues on the exposed face of the helix are thought responsible for DNA binding. Summers *et al.* (1990) have determined the solution structure of a zinc finger domain from

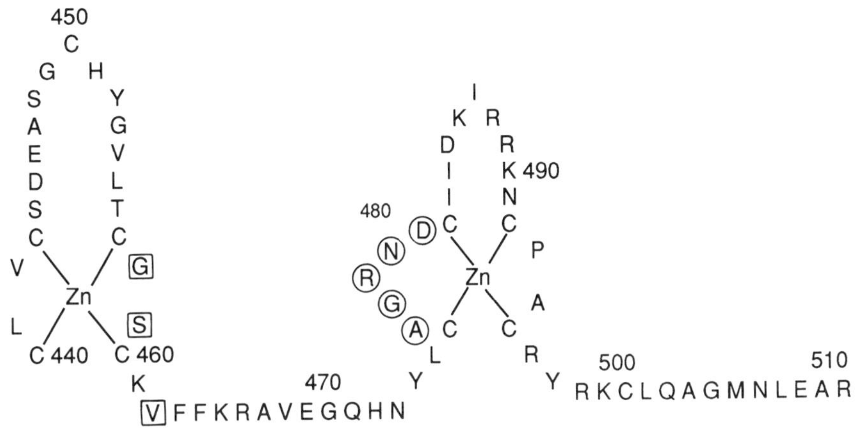

Figure 1.6. Diagram of segment (residues 440–510) of rat glucocorticoid receptor DNA-binding domain. Boxed residues indicate those essential for discrimination between glucocorticoid and oestrogen effects on DNA regulation. (Reproduced with permission from Hard *et al.*, 1990. Copyright 1990 by the AAAS)

the p55 *gag* protein of HIV, the virus thought responsible for AIDS. Others have determined the solution structure of the DNA-binding domain of the glucocorticoid receptor (Lee *et al.*, 1989) which contains two zinc fingers and of the oestrogen receptor (Schwabe *et al.*, 1990).

Recently, the DNA dimer itself has become the target of rational chemotherapy with efforts to design sequence-specific reagents (Moser and Dervan, 1987; Sluka *et al.*, 1987) for potential gene therapy. The reviews of Dervan (1986), Barton (1988) and Cohen (1989) summarize the current status of work in this area. The idealized model of DNA is generally regarded as static, but much experimental data exist which emphasize the dynamic aspects of the DNA dimer and its conformational response to environmental effects (Shakked *et al.*, 1989).

RNA is involved with many essential functions in the body. In protein synthesis, it provides structural and recognition roles in the ribosome, serves as the carrier of the genetic information as messenger RNA, and provides crucial specificity in translation as the carrier of amino acids in tRNA. The crystal structure of glutaminyl-tRNA synthetase complexed with $tRNA^{Gln}$ has been solved (Rould *et al.*, 1989) and provides another case where the basis of nucleic acid-protein recognition can be deciphered. To ensure correct loading of Gln, the enzyme interacts with the anticodon region of the tRNA which is the specificity determinant for amino acid incorporation. Many antibiotics act by interfering with protein synthesis and it is likely that RNA targets are involved in many of these cases.

1.2.3 ENZYMES AS DRUG RECEPTORS

The macromolecules responsible for the catalysis of biological chemical reactions are an obvious target when a disease state is associated with an excess production of a biologically active species. An example of current therapeutic interest is hypercholesteraemia, or the overproduction of the steroid, cholesterol, with an increased incidence of atherosclerosis and stroke or heart disease. Decreasing the circulating levels of cholesterol is associated with lower risk and has been approached by both dietary modification to decrease cholesterol intake as well as inhibition of a key enzyme in cholesterol biosynthesis, HMGCoA reductase. Traditionally, biological catalysis was associated with proteins, but work by Celander and Czech (1991) clearly shows certain RNA molecules (ribozymes) have enzymatic activity. As yet, no clear therapeutic targets have been associated with enzymatic RNAs. The concept of catalytic RNAs to target and selectively hydrolyze specific DNA segments is being developed (Sarver *et al.*, 1990) as a potential therapy for AIDS.

Enzymes are a classic target for therapeutic intervention and numerous well-studied examples exist. Haemoglobin has been of interest because of the allosteric control of oxygen affinity as well as the genetic disease, sickle-cell anaemia, associated with a single change in one amino acid residue. A pioneering and almost classic effort (Goodford, 1977) at rational drug design utilized the crystal structure of haemoglobin and the binding site of DPG, diphosphoglycerate, a natural allosteric regulator as a starting point. Compounds designed with affinity for this site or another site caused significant shifts in the oxygen saturation curves. One such compound is being evaluated clinically in sickle cell anaemia. Others have sought to compete with the aggregation of haemoglobin which leads to the sickling phenomena by binding small molecules competitively. These efforts are reviewed by Perutz *et al.* (1986) where the crystal structure of complexes of several small molecules designed for binding to haemoglobin have been determined.

The inhibition of dihydrofolate reductase, DHFR, which plays an essential role in nucleic acid biosynthesis, supplying one-carbon fragments during the biosynthesis of both purines and pyrimidines, has become almost a case study for methodology development in molecular recognition (Brooks and Fleischman, 1990). In terms of chemotherapy, the inhibition of the biosynthesis of thymidylate (dTMP) appears to be the most important role of DHFR inhibitors (Schweitzer *et al.*, 1990). Detailed crystal structures of DHFR from several species in complex with a variety of ligands have been elucidated (Kuyper, 1989). The structure of the complex has formed the basis for the design of new analogues with higher affinity for the enzyme by at least two pharmaceutical groups (Kuyper, 1989).

Proteolytic enzymes are also popular targets for therapeutic intervention. Inhibitors of angiotensin converting enzyme are used in the treatment of hypertension. In just a few years, over a billion dollars in sales have been achieved for this class of inhibitor due to their efficacy and patient compliance due to the minimal side effects in contrast to the other drugs traditionally used. Unfortunately, the crystal structure of angiotensin converting enzyme has not yet been determined, but much can be deduced about the three-dimensional structure of the active site based on structure–activity data alone (Mayer *et al.*, 1987; Marshall and Cramer, 1988). Aspartic proteinases (Rich, 1986; Davies, 1990) such as renin, also involved in the angiotensin cascade (Vallotton, 1987) and hypertension, and the virally encoded protease from the human immunodeficiency virus which causes AIDS are objects of intense interest in the pharmaceutical industry. The crystal structures of both renin (Sielecki *et al.*, 1989) and HIV protease (Miller *et al.*, 1989) have been determined and are being utilized in the design of inhibitors. Even prior to the solution of the crystal structure of human renin itself, models of the renin structure, built based on sequence homology with other aspartic proteinases whose crystal structures were known, played an important role in rationalizing the design and activity of renin inhibitors (Blundell *et al.*, 1987). One significant difference between the viral aspartic proteinases and those from mammalian origin is that the retroviral enzymes are symmetric dimers (Kay and Dunn, 1990). This offers an opportunity to develop symmetric inhibitors which will be selective for the viral enzyme as was demonstrated recently for HIV protease by the Abbott group (Erickson *et al.*, 1990).

The development of inhibitors by modification of substrates is relatively straightforward with a firm foundation in the concept of the transition state. A recent example can be found for the proteolytic enzyme, neutral endopeptidase (Roques and Beaumont, 1990), which also may have a significant role in hypertension. Practical difficulties arise in finding compounds with high affinity, oral bioavailability and minimal side effects. In this regard, availability of a three-dimensional structure of the therapeutic allows the medicinal chemist to design other classes of compounds to interact with the active site than the class suggested by the substrate in the search for a more ideal drug.

The agents developed for hypertensive therapy offer an example of possible strategies regarding a therapeutic approach. The renin–angiotensin cascade is the target for renin inhibitors to prevent production of angiotensin I, whose subsequent metabolism by angiotensin converting enzyme to release the potent hypertensive, angiotensin II, is the target of the ACE inhibitors. An alternative approach is the development of angiotensin II (AII) receptor inhibitors which preclude binding of AII. Clearly, the more efficient approach is to inhibit the cascade at its beginning because of the catalytic nature of the enzyme targets involved. This would imply use of less drug with concomitantly less risk of side effect. AII receptor inhibitors offer the hope of selectivity in action as evidence has been gained suggesting subtypes of AII receptors with different physiological roles. An alternative approach is the prevention of the metabolism of a natural substance whose beneficial effects would be of utility. Neutral endopeptidase (Roques and Beaumont, 1990) plays a role in the degradation of atrial natriuretic factor, a peptide which lowers blood pressure. In general, it has been most useful to develop a prototype inhibitor to allow pharmacological exploration of the physiological role of the system under study. Serendipitous exploitation of the experimental observations, often clinical, have led to novel and unanticipated uses of compounds targeted for a particular system.

1.2.4 MEMBRANES AS DRUG RECEPTORS

Membranes are significant structural elements, both in defining the boundaries of a cell as well as providing interior compartments within the cell associated with particular functions. Mitochondria, which probably evolved from intracellular symbiotic bacteria, are separated from the cytoplasm by a double membrane, and contain the enzymatic machinery for oxidizing substrates and generating ATP, the common source of chemical energy within the cell. Chloroplasts serve to convert solar energy into chemical energy and have a similar structural arrangement and evolutionary origin. The cell nucleus in eukaryotes is separated from the cytoplasm by a nuclear membrane. Enzymes responsible for degradation within the cell are similarly segregated into vesicles called lysozomes. By compartmentalization, various cellular functions can be isolated and the appropriate machinery targeted for those compartments and concentrated within. In addition, chemical gradients of ions can be established by transmembrane proteins which actively transport ions from one compartment to another. These ionic gradients can subsequently be used to transmit information in the form of nerve impulses, stimulate secretion or contraction as in a heartbeat, or help generate ATP. The integrity of membranes and their transmembrane gradients is a crucial aspect of biological systems and many toxicological or therapeutic agents modulate some aspects of this biology. Because of their physical characteristics, membranes provide a hydrophobic barrier preventing the free diffusion of water-soluble compounds, and elaborate mechanisms have evolved to allow the transmission of a signal from the exterior of a cell into the interior. Two major chemical message systems exist in higher organisms: the hormone system, where chemical messengers are transmitted by the circulation to cells whose membranes have receptors which recognize the chemical signals; the neurotransmitters, where the wiring of the nervous system guarantees delivery of sufficiently high concentrations of the chemical messenger to receptors located on the postsynaptic membrane.

Cell membranes themselves can also function as targets for molecular recognition and biological response. Many general anaesthetics are believed to work by their physical chemical effects when dissolved in membranes (Chiou *et al.*, 1990). Local anaesthetics, on the other hand, are thought to work by direct action on an allosteric site on an integral membrane protein, the nicotinic acetylcholine receptor (Forman and Miller, 1989). Several classes of antibiotics function by forming pores in bacterial membranes. This leads to loss of the electrochemical gradient, osmotic changes and cell lysis in many cases. Examples of this class which have been well studied at the molecular level include gramicidin A and peptaibol antibiotics, such as alamethicin. Gramicidin A is a hydrophobic, linear peptide antibiotic, toxic to Gram-positive bacteria, of 15 residues which acts as a channel for monovalent cations in lipid membranes. A head-to-head dimer of a $\beta^{6.3}$ helix proposed by Urry *et al.* (1971) best fit expectations of length (25–30 Å) and pore size (4 Å) needed for channel activity and ion selectivity. Alamethicin is the best known member of the class of fungal antibiotics, known as peptaibols, whose name comes from α-methylalanine (MeA), or aminoisobutyric acid (Aib) residues and a C-terminal amino acid alcohol such as Phe-ol. The unusual amino acid residue favours helical conformations, and a pore is formed in membranes (see Figure 1.7) under an appropriate membrane potential by aggragation of helical monomers (Marshall and Beusen, 1992). Many toxins such as the peptides, melittin and apamin found in bee venoms, mastoparan from wasp venom and bombolitin from the venom of the bumblebee, form amphipathic helices which have direct effects on planar lipid bilayers, causing transmembrane pores (Bernheimer and Rudy, 1986) (Marshall and Beusen, 1992). Recent data (Higashijima *et al.*, 1990) suggest that they may also exert part of their biological spectrum by direct activation of the G-protein involved in many receptor-activated mechanisms.

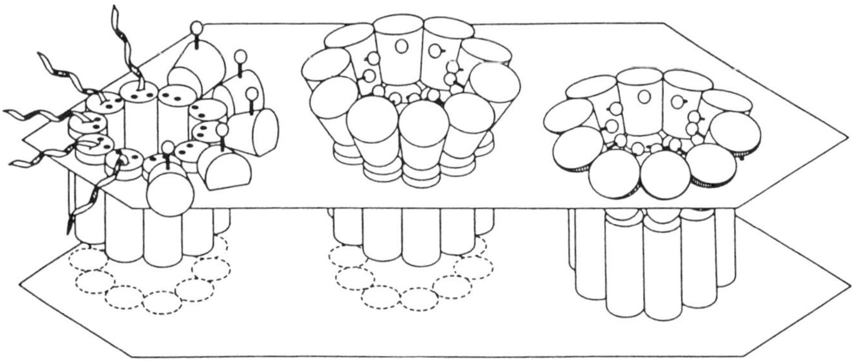

Figure 1.7. Voltage-gating channel model of alamethicin proposed by Fox and Richards (1982). On the left is the structure for the helical bundle partially inserted in the absence of an applied voltage. The C-terminal residues are shown as helical ribbons indicating extended random coil structures, or as cork-shaped α-helical structures partially buried on the surface of the membrane. The middle structure is an intermediate produced by application of voltage and the right view represents the open state which traverses the membrane. (Reprinted with permission from *Nature* vol. **300** p. 329. Copyright © 1982 Macmillan Magazines Ltd)

1.2.5 INTEGRAL MEMBRANE PROTEINS AS RECEPTORS

Because of the difficulties due to molecular size and physicochemical properties for many natural chemical messengers to cross a membrane, cells contain proteins which span the lipid bilayer as constituents of their membranes. The best characterized example of such an integral transmembrane protein is the photosynthetic reaction centre from the purple bacterium, *Rhodopseudomonas viridis*, whose crystal structure has been determined (Deisenhofer and Michel, 1989). Current intense research interest focuses on the true receptors, i.e. those integral membrane proteins which act to transduce the signals of hormones such as insulin, or of neurotransmitters such as acetylcholine and epinephrine. Examples of the explosion in information which has resulted from the ability to clone and express receptors are given by the 5-hydroxytryptamine (serotonin) (Schmidt and Peroutka, 1989; Frazer *et al.*, 1990) and muscarinic acetylcholine (Hulme *et al.*, 1990) families of receptors. These fall into three basic categories based on structure and mechanism of second messenger production. The first (Figure 1.8) are those which are coupled to systems via G-proteins, first recognized in adenylate cyclase-coupled receptors (Dixon *et al.*, 1988). The second class of receptor (Figure 1.9) are those which have a single membrane spanning segment with the molecular recognition domain extracellular and the enzymatic effector, a tyrosine kinase domain, located intracellularly. Occupancy

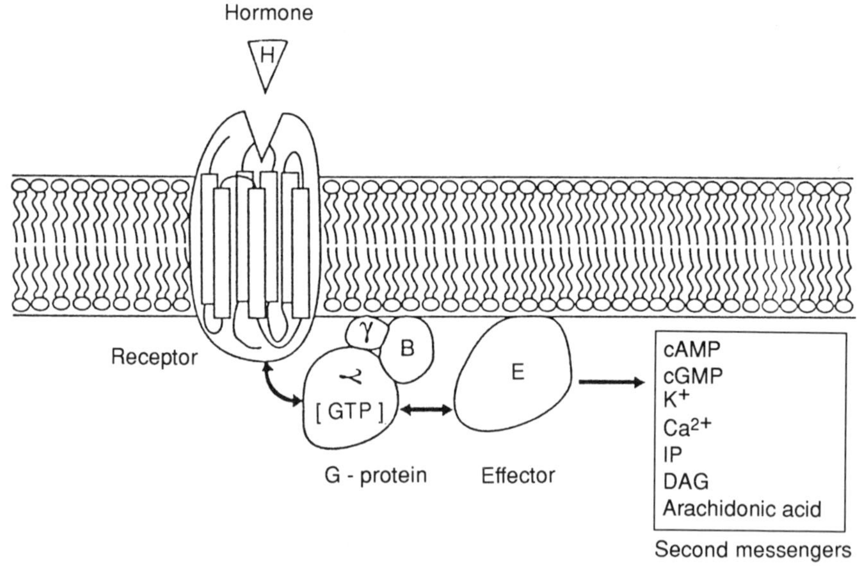

Figure 1.8. Schematic diagram of G-protein coupled receptors with seven helical transmembrane segments containing extracellular recognition site. (Reproduced from Dixon *et al.* (1988), by permission of Academic Press)

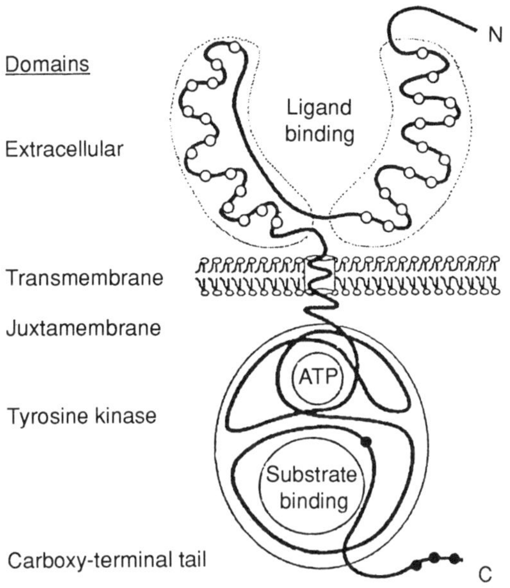

Figure 1.9. Schematic diagram of growth factor receptors which have extracellular hormone receptor domains and intracellular enzymatic domains (Yarden and Ullrich, 1988). (Reproduced, with permission, from the *Annual Review of Biochemistry*, Vol. **57**, © 1988 by Annual Reviews Inc.)

of the binding domain by the appropriate ligand results in activation of the tyrosine kinase intracellularly (Yarden and Ullrich, 1988). The third class of receptor is exemplified by the acetylcholine nicotinic receptor which is a complex of proteins which span the membrane and which gate ions when the binding site is occupied. This receptor has been intensely studied because of the biological hyperexpression of the protein in the electroplax organ of the electric ray, *Torpedo*. While much is known regarding the overall features of this channel (Herz *et al.*, 1989) as shown in Figure 1.10, resolution at the molecular level will be required before the details of transduction can be ascertained. Because of the limited homology of G-protein-linked receptors to bacterial rhodopsin, whose membrane structure is known at low resolution, there have been attempts (Findlay and Eliopoulos, 1990) to predict the structure of several integral membrane receptors. While these are highly speculative, they may have utility when combined with mutational studies (Tota *et al.*, 1991).

Ion channels themselves are targets for therapeutic intervention. These transmembrane protein complexes respond to the transmembrane potential and chemical mediators to regulate the flux of ions across cell membranes. Many toxins act by stimulation or inhibition of ion channels. This includes tetrodotoxin from the puffer fish, *Tetradon tetradon*, responsible for several fatalities

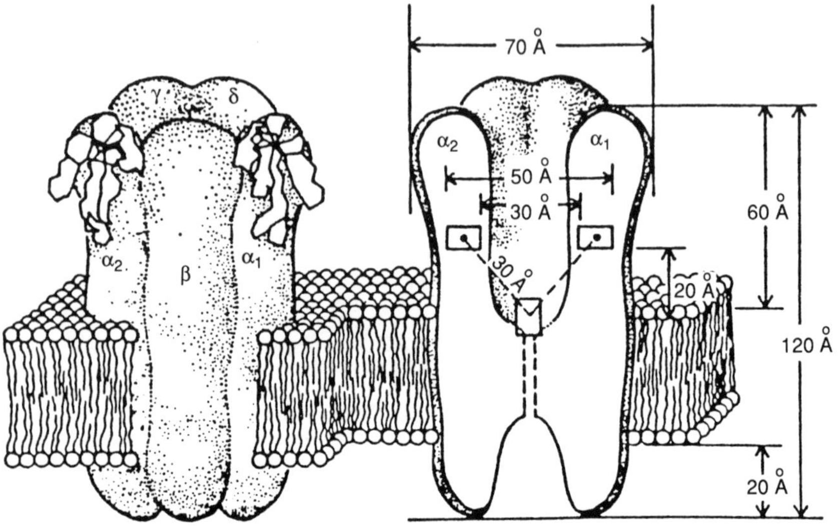

Figure 1.10. Schematic diagram of acetylcholine receptor with two agonist sites indicated on α-subunits. (Reproduced from Herz *et al.* (1989) by permission of The American Society for Biochemistry and Molecular Biology)

each year among Japanese gourmets, as well as capsaicin (Bevan and Szolcsanyi, 1990), the active ingredient of pepper. Cloning of ion channel protein has indicated a wide diversity of subtypes for each cation, and structure-activity studies may be more understandable once this complication has been resolved by cloning and characterization of each subtype. An example of this development is reviewed for the potassium channels by Edwards and Weston (1990). Different opioid receptors are thought (Millan, 1990) to selectively regulate potassiuim and calcium channels. A therapeutically important class of cardiovascular drugs, the 1,4-dihydropyridines appear to work by interaction with calcium channels (Triggle and Rampe, 1989).

Other integral cell surface proteins play essential roles in cell-cell interactions. Often these are glycoproteins and it is becoming increasingly apparent what a significant role these carbohydrate moieties and proteins, lectins (Sharon and Lis, 1989), which recognize them play in many biological phenomena, particularly in targeting the uptake of circulating proteins into cells. Proteinaceous segments of surface membrane proteins also play a significant role in biological recognition. The integrins (Figure 1.11) and other surface proteins involved in cell recognition and adhesion are thought to be important in development, inflammation and tumour metastasis (Albelda and Buck, 1990).

Other compounds classified as traditional hormones do not act on transmembrane receptors. These steroid or amino acid derivatives, such as oestrogen or

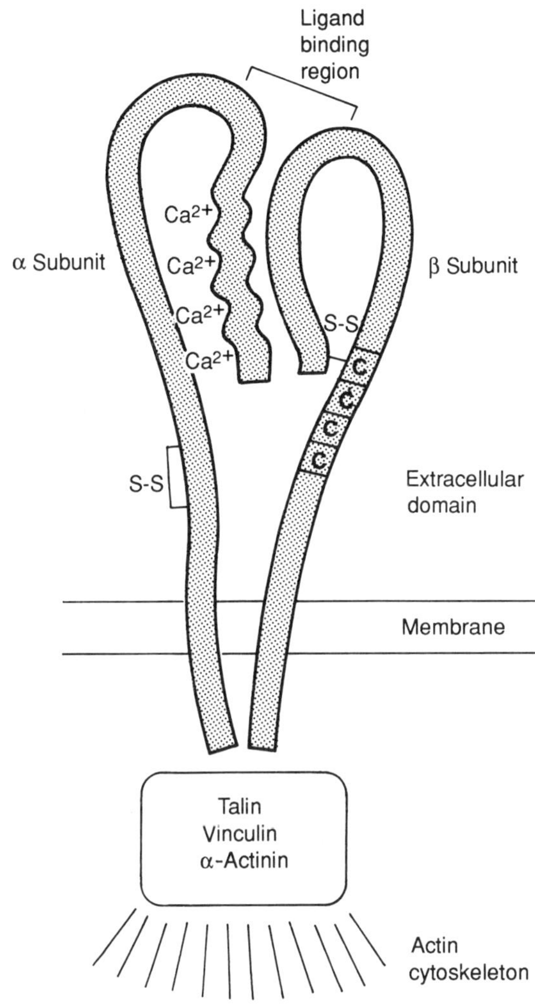

Figure 1.11. Diagram of schematic structure of a typical integrin composed of two integral membrane glycoproteins. (Reproduced from Albelda and Buck (1990) by permission of the publishers)

thyroid hormone as well as the vertebrate morphogen retinoic acid, bind to an intracellular receptor which functions to regulate transcription. The DNA-binding domains show strong homology which can be configured as two zinc-binding fingers with each zinc ion forming a tetrahedral coordination complex with four Cys residues (Evans, 1988). This places the steroid hormone receptor clearly in the category of a protein which modifies eukaryotic transcription as the zinc finger is one of four motifs identified with that function (Struhl, 1989). Binding of the hormone to a separate domain leads to exposure of the DNA-binding domain by a hypothetical allosteric transition.

1.2.6 VIRAL DRUG RECEPTORS

The key to successful rational drug design is selection of a therapeutic target whose function is crucial to the disease state and non-essential for the viability of the host. Because viruses utilize much of the cell's own machinary in their infective state, they do not offer the wide range of therapeutic opportunity available in antimicrobials. The explosion in information regarding the molecular details of the life cycle of retroviruses, stimulated by the therapeutic need for more effective treatment of AIDS, has led to many different approaches to antiviral therapy. An excellent overview of the approaches and the macromolecular targets of the human immunodeficiency virus, the pathogen of AIDS, has been given by Mitsuya *et al.* (1990). The clinically effective drugs, AZT and ddI, show selective effects on reverse transcriptase, a virally encoded enzyme. Besides the development of inhibitors of HIV protease and other viral proteinases (Kay and Dunn, 1990) mentioned above, efforts to develop specific derivatives of nucleic acids which recognize and interact with the viral codon are making significant progress (Buck *et al.*, 1990; Sarver *et al.*, 1990). The crystal structure (Ryu *et al.*, 1990; Wang *et al.*, 1990) of the extracellular fragment of the cell-surface receptor, CD4, for the HIV surface protein, gp120, offers another exciting area for development of a small molecular weight inhibitor of that complex formation. The structure of ribonuclease H from *E. coli*, which is homologous with the enzyme from retroviruses responsible for hydrolysis of RNA complexed with complementary DNA, has recently become available (Yang *et al.*, 1990) and should stimulate development of inhibitors of this enzyme, potential anti-HIV agents. Other therapeutic approaches include inhibition of post-translation modification, myristoylation and glycosylation, of the virally encoded polyproteins which are essential for their proteolytic processing and viral assembly.

The recent availability of the crystal structures of viruses themselves offers new opportunities for therapy (Rossmann, 1989). An interesting case of receptor and drug action (McKinlay and Rossmann, 1989) is the inhibition of the uncoating of the virion of picornaviruses which included two major classes associated with human disease. The first class are the rhinoviruses associated with the common cold and the second class are enteroviruses which cause a spectrum of clinical syndromes. These compounds are extremely hydrophobic in nature and work by binding to a pocket in one of the capsid proteins (Diana *et al.*, 1990). Efforts (Lau and Pettitt, 1989) to understand the details of molecular recognition in this process may increase the effectiveness of new compounds. A conformation change of the protein occurs which is assumed responsible for the increased stability of the viral capsid with resulting loss of viral infectivity due to imprisonment. By mutational analysis, all mutants resistant to these drugs have been mapped to two sites within the drug binding pocket with resistant mutants having larger amino acid side chains than the native, presumably preventing binding of the drugs by steric repulsion.

1.2.7 SUMMARY

In summary, the functional definition of a receptor as 'that molecule in the biological system through which the molecular interaction responsible for the biological effect of the compound under study occurs', may in fact be sufficiently broad to cover the wide variety of known receptors. We have emphasized examples where much of the three-dimensional information characterizing the macromolecular therapeutic target is available. In most cases, however, our chemical knowledge about the receptor is limited to sequence information at best. In these cases, the details of the three-dimensional structure necessary for molecular modelling and design are not yet known and require the use of inductive techniques to deduce properties of the receptor useful for drug development studies.

As in much of biology, if one can conceive of a plausible mechanism of therapeutic intervention, then one can probably discover an example of its use in nature due to the time which evolution has had to exploit its opportunities. This is certainly true for many of our best strategies, e.g. transition-state analogues, mechanism-based inhibitors, etc. For this reason, the study of naturally occurring toxins and poisons has proved a fruitful source of inspiration for the medicinal chemist. With the increased ability to derive three-dimensional information regarding the macromolecular therapeutic target as exemplified in the case of HIV, the coming decade should prove most rewarding for those involved with rational drug design.

1.3 OTHER FUNCTIONS OF MACROMOLECULES

1.3.1 TRANSPORT

Drugs often have considerable hydrophobic character designed into their structures which allow them to cross lipid membrane barriers *en route* to their sites of action. This property leads to a propensity to bind to hydrophobic sites on various macromolecules in the body, often in a relatively non-specific way. This provides a buffer compartment which can play an important role in pharmacokinetics. For example, in the case of the amino acid thyroid hormones, thyroxine and T_3, most of the circulating hormone is bound to proteins in the plasma with only a small percentage in the active soluble form. This requires sophisticated analytical procedures to determine the free, active concentrations of the hormone rather than simply the blood levels for correct diagnostic purposes.

On occasion, serendipitous binding to non-therapeutic target proteins may provide some advantage for one class of drug over another. Transport across biological membranes and within cellular compartments is regulated by sets of transmembrane proteins which often have specific recognition signals, such as glycosylation of particular residues, which might be complicated to utilize directly. Binding of a drug to a protein which is transported can sometimes change the volume of distribution, i.e. alter localization within the body dramatically. This is a problem that is particularly acute with regard to delivery of drugs to the CNS, and much research needs to be done to determine how transport of various hydrophilic molecules such as glucose, or gamma-aminobutyric acid (GABA) into the CNS is mediated. An example of a transport system and approaches to its molecular characterization is given in a summary (Bowery, 1990) of recent work on a GABA transport protein.

1.3.2 METABOLISM

Another area of extreme practical importance is the problem of metabolism and excretion of drugs from the body. The body, particularly in the liver, has very efficient enzymatic machinery, the cytochrome P450 system (Guengerich, 1989), for the oxidation of hydrophobic exobiotics, conjugation to hydrophilic groups and excretion through the bile into the intestines. The ability of a drug to avoid efficient clearance by this system when first absorbed in the gut and passed through the portal circulation is important. Likewise, numerous other macromolecules are involved in catabolism in the body. The difficulty with the use of most peptides, or proteins, as drugs relates to their inability to penetrate and cross lipid membrane barriers as well as their inherent susceptibility to the large array of proteolytic enzymes facing them in the body, particularly on oral administration. Much important medicinal chemistry is concerned with the conversion of a compound with a desirable set of biological activities into a compound retaining the biological activity with appropriate physicochemical and pharmacokinetic properties, i.e. oral bioavailability and prolonged half-life.

1.4 CONCLUSIONS

Molecular recognition, where one of the components is a macromolecule, is a dominant theme in biological systems. The conformation assumed by the macromolecule stabilizes a recognition, or active site, geometry which determines the receptor-bound conformation of small molecular weight ligands, whether substrates, neurotransmitters, hormones, or drugs. Often the molecule to be recognized is a site on another macromolecule which forms the basis of macromolecular association and assemblies. Recent progress in the rapid determination of the chemical and three-dimensional structure of many biologically important macromolecules provides crucial information in the quest for better i.e. more specific, pharmaceuticals. Because a drug is exposed to many different macromolecules in the course of its action besides that to which it was targeted, many compounds which show excellent profiles *in vitro* fail to prove therapeutically effective due to interaction with other macromolecules. As our knowledge-base of molecular structures increases, the possibility to design molecules which are more specific for their targets and less likely to bind to other macromolecules at effective therapeutic concentrations becomes more feasible.

ACKNOWLEDGEMENTS

Support from the NIH (GM24482) during the preparation of this review is gratefully acknowledged.

REFERENCES

Albelda, S. M. and C. A. Buck. (1990). *FASEB J.*, **4**, 2868–80.
Aposhian, V. V. and Aposhian, M. M. (1990). *A. Rev. Pharmacol. Toxicol.*, **30**, 279–306.
Barton, J. K. (1988). *Chem. & Engng. News*, **66**, 30–42.
Berg, J. M. (1990). *J. Biol. Chem.*, **265**(12), 6513-16.
Bernheimer, A. W. and Rudy, B. (1986). *Biochim. Biophys. Acta*, **864**, 123–41.
Bevan, S. and Szolcsanyi, J. (1990). *Trends Pharmacol. Sci.*, **11**(11), 330–3.
Blundell, T. L., Cooper, J., Foundling, S. I., Jones, D. M., Atrash, B. and Szelke, M. (1987). *Biochemistry*, **26**(18), 5585–90.
Bowery, N. G. (1990). *Trends Pharmacol. Sci.*, **11**(11), 435–6.
Boyd, F. L., Cheatham, S. F., Remers, W., Hill, G C. and Hurley, L. H. (1990). *J. Am. Chem. Soc.*, **112**, 3279–89.
Brennan, R. G. and Matthews, B. W. (1989a). *Trends Biochem. Sci.*, **14**, 286–90.
Brennan, R. G. and Matthews, B. W. (1989b). *J. Biol. Chem.*, **264**(4), 1903–6.
Brooks, C. L., III and Fleischman, S. H. (1990). *J. Am. Chem. Soc.*, **112**, 3307–12.
Buck, H. M., Koole, L. H., Van Genderen, M. H. P., Smit, L., Geelen, J. L. M. C., Jurriaans, S. and Goudsmit, J. (1990). *Science* **248**, 208–11.
Celander, D. W. and Czech, T. R. (1991). *Science*, **251**, 401–6.
Chiou, J.-S., Ma, S.-M., Kamaya, H. and Ueda, I. (1990). *Science*, **248**, 583–5.
Cohen, J. S. (1989). *Trends Pharmacol. Sci.*, **10**(11), 435–7.
Cram, D. J. (1988). *Science*, **240**, 760–7.
Davies, D. R. (1990). *A. Rev. Biophys. Biophys. Chem.*, **19**, 189–215.
de Vlieg, J., Berendsen, H. J. C. and van Gunsteren, W. F. (1989). *Proteins*, **6**, 104–27.
Deisenhofer, J. and Michel, H. (1989). *Science*, **245**, 1463–73.
Dervan, P. B. (1986). *Science*, **232**, 464–71.
Diana, G. D., Treasurywala, A. M., Bailey, T. R., Oglesby, R. C., Pevear, D. C. and Dutko, F. J. (1990). *J. Med. Chem.*, **33**, 1306–11.
Dixon, R. A. F., Strader, C. D. and Sigal, I. S. (1988). *A. Rep. Med. Chem.*, **23**, 221–33.

Edwards, G. and Weston, A. H. (1990). *Trends Pharmacol. Sci.*, **11**(10), 417–22.
Erickson, J., Neidhart, D..J., VanDrie, J., *et al.* (1990). *Science*, **249**, 527–33.
Evans, R. M. (1988). *Science*, **240**, 889–95.
Findlay, J. and Eliopoulos, E. (1990). *Trends Pharmacol. Sci.*, **11**, 492–9.
Forman, S. A. and Miller, K. W. (1989). *Trends Pharmacol. Sci.*, **10**(11), 447–52.
Fox, R. O. and Richards, F. M. (1982). *Nature*, **300**, 325–30.
Frazer, A., Maayani, S., and Wolfe, B. B. (1990). *A. Rev. Pharmacol. Toxicol.*, **30**, 307–48.
Furst, P. and Hammer, D. (1989). *Proc. Natl Acad. Sci. USA*, **86**, 5267–71.
Gao, X. and Patel, D. J. (1989). *Q. Rev. Biophys.* **23**(2), 93–138.
Goodford, P. J. (1977). In *Drug Action at the Molecular Level* (Ed. G. C. K. Roberts), Macmillan Press, Ltd, London, pp. 109–26.
Guengerich, P. (1989). *Trends Pharmacol. Sci.*, **10**, 107–14.
Hard, T., Kellenbach, E., Boelens, R., Mahler, B. A., Dahlman, K., Freeman, L. P., Carlstadt-Duke, J., Yamamoto, K. R., Gustafson, J.-A. and Kaptein, R. (1990). *Science*, **249**, 157–60.
Herz, J. M., Johnson, D. A. and Taylor, P. (1989). *J. Biol. Chem.*, **264**, 12439–48.
Higashijima, T., Burnier, J. and Ross, E. M. (1990). *J. Biol. Chem.*, **265**(24), 14176–86.
Hu, J. C., O'Shea, E. K., Kim, P. S. and Sauer, R. T. (1990). *Science*, **250**, 1400–4.
Hulme, E. C., Birdsall, N. J. M. and Buckley, N. J. (1990). *A. Rev. Pharmacol. Toxicol.*, **30**, 633–73.
Hurley, L. F. (1989). *J. Med. Chem.*, **32**(9); 2027–33.
Hurley, L. F. and Boyd, F. L. (1988). *Trends Pharmacol. Sci.*, **9**, 402–7.
Kay, J. and Dunn, B. M. (1990). *Biochim. Biophys. Acta*, **1048**, 1–18.
Klevit, R. E., Wemmer, D. E. and Reid, B. (1986). *Biochemistry*, **25**, 3296–303.
Kuyper, L. F. (1989). In *Computer-aided Drug Design* (Eds T. J. Perun and C. L. Propst), Marcel Dekker, Inc., New York, pp. 327–70.
Lau, W. F. and Pettitt, B. M. (1989). *J. Med. Chem.*, **32**, 2542–7.
Lee, M. S., Gippert, G. P., Soman, K. V., Case, D. A. and Wright, P. E. (1989). *Science*, **245**, 635–7.
Mader, S., Kumar, V., de Verneuil, H. and Chambon, P. (1989). *Nature* **338**, 271–4.
Marmorstein, R. Q. and Sigler, P. B. (1989). *J. Biol. Chem.*, **264**(16); 9149–54.
Marshall, G. R. and Beusen, D. (1992). In *Membrane Electrochemistry*. (Eds M. Blank, I. Vodyanoy and F. Hong), American Chemical Society, Washington DC (in press).
Marshall, G. R. and Cramer III, R. D. (1988). *Trends Pharmacol. Sci.*, **9**, 285–9.
Mayer, D., Naylor, C. B., Motoc, I. and Marshall, G. R. (1987). *J. Comput.-Aided Mol. Design*, **1**, 3–16.
McKinlay, M. A. and Rossmann, M. G. (1989). *A. Rev Pharmacol. Toxicol.*, **29**, 111–22.
Millan, M. J. (1990). *Trends Pharmacol. Sci.*, **11**(2), 70–6.
Miller, M., Sathyanarayana, B. K., Wlodawer, A., Toth, M. V., Marshall, G. R., Clawson, L., Selk, L., Schneider, J. and Kent, S. B. H. (1989). *Science*, **246**, 1149–52.
Mitsuya, H., Yarchoan R. and Broder, S. (1990). *Science*, **249**, 1533–44.
Moser. H. E. and Dervan, P. B. (1987). *Science*, **238**, 645–50.
Mueller, L., Heald, S. L., Hempel, J. C. and Jeffs, P. W. (1989). *J. Am. Chem. Soc.*, **111**, 496–505.
O'Neil, K. T., Hoess, R. H. and DeGrado, W. F. (1990). *Science*, **249**, 774–77.
Otwinowski, Z., Schevitz, R. W., Zhang, R. G., Lawson, C. L., Joachimiak, A., Marmorstein, R. Q., Luisi, B. F. and Sigler, P. B. (1988). *Nature*, **335**, 321–9.
Pedersen C. J. (1988). *Science*, **241**, 536–40.
Pelton, J. G. and Wemmer, D. E. (1990). *J. Am. Chem. Soc.*, **112**, 1393–9.
Perutz, M. F., Fermi, G., Abraham, D. J., Poyart, C. and Bursaux, E. (1986). *J. Am. Chem. Soc.*, **108**, 1064–78.

Qian, Y. Q., Billeter, M., Otting, G., Muller, M., Gehring, W. J. and Wuthrich, K. (1989). *Cell*, **59**, 573–80.
Rafferty, J. B., Somers, W. S., Saint-Girons, I. and Phillips, S. E. V. (1989). *Nature*, **341**, 705–10.
Rebek, J., Jr (1987). *Science*, **235**, 1478–84.
Rich, D. H. (1986). In *Proteinase Inhibitors* (Eds A. J. Barrett and G Salvesen), Elsevier Science Publ., Amsterdam, pp. 179–217.
Roques, B. P. and Beaumont, A. (1990). *Trends Pharmacol. Sci.*, **11**(6); 245–9.
Rossmann, M. G. (1989). *FASEB J.*, **3**, 2335–43.
Rould, M. A., Perona, J. J., Soll, D. and Steitz, T. A. (1989). *Science*, **246**, 1135–42.
Ryu, S.-E., Kwong, P. D., Truneh, A., Porter, T. G., Arthos, J., Rosenberg, M., Dai, X., Xuong, N.-H., Axel, R., Sweet, R. W. and Hendrickson, W. A. (1990). *Nature*, **348**, 419.
Sarver, N., Cantin, E. M., Chang, P. S., Zaia, J. A., Ladne, P. A., Stephens, D. A. and Rossi, J. J. (1990). *Science*, **247**, 1222–5.
Schmidt, A. W. and Peroutka, S. J. (1989). *FASEB J.*, **3**, 2242–9.
Schwabe, J. W. R., Neuhaus, D. and Rhodes, D. (1990). *Nature*, **348**, 458–61.
Schweitzer, B. I., Dicker, A. P. and Bertino, J. R. (1990). *FASEB J.*, **4**, 2441–52.
Shakked, Z., Guerstein-Guzikevich, G., Eisenstein, M., Frolow, F. and Rabinovich, D. (1989). *Nature*, **342**, 456–60.
Sharon, N. and Lis, H. (1989). *Science*, **246**, 227–34.
Sielecki, A. R., Hayakawa, K., Fujinaga, M., Murphy, M. E., Fraser, M., Muir, A. K., Carilli, C. T., Lewicki, J. A., Baxter, J. D, and James, M. N. (1989). *Science*, **243**, 1346–51.
Singh, U. C., Pattabiraman, N., Langridge, R. and Kollman, P. A. (1986). *Proc Natl Acad. Sci. USA*, **83**, 6402–6.
Sluka, J. P., Horvath, S. J., Bruist, M. F., Simon, M. I. and Dervan, P. B. (1987). *Science*, **238**, 1129–32.
Stoddart, J. F. (1987). In *Enzyme Mechanisms* (Eds M. I. Page and A. Williams), Royal Society of Chemistry, London, pp. 35–55.
Struhl, K. (1989). *Trends Biochem. Sci.*, **14**, 137–40.
Summers, M. F., South, T. L., Kim, B. and Hare, D. R. (1990). *J. Am. Chem. Soc.*, **29**, 329–40.
Tabushi, I. (1982). *Acc. Chem. Res.*, **15**, 66–72.
Takusagawa, F., Dabrow, M., Neidle, S. and Berman, H. M. (1982). *Nature*, **296**, 466–9.
Talanian, R. V., McKnight, C. J. and Kim, P. S. (1990). *Science*, **249**, 769–71.
Tota, M. R., Candelore, M. R., Dixon, R. A. F. and Strader, C. D. (1991). *Trends Pharmacol. Sci.*, **12**(1), 4–6.
Triggle, D. J. and Rampe, D. (1989). *Trends Pharmacol. Sci.*, **10**, 507–11.
Umesono, K. and Evans, R. M. (1989). *Cell*, **57**, 1139–46.
Urry, D. W., Goodall, M. C., Glickson, J. D. and Mayers, D. F. (1971). *Proc. Natl. Acad. Sci. USA*, **68**, 1907–11.
Vallotton, M. B. (1987). *Trends Pharmacol. Sci.*, **8**, 69–74.
Vinson, C. R., Sigler, P. B. and McKnight, S. L. (1989). *Science*, **246**, 911–16.
Waltho, J. P. and Williams, D. H. (1989). *J. Am. Chem. Soc.*, **111**, 2475–80.
Wang, A. H. -J., Ughetto, G., Quigley, G. J., Hakoshima, T., van der Marel, G. A., van Bloom, J. H. and Rich, A. (1984). *Science*, **225**, 1115–21.
Wang, J., Yan, Y., Garrett, T. P. J., Liu, J., Rodgers, D. W., Garlick, R. L., Tarr, G. E., Husain, Y., Reinherz, E. L. and Harrison, S. C. (1990). *Nature*, **348**, 411.
Yang, W., Hendrickson, W. A., Crouch, R. J. and Yoshinori, S. (1990). *Science*, **249**, 1398–405.
Yarden, Y. and Ullrich, A. (1988). *A. Rev. Biochem*, **57**, 443–78.

2 Ligand Fitting Methodology

ALWYN S. GILBERT AND JOHN N. CHAMPNESS
The Wellcome Research Laboratories, Beckenham, Kent, UK

2.1 INTRODUCTION

Given a sufficiently detailed and accurate three-dimensional description of the functional region of a biological macromolecule, it is theoretically possible to design a small molecule that can interact with it and modify its function. Such modification may alter the biological properties of the system in a way that is therapeutically useful. The initial design criteria are merely that the drug must fit into the relevant part of the macromolecule and be capable of displaying a useful degree of binding to it, i.e. it must be a ligand.

Put simply like this, ligand design would seem to be a fairly straightforward, if perhaps tedious, process. Unfortunately very few macromolecules have been amenable to yielding 3-D structural information and even where 'pictures' can be obtained the accuracy in placing chemical groups and atoms is often too poor

The Design of Drugs to Macromolecular Targets. Edited by C. R. Beddell
© 1992 John Wiley & Sons Ltd

to allow the interactions between drug and macromolecule that are necessary for binding, for example hydrogen bonds, to be modelled with confidence. An additional uncertainty is that macromolecules are flexible and may exhibit conformational changes on small-molecule binding. Furthermore, structural information available from a complex with one ligand may exhibit a conformation that is not relevant for interaction with a small molecule of rather different structure. Of course, the ability to bind is only a necessary, not a sufficient, condition for a chemical to be useful as a medicine as many other factors are important such as transport, host metabolism, etc., but they lie outside the scope of this discussion.

These are significant impediments to successful drug design; nevertheless by combination of the available data on the macromolecule under study with knowledge of how chemical groups interact with each other, and with imaginative compound design, it is possible to propose small molecules that have a reasonable chance of binding in an appropriate manner. Fine tuning of the structure can then be carried out by other methods such as classical quantitative structure activity relationships (QSAR). This potential for lead generation is perhaps the most important but also the most underestimated aspect of 3-D modelling.

Our purpose in this chapter is to review the present state of the art in assessing how well a putative drug molecule can bind to a macromolecular site using the techniques of 3-D modelling, or as it is often known, small molecule-receptor docking. We will pay attention to what is known or can be inferred about the relevant molecular interactions and in addition will survey the various computer tools that are now available for visualization of the molecules concerned and estimation, both qualitative and quantitative, of the energies of interactions. We will not go into great detail on any particular aspect as most of what we will discuss is covered in greater depth elsewhere in this book; rather our intention is to provide an overview of the basic principles that need to be understood by the 3-D modeller and some of the tools now available for this type of work. But because many of the assumptions that have to be made are crucially dependent on the accuracy of the structural data we will start out with a short discussion on how such data are obtained and how much reliance can be placed upon the models derived from them.

2.2 STRUCTURAL STUDIES ON PROTEIN MACROMOLECULES

Most useful data have been derived from X-ray crystallographic studies; we know of only half a dozen neutron diffraction studies on large biological molecules, while NMR spectroscopy has only recently begun to provide the appropriate information. Certain disadvantages of X-ray studies are well known, for instance the scattering intensity is proportional to atomic number so hydrogens are difficult to observe.

The X-ray diffraction pattern is related to the three-dimensional arrangement of the repeating unit of atoms in the crystal by the process of Fourier transformation. The whole assembly contributes to each point in the pattern and for biological macromolecules such as proteins the necessity of determining the phase of each part of the pattern (only the amplitude can be directly measured) is usually dealt with by multiple isomorphous replacement using heavy metal atoms. In some instances endogenous scatterers have also been used and, alternatively, where homology exists, a structure can be determined by 'molecular replacement' with one already known, suitably orientated and located. Once an approximate structural solution is obtained it may be further refined against the restraints of the X-ray data and known structural parameters. These methods allow the calculation of a 3-D map of the electron density which, if the resolution is sufficiently good, provides an interpretable likeness of the molecule. For an authoritative textbook on the topic of crystallography of macromolecules the reader is referred to Blundell and Johnson (1976).

Whilst in the case of small molecules resolutions of better than 1 Å are achievable, presenting the atoms as individual spheres of density, proteins frequently yield resolution no better than 2 Å, and since most covalent bonds are shorter than this the atoms appear blurred together so that, for example, a benzene ring shows up as a lumpy doughnut. The benzyl group is one of the easier chemical groups in a protein to identify because the heavier side chain atoms are all of one type. But carbons show very similar electron density to oxygens (and nitrogens) so that, for example, a valine side chain can be mistaken for threonine. Thus the initial model has to be constructed using the known sequence of the protein and any other chemical information to hand. This is usually done by means of a visual fit of the macromolecule's skeletal structure to the electron density map. The software FRODO (Jones, 1978) was developed to provide a means of doing this with a computer-graphics workstation. Adjustments are subsequently made to remove incompatabilities between the model and the electron density distribution, and an interactive process of modelling alternating with cycles of refinement as indicated above generally achieves a sensible and useful result.

There is, however, always significant uncertainty in the precise positioning of atom centres and of course hydrogens are virtually invisible. It is important to note that the positioning is always a compromise, often yielding bonds and bond angles that are obviously wrong but difficult to correct. At 2 Å resolution it is possible for atoms to be generally misplaced by up to 0.3 Å and if this error is repeated on both sides of a binding site so as to apparently open it up, the designer may well be led into proposing molecules that in fact are too large and therefore will not bind. Furthermore, larger local errors may occur. Ways of assessing and minimizing these errors have recently been reviewed by Branden and Jones (1990).

2.3 MOLECULAR INTERACTIONS

In order to compete with the natural substrate a drug must make a number of specific interactions with atoms of the protein at the active site. It can do this by forming a variety of non-covalent bonds which taken together must be sufficiently favourable to overcome any steric repulsions, the entropic penalty of freezing translational and rotational degrees of freedom and the free energy of desolvation. It is typically desirable to aim for a binding constant in the region of 10^9 (in mole terms), though many drugs are efficacious while binding less strongly, and this requires that the free energy of transfer from solution to bound complex, ΔG, should be about -12.5 kcal mol^{-1}.

The cost of freezing the translational and overall rotational freedom of a small molecule or ligand has been estimated by Andrews *et al.* (1984) from the work of work of Page (1977) at about $+14$ kcal mol^{-1} at 300 K. A C—C single bond between sp^3 hybridized carbons can be considered to possess three degrees of freedom when freely rotating but only one when constrained, which results in an entropy penalty of $\log_n 3$ e.u. and thus a free energy of about $+0.6$ kcal mol^{-1} when binding at 300 K. The overall situation in design terms can therefore be represented as below

$$\Delta G_{\mathrm{NC}} + \Delta G_{\mathrm{DS}} + \Delta G_{\mathrm{DF}} < -26.5 \text{ kcal mol}^{-1} \quad (1)$$

where NC refers to non-covalent interactions, DS refers to desolvation and DF refers to internal degrees of freedom.

Note that contributions to desolvation and restricting internal degrees of freedom are made by both ligand and protein, the desolvation energy of the former being given approximately by RT $\log_n P$ where P is the octanol/water partition coefficient which can be calculated by the method of Hansch and Leo (1979).

The non-covalent interactions can be divided into four types; hydrogen bonds, ionic bonds, forces between dipoles and the interplay of induced dipoles usually and loosely termed van der Waals forces. The energies involved are rather variable and therefore not easy to estimate. However, before discussing each type in turn we will describe an empirical study by Andrews (Andrews *et al.*, 1984; Andrews, 1986) which gives a rough guide to the free energies of some of these interactions.

Essentially Andrews *et al.* utilized a database of 200 drug molecules and ligands with their known or derived binding constants and performed multiple linear regression to calculate the coefficients of binding of ten common functional groups. Desolvation energies were not separately considered and the regression equation was simply

$$\Delta G_{\mathrm{B}} = \sum (n_{\mathrm{x}} \Delta G_{\mathrm{x}} + n_{\mathrm{DF}} \Delta G_{\mathrm{DF}}) + 14 \text{ kcal mol}^{-1} \quad (2)$$

where ΔG_B is the overall free energy of binding as calculated from the binding constant; ΔG_x is the contribution to the free energy of binding of chemical group x; n_x is the number of such groups in the molecule; ΔG_{DF} is as defined for equation (1); n_{DF} is number of freely rotating bonds in the molecule; and the figure of 14 kcal mol^{-1} refers to the freezing of translational and overall rotational freedom of the ligand as already described. The summation is over all group types.

In summary, Andrews *et al.*, concluded that ionic bonds, H-bonds and van der Waals bonds contribute in the region of (with desolvation taken into account) −10, −3 and −1 kcal mol^{-1} respectively to the binding free energy while each internal degree of freedom contributes +0.7 kcal mole^{-1}, a value close to the rough calculation for a C—C bond given above. It should be stressed that these are average values for this particular set of 200 ligands; some drugs evidently fit 'better' into their receptors than others, which for instance might not form all the H-bonds for which they have functional groups. In particular cases, however, the contribution from individual functional groups may be very much greater than expected, as shown by Wolfenden and Kati (1991) for hydroxyl groups in transition state analogues. It should also be noted that the ΔG_x can be expected to encompass a variety of terms, not only for instance enthalpic contributions from the formation of H-bonds but also entropic contributions where water molecules gain translational freedom by being displaced from protein surface.

2.3.1 HYDROGEN BONDS

H-bonds are formed between hydrogens attached to an electronegative donor atom (e.g. O, N) and an electronegative or basic acceptor (e.g. carbonyl oxygen). Either donor or acceptor can be ionized, but if both are, the bond could be considered a special case of ionic bonding. Both donor and acceptor can be involved in more than one H-bond at a time though unshared H-bonds are most common. Charged groups seem to be better at donating or accepting multiple H-bonds. With regard to proteins the most numerous donor and acceptor groups are of course N—H and O=C respectively, but several of the amino acid side chains possess H-bonding groups, for instance $—NH_3^+$ (from Lys), $—CO_2^-$ (from Asp, Glu) and —OH (from Ser, Thr, Tyr). Water molecules attached to such groups are often available for further H-bonding with small molecules. A review of H-bonds in proteins has been given by Baker and Hubbard (1984).

The geometries of H-bonds are quite variable as shown in a survey of data on organic crystals by Taylor and Kennard (1984). Completely linear bonds, i.e. D—H ... A, $\phi = 180$, appear to be quite rare though the authors conclude that there is an energetic preference for linearity, the mean value of the departure from linear of about 20° being a consequence of spatial factors. However, the acceptor angles show considerable deviation from the head-on configuration, for instance the distribution of in-plane N(H)...O=C angles show a maximum with the donor approaching along the direction of the carbonyl sp^2 lone pairs, while examples can be found of the out-of-plane angle being almost 90°, though the distribution of these angles peaks at 0°. This geometrical flexibility is no doubt as useful to nature as it is to the designer.

The lengths of H-bonds also show some variation with O(H)...O and O(H)...N at about 2.7 Å and N(H)...O and N(H)...N at nearer 3 Å. The more electronegative the acceptor and the more electropositive the hydrogen the shorter and stronger the bond in every case. It is very uncommon for an H-bond to be less than 2.5 Å while it is usually considered that 3.5 Å is the limit beyond which the energy of interaction is so small that no bond exists.

A recent and comprehensive study by Abraham *et al.* (1989) has examined the H-bonding ability of various chemical moieties in 1,1,1-trichloroethane (TCE) solution. Both simple and more complex groups have been studied and the work provides a useful fund of ideas to the modeller. Amide carbonyls are found to be potent H-bond acceptors while simple ketones, ester carbonyls and aliphatic ethers are less strong. Amide N—H is a weaker donor than aliphatic O—H and much weaker than phenolic hydroxyl, however N—H (urea) is as strong as a phenol. Aromatic ether oxygens have a low capacity for H-bonding but should not be ignored, for placing one next to an H-bond donor on the macromolecule is certainly better than using a methylene if there are no other alternatives.

H-bond strengths are very much dependent on the polarity of the solvent and the use of 1,1,1-TCE is appropriate as it probably mirrors the situation found in protein interiors rather more closely than do the very low polarities of other chlorinated solvents that have been used in previous studies. Because the electrostatic component of H-bonding seems to be dominant for the type of H-bonds most relevant to drug design, the energy of interaction decreases with rising solvent polarity. Thus the modeller should expect that H-bonds, particularly those that are charge assisted, will be much weaker than 'normal' if they are formed away from the hydrophobic interiors of proteins and surrounded by solvent water.

2.3.2 IONIC BONDS

Many examples of ionic bonds occur in drug–protein complexes. Examples of ion pairs include drug carboxylate or phosphate groups bound to ionized arginine or lysine side chains of the protein, ionized drug nitrogen atoms binding to protein carboxylates, and metal ions coordinated to protein side chains or bridging between protein and ligand.

2.3.3 FORCES BETWEEN DIPOLES

Permanent dipoles can interact with other permanent dipoles or with point charges. The most common dipoles in proteins are those formed by the amide linkage in the main chain, and where the configuration is helical these line up to reinforce each other, so generating a large overall dipole moment. While the presence at either end of certain helices of ionized amino acid side chains can reduce the dipole moment, this is probably not often the case so that serious consideration must be given to the possibility of significant interactions where the binding site is near an α-helix and the ligand has a large dipole or charged atom. Dipoles certainly do play a role in drug binding as, for instance, Young *et al.* (1985) have shown by a QSAR study of the antagonist activity of cimetidine analogues towards the histamine H_2 receptor.

2.3.4 VAN DER WAALS FORCES

These forces arise from the interplay of dipoles induced in the electron clouds of non-bonding atoms when in close proximity. They are the weakest but most ubiquitous of interactions between molecules and at long distances are attractive with maximum binding when the atoms are separated by the sum of their so-called van der Waals radii. On closer approach a repulsive element due to the overlap of non- or anti-bonding electron clouds rapidly takes over so that forcing two hydrogens only 0.5 Å closer than optimum requires several kcal mol^{-1} of energy. Set in the context of the binding energies expected from other interactions, this factor places a premium on the good fitting of the ligand, and highlights the general problem of uncertainties in the X-ray crystal structure that has been discussed earlier. Minimizing this drawback, however, is the likelihood that all proteins are flexible to a certain degree and the designer must therefore use his judgement in situations involving overtight fits.

The work of Andrews (Andrews *et al.*, 1984; Andrews, 1986) gives a good idea of how much binding energy can normally be obtained from the more common molecular components such as C, O etc.; however, the 'normal' case probably relates to groups situated well within the fairly hydrophobic clefts in proteins where X-ray crystallographic studies find most drug molecules binding. Where a ligand binds to the surface of a protein the opportunity for attractive van der Waals interactions with the macromolecule will be far less. By contrast, where a ligand's shape is highly complementary to a cleft so that it fits tightly, then significant van der Waals interaction can be made and this may be contributory to the high binding energies shown by some molecules which have few or only single H-bonding groups such as the benzodiazepines.

2.4 QUANTITATIVE ESTIMATIONS OF BINDING ENERGIES

It would obviously be desirable to have a more pertinent method of estimating the binding energy of a ligand other than the rather general values quoted in the previous section: it would be very helpful to be able to calculate the energy of interaction of the ligand with the macromolecule as modelled. At its simplest this involves treating all molecules as rigid bodies, at its most complex all molecules are treated as flexible, their atoms able to translate under certain constraints, and the most complete simulation that can be practically envisaged would also include solvent molecules.

With ever widening access to modern, more powerful computers, it is at least possible to attempt the more complex calculations. However, it is doubtful if the results are sufficiently reliable, given the approximations that are necessarily involved in calculating the various interacting and competing forces. In a macromolecule/ligand complex, processes elsewhere in the system may mask the presence of the ligand, the effect of cumulative errors throughout the large structure swamping the relatively small component of the energy that is due to ligand binding. Restricting the simulation to the binding site itself, however, carries the dangers that conformational changes in the protein induced by the presence of the ligand cannot properly be allowed for. One way around those problems is, of course, not to try to calculate the absolute energy changes, but to examine the differences when simulating a number of similar ligands, thus cancelling out the effects of large errors which, it can be assumed, will remain constant if changing the ligand results in only a small perturbation of the overall system.

In this section we briefly describe and review methods and examples of such energy calculations for simulating the binding process. Essentially the methodology can be grouped under the headings of molecular mechanics and molecular dynamics; programmes based on the former are the most widely used and available while the latter technique is of more recent origin requiring rather greater computer resources.

2.4.1 MOLECULAR MECHANICS

The outcome of using this methodology is a figure for the potential energy of the whole assemblage with regard to the atomic positions as defined by the modeller and the covalent and non-covalent forces between them.

For very small displacements about equilibrium bond distances and angles, the force field of covalent bonds can be considered as harmonic, thus the potential energy is given by

$$E = mK(x - x_0)^2 \tag{3}$$

where K is the harmonic force constant, m the reduced mass of the atoms involved and x and x_0 the defined value and the equilibrium value respectively of the bond coordinates. This approximation has been extensively employed to allow the calculation of the normal modes of vibration of single molecules in order to simulate infra-red and Raman spectra under the name of normal coordinate analysis, but in the situation of ligand binding models, displacements from equilibrium distances may be large enough to require more realistic descriptions of bond force fields such as the Morse potential. Then for the case of changes in bond length

$$E = D(1 - \exp(-mK(x - x_0)))^2 \tag{4}$$

where m, K and x, x_0 (now referring to bond distances) are as before and D is the depth of the potential well. It may also be necessary to utilize cross-terms to allow for changes in the force constant K induced by adjacent bond stretching or angle bending.

Van der Waals forces are typically described by a version of the well known Lennard-Jones (1924) potential giving

$$E = c\left[\left(\frac{r_0}{r}\right)^{12} - 2\left(\frac{r_0}{r}\right)^{6}\right] \tag{5}$$

where c is a constant that contains force terms, r is the separation of the two non-bonded atoms as placed by the modeller and r_0 is the separation of lowest potential energy (i.e. the sum of the van der Waals radii of the atoms). An alternative potential equation with a softer repulsive component to take account of the deformability of C—H groups has been given by Buckingham and Utting (1970).

Similar types of potential function are used to model hydrogen bonding even though there is an electrostatic component to the interaction, while ionic bonds can be described as the attraction between point charges, the energy being given by

$$E = \frac{q_i q_j}{D r_{ij}} \tag{6}$$

where q_i and q_j are partial electronic charges on atoms i and j and r_{ij} is the distance between them. Thus purely electrostatic interactions can be seen to operate over a larger range than either H-bonds or van der Waals forces. The values of the partial charges for any atoms in a given molecule can be estimated by semi-empirical methods such as that by Del Re (1958) or Gasteiger and Marsili (1980). D in the above equation is the dielectric constant of the medium in which the interaction takes place and it also appears as the denominator in the equation for forces between dipoles and between dipoles and ions. This parameter can present a problem; the forces it is involved in mediating are large it should be set to as this changes by more than an order of magnitude going from water to the interior of a protein.

A well known programme suite including molecular mechanics capability is SYBYL (Tripos Associates), and like other sophisticated packages it contains options allowing the user, for example, to choose different methods of assigning atom partial charges. Such programmes also provide facilities for construction of molecules either atom by atom or from fragment libraries and such constructs must often be modified by the programme to approach the minimum potential energy configuration as a check that they have not been built to dimensions (in particular valence and torsional angles of rings) that are inappropriate to the force fields.

We have already made some general remarks about the problems of energy calculations on complete ligand/macromolecule assemblages that are associated with the specification of the physical system, but even were these to be surmounted there are still severe computational difficulties to be overcome. These arise because of the necessity, if a binding energy is to be calculated, of finding the potential energy minima of the system with and without interaction between ligand and protein. Not only is there as yet no infallible method of minimization available but it is not generally possible to distinguish between the global and local energy minima. In addition, molecular mechanics cannot calculate the entropy change during binding. At best, therefore, molecular mechanics is sensibly restricted to providing comparisons between different ligands and their local interactions, i.e. relative and partial binding energies.

While the methodology would thus seem to have a rather narrow field of applicability it can nevertheless provide useful information. Most ligand–macromolecular modelling packages provide simple energy calculation modules that can give warning, for instance, on unduly tight fits (the repulsive parts of the Lennard-Jones or Buckingham potentials).

It is possible to go further than this; Goodford (1985) has developed a software package (GRID) that calculates the interaction energy of a number of potential binding groups (e.g. carbonyl, amide nitrogen, etc.) throughout the binding site and displays these on a three-dimensional grid, thereby providing the modeller with an indication of what constituents should be built into a ligand and their approximate location. Hydrogen-bond potentials and associated energy contours have been described by Boobbyer *et al.* (1989). A different approach has been developed by Kuntz *et al.* (1982); here only van der Waals and steric interactions are considered and a rigid ligand is moved around the binding site until the optimum position and orientation is found. A development of this method by Des Jarlais *et al.* (1990), wherein the binding site is scanned to give a 'negative image' of itself that can then be used to screen a database of compounds for complementarity of shape, has successfully identified a novel inhibitor of the HIV-1 protease enzyme.

2.4.2 MOLECULAR DYNAMICS

This technique uses the same force fields as molecular mechanics to obtain the instantaneous forces on each atom; then by successive integration from acceleration through velocity to distance travelled, attempts to simulate the trajectories of each part of the molecular assemblage.

The expectation embodied in the approach is that by mimicking the 'natural' dynamic nature of molecules they will 'settle down' eventually into a series of states near to the equilibrium, i.e. the minimum energy configuration, so that realistic energies can be obtained. The dynamics process thus replaces the artificial energy minimizations carried out in simple molecular mechanics and if it is run at a suitable 'temperature' the molecular assemblage should not get trapped in local energy minima. An additional benefit is that, in principle, diffusion can be simulated, i.e. instead of the modeller manually docking the ligand into the binding site it can be allowed to move there of its own accord. This involves putting together protein, ligand and a number of solvent molecules into a 'box', so that solvent cannot evaporate away, then allowing trajectories to develop while monitoring the overall energy of the system. The process proceeds in a number of steps, a set of new trajectories being calculated for each atom at each step, this requiring very short simulated time intervals, generally fractions of picoseconds.

It should be pointed out that simulation of molecular dynamics is still very much in its infancy. It has limitations; no one has yet taken a linear protein chain and folded it to produce secondary and tertiary structure, a task which in principle the technique should be able to accomplish. So far its applications relevant to modelling have been in comparing the binding of similar ligands to enzymes using so-called 'free energy perturbation methods', but because it has been possible to simulate the whole of the binding process some interesting conclusions have been drawn as to the sources of binding energy differences. Thus Wong and McCammon (1986) have found that the selectivity of some benzamidine analogues in binding to trypsin is a consequence of desolvation effects while Fleischman and Brooks (1990) have come to a similar conclusion with regard to trimethoprim and analogues binding to dihydrofolate reductase. These results are a timely warning to modellers not to concentrate exclusively on the direct interactions between ligand and binding site. These studies are somewhat involved, but molecular dynamics can be of some use without resorting to complex detail. For instance, a simple use of the technique is merely to observe the behaviour of a modelled compound in the active site (usually held rigid and without solvent except for crystallographically observed waters) to see whether or not it can exhibit an alternative and otherwise overlooked mode of binding. Novel molecules that have been found to be biologically active from random screening can also be handled to get some idea of how and where they are capable of interacting.

2.5 VISUALIZATION METHODS

By far the most convenient method of displaying molecular models is to use computer graphics. The last decade has seen a great expansion in the use of molecular graphics hardware and software systems both in industry and universities and the field has been extensively reviewed, for example by Cohen *et al.* (1990).

It is a relatively simple matter for computer software to deal with atom position data if it is provided in terms of *x*, *y*, *z* coordinates and access to the results of a wide range of crystallographic studies on macromolecules and their ligands, and some NMR studies too, is afforded by the ready availability on subscription of such data from the Brookhaven Protein Data Bank (Bernstein *et al.*, 1977; Abola *et al.*, 1987).

A range of interactive computer graphic display tools is required in order for a modeller to utilize the data. Essentially it must be possible to view the overall picture from any angle and scale by means of *x*, *y* and *z* rotation and zoom facilities and to be able to move and rotate the putative ligand independently in any direction and about any axis. In addition, it must be possible to manipulate the ligand's torsion angles so that any desired conformation can be tested. And of course some type of 'construction kit' must be available so that the putative ligands can be built and modified. A facility to label atoms on the computer screen and measure distances between atoms is also needed.

A computer screen can only give a two-dimensional projection of the macromolecule. A crude method of adding extra dimensionality is for the display software to provide depth cueing whereby those parts of the 'picture' that are far back are reduced in intensity on the screen but a far better way is to generate stereo pairs, that is two slightly different views that correspond to left and right eye perspectives. These are displayed in regular rapid succession. The left and right eye views have, of course, to be shown only to the appropriate eye. In practice this involves synchronizing the graphics screen with an interposing physical device. Langridge and Klein (1990) described a cylindrical device driven by a stepping motor in which appropriate placed slits allow alternatively one eye then the other to view the screen. When the cylinder is in phase with the alternating screen, a true stereo image is seen. In the authors' own laboratory and elsewhere use has been made of special spectacles consisting of two liquid crystals each of which is alternately opaque and translucent under electrical control in phase with the alternating screen. Most recently hardware manufacturers have started to supply large liquid crystal screens which fit over a graphics terminal and are synchronized with the image to transmit alternatively left or right circularly polarized light. The experimenter wears spectacles with circularly polarizing lenses to observe the picture in stereo. The advantage here over the previously described methods is that the experimenter has neither to sit in one place nor to wear spectacles wired to a control box.

There are other aids to modelling or docking ligands into the active site of macromolecules and one or two 'automatic' ones have been briefly described in the previous section. A most important facility is the capacity to visualize the van der Waals surfaces of atoms and groups, i.e. a representation of the extent of the electron density around an atom so that the modeller can get a rapid impression of how the ligand fits into the active site and what steric problems might arise. Much graphics technology has been devoted to representing molecules as physical CPK-type models; this renders the rear-most parts of the molecule invisible from the front. The van der Waals surface is more conveniently realized by the display of the surface as a fairly open three-dimensional contour map so that atoms behind those at the front of the picture can still be seen. It is very helpful if the surfaces are colour-coded, e.g. red for an H-bond acceptor atom (e.g. carbonylic oxygen), blue for an H-bond donor (e.g. an N—H group), green for a 'hydrophobic' group (e.g. methyl), and purple for a polar group (e.g. a hydroxyl or side chain amide group) whose orientation is not known owing to experimental uncertainties of the hydrogen positions. Note, however, that these surfaces will only be an approximation to the real state of affairs because among many of the uncertainties so far described the positions of hydrogens on a protein often cannot be determined. A further problem with the van der Waals surface algorithm is that it draws regions of surface which are in fact inaccessible even to the smallest ligand structure.

The surface of solvent accessibility algorithm (Richards, 1977; Connolly, 1983) has been developed to overcome this disadvantage. The algorithm allows for a water molecule of set van der Waals radius to roll over the molecular surface represented by the envelope of the van der Waals surfaces of all the atoms in the molecule. A dot is drawn at regularly placed contact points. At positions of the probe where there is more than one point of contact with the molecule a dot is placed instead on the inner surface of the probe molecule between those contact points. The resulting dot surface has a smoothly varying shape with no inner recesses or kinks into which a solvent molecule could not fit.

Despite the depiction of these surfaces as transparent, drawn with dots or chicken wire, their presence can unduly complicate the modeller's view. Figure 2.1(a) (see colour section) shows a view of the HIV-1 protease structure solved by Miller *et al.* (1989) coordinates being obtained from the Brookhaven Data Bank (Abola *et al.*, 1987). The water molecule depicted is situated in the active site region; it donates a hydrogen bond to a carbonyl oxygen of the hexapeptide inhibitor MVT101 and accepts a hydrogen bond from the peptide NH of the Ile 50. This view has incorporated depth-cueing, front and back clipping planes and colour-coded van der Waals surfaces in an effort to depict one feature of a multi-faceted intermolecular interaction. It has had to be carefully tailored.

A very much more useful surface representation has been introduced by C. D. Barry (unpublished). This is the 'double' or 'added' radius surface whereby the contours around the atoms and groups of the macromolecule serve to represent both macromolecule and ligand. This allows the ligand to be depicted purely as a stick model and very considerably simplifies the display. To obtain the optimum van der Waals interaction merely requires a ligand atom (the end-point of a vector) to be positioned at or close to the surface. Penetration of the surface indicates the probability of steric repulsions unless the atom is donating or accepting an H-bond from the macromolecule.

Figures 2.1(b) and 2.1(c) (see colour section) illustrate the ability of such a surface to display underlying molecular properties. The colour-code here and throughout is as indicated above, and the software modified by one of us (J.N.C.) to display the contour surfaces is FRODO (Jones, 1978). The same protein is depicted in each case, HIV1-protease, and the same region of structure, residues 25–29 abutting the active site. Figure 2.1(b) shows the protein surface adjacent to where a hexapeptide inhibitor (MVT101) binds; Figure 2.1(c) shows the same surface in the absence of inhibitor. The two surfaces illustrate quite major differences in the property profile of the macromolecule attendant upon quite a small rearrangement in this particular region of the structure, notably the carbonyl group of residue 27. This is not to be confused with changes of a grosser nature that occur elsewhere upon binding of ligand, notably in the flap region.

The double radius surface is calculated by adding on a distance equal to a methyl radius to an atomic van der Waals radius. This is obviously a compromise so that constituents of the ligand that possess larger radii, for example sulphur, must be kept from actually touching the surface. It is obviously not an appropriate treatment in cases where it is important to take accurate hydrogen positions into account but the technique has great value so far as ligand/macromolecule interactions are concerned, as hydrogen positions are normally missing from the experimentally determined structural data. Moreover, the solvent inaccessibility problem is solved as the added radius envelope requires the interacting ligand molecule to be represented merely as a skeleton.

2.6 EXAMPLES

The remaining part of this chapter will be devoted to discussing some examples that illustrate the general approach to drug design set out in the previous sections. They will not be considered at length as at least some of the examples are considered in greater detail elsewhere in this book (for instance see Chapters 3 and 4).

2.6.1 ACE AND RENIN

One of the early examples in which the importance of macromolecule structural information was recognized is angiotensin converting enzyme (ACE). This enzyme has a role in blood pressure regulation through its action in cleaving a dipeptide from the C-terminus of the substrate angiotensin I. Inhibition of the enzyme results in relief of hypertension. Cushman *et al.* (1977) recognized similarities in the properties of ACE to those of pancreatic carboxypeptidase A whose structure is known. By postulating certain essential functional groups common to both enzymes and by reference to known inhibitors of carboxypeptidase, Cushman *et al.* synthesized and studied the structure–activity relationship of a range of carboxyalkanoyl and mercaptoalkanoyl derivatives of proline. From this work captopril (**1**)

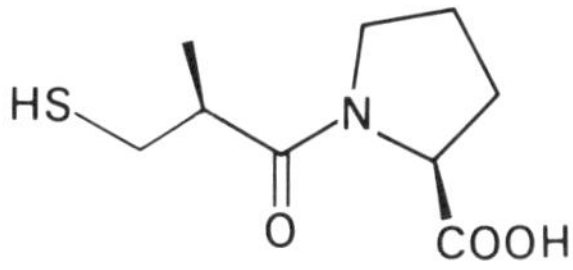

(**1**)

emerged as a potent ACE-inhibitor exemplifying, in the words of Cushman *et al.*, 'the great heuristic value of an active-site model in the design of inhibitors, even when such a model is a hypothetical one'.

Little use has been made to date of the wealth of structure homology information inherent in the overlap of the three-dimensional structure information of the Protein Data Bank (Abola *et al.*, 1987) and the knowledge of sequences of thousands of proteins many of which will have biochemical properties akin to solved three-dimensional structures. Interest has centred on the enzyme renin which cleaves angiotensin I from angiotensinogen (see for example Arrowsmith *et al.*, 1989). The homology of renin with a fungal aspartic protease endothiapepsin whose structure is known (Sibanda *et al.*, 1984) was used to model a hypothetical renin active-site structure. The model served as an ideas generator in the context of certain amino alcohols, a class of aspartyl protease inhibitors which have similarities to the relevant sequence of the renin substrate angiotensinogen, but contain a non-hydrolysable bond at the point where substrate would be cleaved. In contrast to other classes of isostere, amino alcohols proved to be potent inhibitors, even as relatively small molecules, mimicking, for example, a tetrapeptide sequence. Arrowsmith *et al.* (1989) reported a proposed mode of binding for one such compound, BW625C (**2**)

$$\text{H.Leu.}AA\text{.Val.Ile.Phe.OCH}_3$$

(**2**)

in which *AA* represents —CH(OH).CH_2.NH— (*R*), replacing the normal peptide linkage. According to their model, the four side chain moieties interacted with four identifiable subsites (S1, S1′, S2′, S3′) of the renin enzyme. Poor occupancy of S1 by the isobutyl moiety of the leucine in the postulated interaction led to proposal and synthesis of other amino alcohols. One of these, in which the isobutyl was replaced by a phenylpropyl group, showed at least an order of magnitude improvement in binding. This compound was subsequently further improved upon by continuing modelling and QSAR studies. An indirect means of testing the binding hypothesis for BW625C was afforded when its *S* stereoisomer, BW624C, was observed by X-ray crystallography in complex with endothiapepsin (Cooper *et al.*, 1988). The side chain occupancy for the subsite S1 was indeed as postulated; however, certain other interactions of BW624C with endothiapepsin, notably the N-terminus and the hydroxyl group, were different from those proposed for BW625C with renin.

2.6.2 HAEMOGLOBIN

This was the first macromolecule to yield to a successful ligand design exercise. The work was carried out without the help of computer graphics by utilizing models designed by Beddell *et al.* (1973) and built in physical form by Norrington (1974) from coordinates supplied by Perutz *et al.* (1968) of an X-ray crystallographic study of haemoglobin.

The ligands were designed to mimic the effect of the diphosphoglycerate (DPG) anion through a shift to the right in the overall oxygen saturation curve of haemoglobin, a decrease in the affinity of haemoglobin for oxygen. DPG binds to, and stabilizes, the deoxy form of haemoglobin; if it is absent the protein can undergo a long range conformational rearrangement that enables it to bind oxygen at another site. Beddell *et al.* (1976) were able to fit models of a series of bibenzyl dialdehydes to the DPG binding site that made a number of interactions which were considered sufficient to give a significant degree of binding. Schematic representations of the known mode of binding of DPG as shown by Arnone (1972) and the postulated mode of binding of three compounds is shown in Figure 3.3, (see Chapter 3) where a more detailed description of the work can be found. When synthesized, these compounds were indeed found (Beddell *et al.*, 1976) to have the desired effect, though only compound **3** of Chapter 3 proved to be as potent as DPG.

DPG is known (Beddell *et al.*, 1979) to have a free energy of binding of about $-7\ \text{kcal mol}^{-1}$. It is interesting to note that this is several kilocalories less strongly bound than might be expected on the basis of Andrews' (Andrews *et al.*, 1984; Andrews, 1986) work suggesting that it fits less well than 'average' in terms of Andrews's database of drug molecules.

Synthetic compounds **1**–**3** of Chapter 3 were each designed to make two 'reversible' covalent bonds with valine residues of human haemoglobin. Beddell *et al.* (1979) were able to estimate the contribution to the binding free energy from these bonds by comparing the binding of DPG and compounds **1** and **3** in Chapter 3 to a number of mutant and animal haemoglobins chosen so that one or more key residues in the active site were different, thereby modifying in each individual case the pattern of H-bonding or covalent interactions. Multiple linear regression analysis using a three term equation (H-bond/ionic, covalent, constant) then allowed them to conclude that each H-bond/ionic interaction contributed -0.75 kcal mol^{-1} and each 'reversible' covalent bond about -1.6 kcal mol^{-1}. Some 86% of the variation in binding constant is explained by the regression analysis, suggesting that there is little contribution from the aromatic rings of compounds **1** and **3**, which is understandable given the hydrophilic environment of the site and the positioning of the ligands which do not allow the promotion of attractive van der Waals forces. The assumption made here, of course, is that each interaction takes place in the same way for each ligand and macromolecule.

The magnitudes of these values seem small, particularly the contribution to binding made by the H-bonds, and therefore some other source of binding is indicated. The regression analysis delivers a constant of -2 kcal mol^{-1} and if 14 kcal mol^{-1} is subtracted to allow for the loss of the ligand's translational and rotational entropy, together with 1–2 kcal for the freezing of internal conformational freedom (see Section 2.3 above), then at least some -16 kcal mol^{-1} of binding free energy must be accounted for. A possible explanation is that a large part of this derives from the hydrogen/ionic bond between lysine NH_3^+ and the ligand carboxylate moiety which can be formed between all the ligands and all the haemoglobins in this study and which on the basis of Andrews' (Andrews *et al.*, 1984; Andrews, 1986) work might be expected to generate in the region of -10 kcal mol^{-1}. Against the background of one strong interaction a number of weaker H-bonds are presumably either the result of non-perfect positioning of the ligands or presence of solvent.

Compounds **1**–**3** (Chapter 3) were not intended to be used as drugs but rather to validate the technique of 3-D modelling. Further work by Beddell *et al.* (1984) has produced ligands that bind to and stabilize the oxy conformation of human haemoglobin, thereby shifting the overall oxygen saturation curve to the left. One of these compounds (BW12C) has been considered as a potential anti-sickle-cell disease agent and Figure 3.7(b) (see Chapter 3) is a schematic representation of this molecule as modelled binding to the α-chain terminal amino region of the macromolecule. The main features to be noted are the formation of a Schiff's base by the free amino group of the protein, while the carboxylate moiety of BW12C forms a strong charge-assisted H-bond with the corresponding ionized protein amino group of this symmetrical binding site. The alkyl chain connecting the two functional groups of BW12C together is designed to produce an attractive van der Waals interaction with hydrophobic areas of the protein formed by proline side chains.

2.6.3 DIHYDROFOLATE REDUCTASE

Probably more is known about the three-dimensional structure and modes of drug and cofactor binding to dihydrofolate reductase than to any other protein.

The most important inhibitor of the bacterial enzyme is the drug trimethoprim (TMP) (**3**).

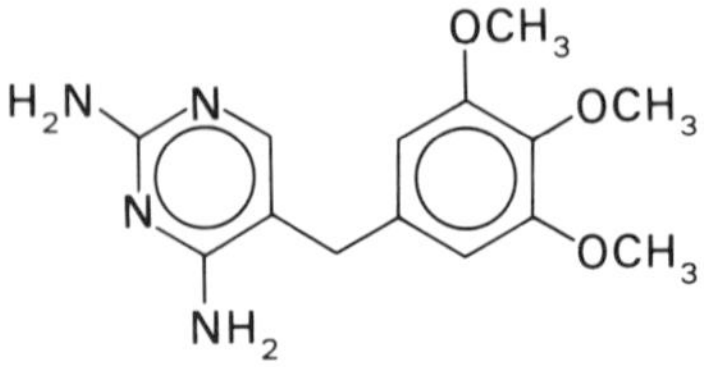

(**3**)

It was not a product of logical design, but was discovered by screening; nevertheless, considerable effort has been made by many groups of workers to improve upon it. Figure 2.2(a) (colour section) shows the molecule in the active site as determined by an X-ray crystallographic study on a ternary complex of the enzyme with drug and NADPH bound (Champness *et al.*, 1986b). The figure illustrates a skeletal model of the ligand and enzyme. The pyrimidine ring of the drug makes a number of H-bond interactions with the protein, particularly with Asp 27. Of particular interest, however, is the trimethoxybenzyl ring which is indicated from binding constant comparisons with pyrimidine alone to contribute several kcal per mole of binding free energy. This is not unexpected as its position within a hydrophobic cleft of the protein enables many attractive van der Waals interactions to be made.

Modifications to the benzyl ring by replacing one or more of the trimethoxy groups by larger hydrophobic groups have been found by Roth *et al.* (1987) to give little improvement in binding energy and in some cases can actually decrease it. This is generally consistent with the model as shown in Figure 2.2(a), as such substituents should either encounter steric blocking or protrude well into solvent.

Kuyper and co-workers (1985), however, decided to employ a different tactic: on the basis of modelling that utilized the shallow extension of the binding cleft, a channel running from the *meta*-methoxy group towards Arg 57, they designed and synthesized TMP derivatives in which one of the *meta*-methoxy groups is replaced by a group of the form $O(CH_2)_nCOO^-$ in order to make a charge-assisted H-bond with the side chain of Arg 57. As indicated by X-ray crystallographic studies of compound and enzyme in binary complex (lacking NADPH), the proposed interaction does indeed take place. Difference Fourier maps (Champness *et al.*, 1986a) clearly showed the electron density due to the side-chains ($n = 2$ and 5 respectively) of two of the compounds investigated, and were sensitive enough also to indicate the slight relocation of the benzyl group of TMP that resulted. The difference map for the $n = 5$ compound versus TMP is reproduced in Figure 2.2(b), and Figure 2.2(c) (see colour section) shows the crystallography-derived coordinates of the compound and certain active-site residues set in the context of the double-radius surface, which highlights the properties of the surrounding active site. The benzyl group of the inhibitor is shown sandwiched between isoleucyl and leucyl side chains. The channel is largely lipophilic in character except for where the Arg 57 side chain is located, characterized on the surface by a small blue patch. The increase in binding energy achieved, about $-1.7\ \text{kcal mol}^{-1}$, is not large but the probable reasons are evident from the model. First, the carboxylate–arginine H-bond is sufficiently near the surface to be influenced by solvent access, and thus the high dielectric constant of water can be expected to reduce the interaction strengths well below the $-10\ \text{kcal mol}^{-1}$ that might be achieved if it was elsewhere. Secondly, the entropic penalty of freezing the conformational freedom of the long alkoxy chain will cancel out most or all of the attractive van der Waals interactions. Nevertheless, a significant improvement has been obtained.

2.6.4 HAEMAGGLUTININ

Multiple copies of the protein haemagglutinin are embedded in the lipid membrane of the influenza virus particle and are implicated in the cell-infection process through adhesion of a particular region of the protein to sialic-acid containing regions in the cell wall. In recent years the X-ray structure of haemagglutinin complexed with sialyllactose containing *N*-acetylneuraminic acid (NANA)

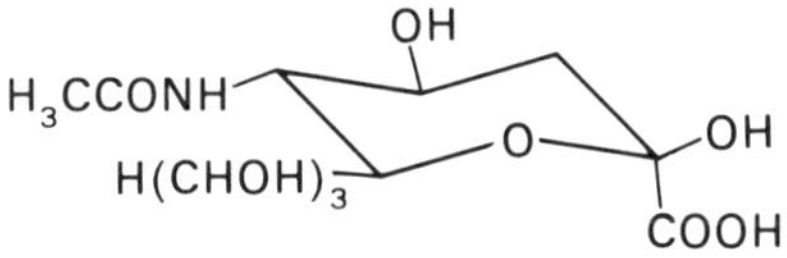

(4)

has become available (Weis *et al.*, 1988) and use has been made of the information to propose molecules that might interact with the relevant region of the protein, thus inhibiting haemagglutination and hence infection (Schrell and Whitesides, 1990: S. R. Challand, C. R. Beddell and J. N. Champness, unpublished). Figure 2.3 (colour section) shows a double radius surface which has been computed from the protein structure crystallographically determined by Weis *et al.* The surface indicates the properties of the underlying macromolecular structure. The NANA moiety is reported by Weis *et al.* to interact with a shallow binding region dominated by Glu 190 at the back (in this view), Trp 153 on the right, and main chain 135–137 at the front, with the *N*-acetyl amino group donating a hydrogen bond to 135 carbonyl and the glycerol making contact with Glu 190. The interaction is fairly weak and it seems probable that it is potentiated via the contact made by multiple copies of haemagglutinin on the virus with several sialic acid residues on the cell wall. A mapping of the sialic acid binding site has been carried out by Wade (1990) to investigate the most favourable locations of particular substructural probes, for example methyl, amide or hydroxyl groups, using the program GRID (Goodford, 1985; Boobbyer *et al.*, 1989) to compute sites of minimum interaction energy.

Challand and Beddell (unpublished) proposed a series of novel compounds to make the sialic acid interactions indicated above. An initial suggestion, a 1,6-substituted naphthalene

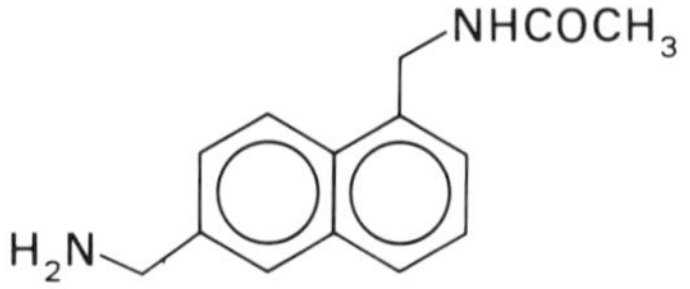

(**5**)

was postulated to bind with its methylamine group interacting with Glu 190 and the NH of the amide group interacting with the carbonyl oxygen of 135 (Figure 2.3). This compound and related naphthalenes were synthesized and tested for activity indirectly via an erythrocyte agglutination assay. Interestingly, whilst a few of the target compounds showed some weak activity, certain non-target intermediates were also effective. It seemed the distance between the primary amine and the acetyl group was an important determining factor for activity. The study was not continued, but it highlighted the importance of assaying not only the carefully tailored target molecules but also related compounds having similar elements of substructure, such as synthetic intermediates. This in turn emphasizes the importance of examining molecular libraries of compounds, such as corporate databases, for compounds reasonably similar to the target structure.

2.7 AFTERVIEW

We have attempted to give a succinct description of the principles and practice of designing and fitting small molecules to binding sites of proteins. A few examples of real applications have been briefly presented and discussed.

The subject has been given some exposure in the 'popular press', a fairly recent instance being 'Computers picture the perfect drug' by Ripka (1988). Such articles have an upbeat and optimistic tone and often pay scant regard to the uncertainties and approximations inherent in the methodology that we have alluded to above. Whilst information concerning 3-D macromolecule structure has an obvious part to play in ligand design, alongside other compound design technology such as QSAR or pharmacophore mapping, oversimplistic presentation of its potential has led to unrealistic expectations. Not only has binding of a compound to a target macromolecule to be optimized, as described here, but often, to achieve species selectivity, hence therapeutic index, its potential to bind to another related macromolecule has to be minimized. In addition, the problems of bioavailability and toxicity have to be separately evaluated elsewhere.

Another reason for disappointed expectations hitherto is that many modelling exercises have waited for a crystallographic study of protein and ligand/candidate drug, the latter having been derived by other methods, mostly for historical reasons. This means that 3-D modelling has often been reduced to the minimal task of subsequently optimizing a given ligand by minor modifications which, given the uncertainties relating to the 3-D structures of macromolecules, might better have been carried out using the methods of QSAR. Many known inhibitors (ligands) of enzymes are peptides and such substances generally suffer from the drawback of having poor potential as medicines because of lack of bioavailability, metabolic degradation, etc. Thus the methodology of macromolecule fitting has tended to be used late in the day and rather than being seen as an ideas-generation procedure has been attempted as an after-the-event tuning process.

In our view far too little use has been made of the opportunity to design novel chemical structures as ligands. One might cite the examples given above, also the work on thymidylate synthase (Appelt *et al.*, 1991) and on HIV-1 protease (Erickson *et al.*, 1990) as some of the few exceptions. This approach, of course, requires some degree of imagination and often brings into focus the problem of difficult chemical synthesis, though with medicinal chemists integrated into the design process impossible synthetic problems should generally be avoidable. An essential extension to the design process is to regard a newly designed compound as a template on which a variety of chemical structures can be accommodated and then to conduct a screening exercise using, for example, the MACCS (Molecular Design Ltd) software suite to search for substructure against a large database, for example a corporate compound library. Thus the end result of a modelling exercise might be regarded to be not a specific molecule but rather a variety of related chemical structures which can then be selected on the basis of relationship to known compounds or ease of chemical synthesis.

Many recent developments in the field have concerned energy calculations or minimizations using the techniques of molecular dynamics with the possibility of completely automated docking facilities. Such developments can only represent a part of what is required for 3-D modelling to fulfil its promise, as in our experience the greatest impediment is the hurdle of imagining novel chemical structures that will both bind and be available for biological testing.

REFERENCES

Abola, E. E., Bernstein, F. C., Bryant, S. H., Koetzle, T. F. and Weng, J. (1987). In *Crystallographic Databases—Information Content, Software Systems, Scientific Applications* (Eds F. H. Allen, G. Bergerhoff and R. Sievers), Data Commission of the International Union of Crystallography, Bonn/Cambridge/Chester, pp. 107–32.

Abraham, M. H., Duce, P. P., Prior, D. V., Barratt, D. G., Morris, J. J. and Taylor, P. J. (1989). *J. Chem. Soc., Perkin Trans 2*, 1355–75.

Andrews, P. R. (1986). *Trends Pharmacol. Sci.*, **7**, 148–51.

Andrews, P. R., Craig, D. J. and Martin, J. L. (1984). *J. Med. Chem.*, **27**, 1648–57.

Appelt, K., Bacquet, R. J., Bartlett, C. A., Booth, C. L. J., Freer, S. T., Fuhry, M. A. M., Gehring, M. R., Hermann, S. M., Howland, E. F., Janson, C. A., Jones, T. R., Kan, C.-C., Kathardekar, V., Lewis, K. K., Marzoni, G. P., Matthews, D. A., Mohr, C., Moomaw, E. W., Morse, C. A., Oatley, S. J., Ogden, R. C., Reddy, M. R., Reich, S. H., Schoettlin, W. S., Smith, W. W., Varney, M. D., Villafranca, J. E., Ward, R. W., Webber, S., Webber, S. E., Welsh, K. M. and White, J. (1991). *J. Med. Chem.*, **34**, 1925–34.

Arnone, A., (1972). *Nature*, **237**, 146–49.

Arrowsmith, R. J., Harris, C. J., Davies, D. E., Morton, J. A., Dann, J. G. and Champness, J. N. (1989). In *Molecular Recognition: Chemical and Biochemical Problems* (Ed. S. M. Roberts), Royal Society of Chemistry, London, p. 112–22.

Baker, E. N. and Hubbard, R. E. (1984). *Prog. Biophys. Molec. Biol.*, **44**, 97–179.

Beddell, C. R., Goodford, P. J., Norrington, F. E. and Wilkinson, S. (1973). *Br. J. Pharmac.*, **48**, 363P–4P.

Beddell, C. R., Goodford, P. J. Norrington, F. E., Wilkinson, S. and Wootton R. (1976). *Br. J. Pharmac.*, **57**, 201–9.

Beddell, C. R., Goodford, P. J.,. Stammers, D. K. and Wootton, R. (1979). *Br. J. Pharmac.*, **65**, 535–43.

Beddell, C. R., Goodford, P. J., Kneen, G., White, R. D., Wilkinson, S. and Wootton, R. (1984). *Br. J. Pharmac.*, **82**, 397–407.

Bernstein, F. C., Koetzle, T. F., Williams, G. J. B., Meyer, E. F., Brice, M. D., Rodgers, J. R., Kennard, O., Shimanouchi, T. and Tasumi, M. (1977). *J. Mol. Biol.*, **112**, 535–42.

Blundell, T. L. and Johnson, L. N. (1976). *Protein Crystallography*, Academic Press, New York.

Boobbyer, D. N. A., Goodford, P. J., McWhinnie, P. W. and Wade, R. C. (1989). *J. Med. Chem.*, **32**, 1083–94.

Branden, C. I. and Jones, T. A. (1990). *Nature*, **343**, 687–9.

Buckingham, A. D. and Utting, B. D. (1970). *A. Rev. Phys. Chem.*, **21**, 287–316.

Champness, J. N., Kuyper, L. F. and Beddell, C. R. (1986a). In *Topics in Pharmacology*, **3** (Eds A. S. V. Burgen, G. C. K. Roberts and M. S. Tute) Elsevier Science Publishers, Amsterdam, p. 336–62.

Champness, J. N., Stammers, D. K. and Beddell, C. R. (1986b). *FEBS Lett.*, **199**, 61–7.

Cohen, N. C., Blaney, J. M., Humblet, C., Gund, P. and Barry, C. D. (1990). *J. Med. Chem.*, **33**, 883–94.

Connolly, M. L. (1983). *Science*, **221**, 709–13.

Cooper, J. B., Foundling, S. I., Blundell, T. L., Arrowsmith, R. J., Harris, C. J. and Champness, J. N. (1988). In *Topics in Medicinal Chemistry*, 4th RSC-SCI Med. Chem. Symposium (Ed. P. R. Leeming), Royal Society of Chemistry, London, p. 308–13.

Cushman, D. W., Cheung, H. S., Sabo, E. F. and Ondetti, M. A. (1977). *Biochemistry*, **16**, 5484–91.

Del Re, G. (1958). *J. Chem. Soc.*, 4031–40.

Des Jarlais, R. L., Siebel, G. L., Kuntz, J. D., Ortiz de Montellano, P., Furth, P. S., Alvarez, J. C., De Camp, D. L., Babé, L. M. and Craik, C. S. (1990). *Proc. Natl. Acad. Sci., USA*, **87**, 6644–8.

Erickson, J., Neidhart, D. J., VanDrie, J., Kempf, D. J., Wang, X. C., Norbeck, D. W., Plattner, J. J., Rittenhouse, J. W., Turon, M., Wideburg, N., Kohlbrenner, W. E., Simmer, R., Helfrich, R., Paul, D. A. and Knigge, M. (1990). *Science*, **249**, 527–33.

Fleischman, S. H. and Brooks, C. L. (1990). *Proteins Struct. Funct. Gen.*, **7**, 52–61.
Gasteiger, J. and Marsili, M. (1980). *Tetrahedron*, **36**, 3219–28.
Goodford, P. J. (1985). *J. Med. Chem.*, **28**, 849–57.
Hansch, C. and Leo, A. (1979). *Substituent Constants for Correlation Analysis in Chemistry and Biology*, Wiley and Sons, Chichester.
Jones, T. A. (1978). *J. Appl. Cryst.*, **11**, 268–72.
Kuntz, I. D., Blaney, J. M., Oatley, S. J., Langridge, R. and Ferrin, T. E. (1982). *J. Mol. Biol.*, **161**, 269–88.
Kuyper, L. F., Roth, B., Baccanari, D. P., Ferone, R., Beddell, C. R., Champness, J. N., Stammers, D. K., Dann, J. G., Norrington, F. E. A., Baker, D. J. and Goodford, P. J. (1985). *J. Med. Chem.*, **28**, 303–11.
Langridge, R. and Klein, T. E. (1990). *Comprehensive Medicinal Chemistry* (Ed. C. Hansch), Vol. 4, Pergamon Press, Oxford, p. 413–29.
Lennard-Jones, J. E. (1924). *Proc. R. Soc. Lond. Ser. A*, **106**, 463–77.
Miller, M., Schneider, J., Sathyanarayana, B. K., Toth, M. V., Marshall, G. R., Clawson, L., Selk, L., Kent, S. B. H. and Wlodawer, A. (1989). *Science*, **246**, 1149–52.
Norrington, F. E. (1974). *Lab. Pract.*, 191–2.
Page, M. I. (1977). *Angew. Chem. Int. Ed. Engl.*, **16**, 449–59.
Perutz, M. F., Muirhead, H., Cox, J. M. and Goaman, L. C. G. (1968). *Nature*, **219**, 131–9.
Richards, F. M. (1977). *A. Rev. Biophys. Bioengng.*, **6**, 151.
Ripka, W. (1988). *New Sci.*, Issue 1617 (June 16), 54–8.
Roth, B., Rauckman, B. S., Ferone, R., Baccanari, D. P., Champness, J. N. and Hyde, R. M. (1987). *J. Med. Chem.*, **30**, 348–56.
Schrell, A. and Whitesides, G. M. (1990). *Liebigs Ann. Chem.*, **1111–4**.
Sibanda, B. L., Blundell, T. L., Hobart, P. M., Fogliano, M., Bindra, J. S., Dominy, B. W. and Chirgwin, J. M. (1984). *FEBS Lett.*, **174**, 102–11.
Taylor, R. and Kennard, O. (1984). *Acc. Chem. Res.*, **17**, 320–6.
Wade, R. C. (1990). In *Use of X-ray Crystallography in the Design of Antiviral Agents* (Eds W. G. Laver and G. M. Air), Academic Press, San Diego, p. 61–74.
Weis, W., Brown, J. H., Cusack, S., Paulson, J. C., Skekel, J. J. and Wiley, D. C. (1988). *Nature*, **333**, 426–31.
Wolfenden, R. and Kati, W. M. (1991). *Acc. Chem. Res.*, **24**, 209–15.
Wong, C. F. and McCammon, J. (1986). *J. Am. Chem. Soc.*, **108**, 3830–2.
Young, R. C., Durant, G. J., Emmett, J. C., Ganellin, C. R., Graham, M. J., Mitchell, R. C., Prain, H. D. and Roantree, M. L. (1985). *J. Med. Chem.*, **29**, 44–9.

3 Compounds Designed to Bind to Haemoglobin

R. WOOTTON
Wellcome Research Laboratories, Beckenham, Kent, UK

3.1 INTRODUCTION

Perhaps the first question to be addressed is why anyone would want to design compounds to bind to haemoglobin (Hb) and the obvious answer 'because it's there' is in fact one good reason. There it certainly is, readily accessible in large amounts in a form pure enough for most purposes and without the need to sacrifice the donor. Hb is one of the most studied of all proteins and has frequently been used as the first biological testing ground for many new ideas and methods, particularly new physical techniques; this, despite the fact that it is certainly not the simplest of proteins in its detailed structure and function.

The Design of Drugs to Macromolecular Targets. Edited by C. R. Beddell
© 1992 John Wiley & Sons Ltd

There are, however, other good reasons for wanting to discover compounds which bind to Hb and affect its functional behaviour. Perhaps the most obvious of these, and one that has certainly received a great deal of attention, is the requirement for a drug to alleviate the severe debilitating symptoms of a genetically inherited disorder known as sickle cell disease (Dean and Schechter, 1978). In this disease, normal haemoglobin, HbA, is replaced by a mutant haemoglobin, HbS. HbS is very similar to HbA in all respects except that, due to the presence of a valine residue at position 6 of the β chains (see later) it is less soluble when the Hb is deoxygenated. It then precipitates inside the red cell, forming long fibrous strands which distort the normal flexible discoid shape and lead to the rigid, characteristic 'sickle cell' shape. These sickled cells are much less readily transported through fine capillaries and may form blockages leading to the painful crises associated with the disease. These crises lead to tissue damage due to lack of oxygen supply and eventually premature death.

Two potential methods of inhibiting the sickling of deoxyHbS fall within the scope of this chapter in that they involve compounds which bind directly to Hb. The structure of the HbS fibres is known in detail (Benesch *et al.*, 1982) and the best approach would be to design compounds to bind to HbS at the contact region(s) between adjacent molecules and thereby inhibit aggregation. Such compounds, direct anti-gelation agents, are discussed in Section 3.5. The other approach relies on the fact that it is only HbS in its deoxygenated state which shows a reduced solubility; oxyHbS has the same solubility as oxyHbA. It might be expected then that a compound which increases the oxygen affinity of Hb and thus maintains a greater proportion of oxyHbS in the circulation would exert an anti-sickling action. This approach is discussed in Section 3.4—left-shifters.

There are situations where physiological problems arise, of course, due to a deficiency in the amount of oxygen reaching the tissues in various disease states. Only about a quarter of the oxygen taken up as the blood passes through the lungs is released to the tissues and there is therefore considerable scope for increasing this amount. The design of compounds towards this purpose is discussed in Section 3.3—right-shifters. Section 3.2 presents a brief summary of the structure and mechanistic aspects of Hb function necessary for the discussion of compound design and action.

3.2 STRUCTURE AND FUNCTION

In order to understand the rationale behind the design of compounds to bind to Hb it is necessary to know the relevant structural details of the design sites, and how binding at these sites affects other parts of the structure. This then requires a knowledge of how these structural features relate to the mechanisms involved in oxygen binding and the regulation of oxygen binding by allosteric interactions both between oxygen binding sites (homotropic allosteric effects) and between these sites and binding sites for naturally occurring effectors (heterotropic effects). Such topics have been the subject of many books and reviews too numerous to mention, but a very good treatment of allosteric effects in Hb has been written by Imai (Imai, 1982) whose group have made substantial contributions towards the present understanding of mechanistic details. This would not have been possible without the detailed three-dimensional structural information from the X-ray work of Perutz (Perutz *et al.*, 1968; Bolton and Perutz, 1970) and others, and the relationship between the structural detail and the precise mechanism of oxygen binding is the subject of two key papers (Perutz, 1976; Baldwin and Chothia, 1979).

3.2.1 BASIC STRUCTURE

Hb is a tetramer composed of two pairs of chains, the α chains containing 141 amino acid residues and the β chains containing 146 residues. The chains contain many differences in amino acid sequence yet their folding to give the tertiary structure results in a very similar overall shape for both chains, which is also shared by the related protein, myoglobin. Each chain contains a haem group, a protoporphyrin ring coordinated to an iron atom in the divalent, ferrous state. The iron atom is coordinated to the four nitrogen atoms of the porphyrin ring and to a histidine residue known as the proximal histidine which is residue 87 in the α chains and 92 in the β chains numbering from the amino terminal end of the chains. The sixth coordination position is free to enable the reversible binding of an oxygen molecule, and the tetramer therefore binds four molecules of oxygen when fully oxygenated.

3.2.2 COOPERATIVITY OF OXYGEN BINDING

Figure 3.1 shows a typical oxygen saturation curve (a) for whole blood, i.e. a plot of the percentage of oxygen bound by Hb against the partial pressure of oxygen. Whereas such ligand binding curves are often defined by a hyperbolic shape, for oxygen binding to Hb the curve is a pronounced sigmoid shape and looks more like the classical pharmacological log dose/response curve with which it should not be confused. The shape of this curve is indicative of a positive cooperativity of oxygen binding (Monod *et al.*, 1965); with no oxygen bound the iron atoms have a relatively low affinity for oxygen, but as successive oxygens bind the structure of the tetramer is altered such that the affinity of the remaining unoccupied iron atoms for oxygen is increased. The precise structural changes that occur during oxygen binding and lead to the cooperativity are now well understood and are discussed in Section 3.2.3.

The sigmoidal shape of the oxygen saturation curve of whole blood is crucial for the efficient delivery of oxygen to the tissues to fuel aerobic metabolism. The P_{50} of whole blood, i.e. the partial pressure of oxygen at which the Hb is 50%

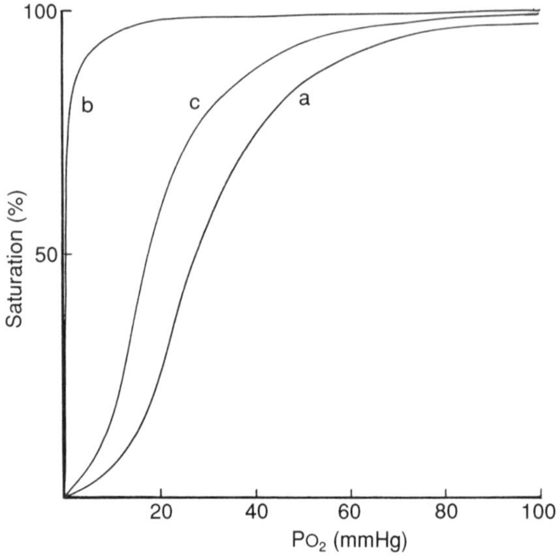

Figure 3.1. Oxygen saturation curves expressed as the percentage of oxygen bound per Hb tetramer against the partial pressure of oxygen for whole blood under various conditions. Curves a and c represent measured curves using a Hem-O-Scan (Aminco, Maryland, USA) for fresh human blood and blood stored overnight at 37°C respectively. Curve b is the theoretical curve expected for a haemoglobin lacking cooperativity

saturated with oxygen, is about 27 mmHg (3.6 kPa). The Po_2 in the lungs is about 100 mmHg (13.3 kPa) and the Po_2 of the venous blood returning to the lungs is about 40 mmHg (5.3 kPa). Inspection of curve a shows that on each circulation about 22% of the oxygen carried by the haemoglobin is released. Curve b shows the oxygen saturation curve that would result if there were no cooperativity of oxygen binding. Clearly this curve is hopelessly inadequate for oxygen delivery since less than 1% of the oxygen would be given up under normal physiological conditions.

The affinity of Hb for oxygen and thus the position of the oxygen saturation curve on the Po_2 axis is subject to control by a number of factors including temperature, pH and Pco_2. Physiological changes in these quantities caused by an increased workload occur rapidly and generally lead to a reduction in oxygen affinity, resulting in a right-shift of the oxygen saturation curve and a beneficial increase in oxygen delivery. However, in 1967 two independent groups (Benesch and Benesch, 1967; Chanutin and Curnish, 1967) discovered a role for a major byproduct of glycolysis in the red cell, 2,3-diphosphoglycerate (DPG) whose function was unknown for more than 40 years. These groups showed that DPG is also a regulator of oxygen affinity and provides a less rapid, but more permanent, means of promoting increased oxygen delivery when necessary, for example in some anaemias and in the hypoxic stress of moving to high altitude. Curve c in Figure 3.1 is an oxygen saturation curve for blood left overnight at 37°C and shows a considerable left-shift from the normal curve due to a loss of DPG by glycolytic metabolism on storage. The presence of normal amounts of DPG thus maintains the normal oxygen delivery of blood but increased DPG levels would lead to a further right-shift of the curve and an increase in oxygen delivery capacity.

3.2.3 THE ALLOSTERIC MECHANISM

As mentioned earlier, the mechanism of the cooperativity of oxygen binding is now known in molecular detail and the same is true for the mechanism of action of natural allosteric regulators of oxygen affinity. The basis of the mechanism lies in the existence of distinct quaternary conformations for the Hb tetramer when it is deoxygenated (deoxyHb) and when it is fully oxygenated (oxyHb). Although composed of four separate chains which are arranged in the tetramer in a roughly tetrahedral arrangement, from the mechanistic point of view the dimer formed from one α chain and one β chain is the important unit and regarding each dimer as a hemisphere with an open interface between them provides a useful if rather simplistic picture. This is portrayed in a very simple form in Figure 3.2. The main change that occurs in going from deoxyHb to oxyHb is a rotation of one $\alpha\beta$ dimer unit relative to the other coupled with a small translation to bring the dimers closer together in oxyHb (Baldwin and Chothia, 1979). This movement brings a different set of residues into alignment in the two conformations and gives rise to a different set of H-bonds, van der Waals contacts and electrostatic interactions. Of particular importance are the electrostatic interactions or 'salt bridges' in the deoxy conformation since these are largely responsible for the stabilization of this conformation with its low oxygen affinity relative to the oxy form.

In the oxy conformation, which predominates when three molecules of oxygen are bound, there is no hindrance to oxygen binding and the oxygen affinity is that expected for an iron atom coordinated in the particular haem group environment; the monomeric oxygen binding protein myoglobin shows a similar oxygen affinity. For this reason the oxy conformation is also referred to as the R (relaxed) form in the nomenclature of the Monod *et al.* (1965) allosteric two-state model. In the deoxy conformation the position of the iron relative to the porphyrin ring and of the whole haem group relative to the globin chain are different and there is steric hindrance to oxygen binding resulting in a lower oxygen affinity (*ca.* 250 times lower though this is variable depending on conditions). The tetramer is constrained to adopt this low oxygen affinity conformation by the strengths of the salt bridges between the subunits at the interface between the $\alpha\beta$ dimers and this conformation corresponds to the T (tense) conformation according to the Monod *et al.* (1965) model. When an

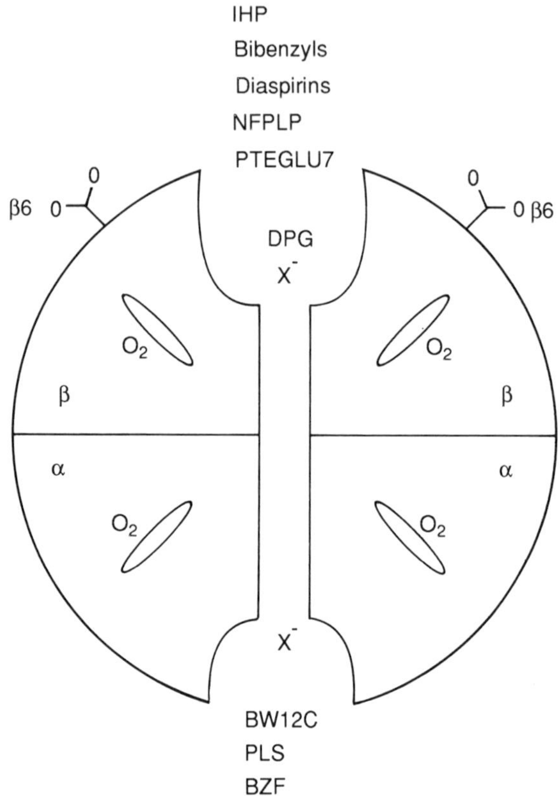

Figure 3.2. Schematic representation of the Hb tetramer showing the rather open interface between the $\alpha\beta$ dimers and the clefts at the β terminus where many right-shifting compounds bind and at the α terminus where predominantly left-shifting compounds bind

individual subunit becomes oxygenated the structural changes resulting in the local haem environment are relayed through the protein chains to the salt bridges at the interface between the dimers and these are weakened such that the T state is destabilized relative to the R state. The next oxygen therefore has a smaller energy barrier to overcome in binding and its affinity is correspondingly higher. This weakening of the salt bridges continues as more oxygens bind until, when three molecules are bound, the T form is less stable than the R form and the R form predominates so that the fourth oxygen binds with its intrinsic unhindered affinity.

The particular salt bridges involved in stabilizing the T state and in constraining its low oxygen affinity have been identified and involve, in particular, the terminal carboxyl and amino residues of the α and β chains. The fold of the globin is such that in the tetramer the amino terminal valine residue (Val 1α_1) of one α chain lies in close proximity to the carboxy terminal arginine residue (Arg 141α_2) of the opposite α chain and *vice versa* and the amino group is linked via an inorganic anion to the guanidinium chain of the arginine. In the β chains the amino terminal valine (Val 1β) and the carboxy terminal histidine (His 146β) are not quite so close and, in fact, the carboxyl group forms a direct salt bridge with a lysine amino group (Lys 40α) of the opposite α chain. The terminal amino group is involved in binding an inorganic anion but its main role in salt bridge formation arises from its involvement in binding the natural effector DPG.

3.2.4 THE MECHANISM OF THE DPG EFFECT

The observation that DPG produces its right-shifting effect on isolated Hb by binding specifically to deoxyHb in a 1:1 molar ratio and that the intact tetramer is necessary for DPG binding led the Beneschs to postulate that it binds to cationic residues at the interface between the $\alpha\beta$ dimers (Benesch and Benesch, 1969). This was confirmed by Arnone (1972) who showed crystallographically that it binds by ionic interactions to the amino terminal groups of the β chains and to residues His 2, His 143 and Lys 82 of the β chains. This is shown schematically in Figure 3.3. Out of a possible eight interactions seven are formed; the carboxylate group of DPG is unable to interact with both Lys 82 residues at the same time and flips between these. In the oxy form the movement of one $\alpha\beta$ dimer relative to the other closes up this DPG binding cleft, resulting in the loss of this specific binding site and a much lower affinity of DPG for the oxy conformation. Figure 3.3 also shows that DPG forms 'salt bridge' links between residues of opposite β chains and thus adds to the salt bridges constraining the low oxygen affinity of the deoxy form. Because of this, DPG has a dual mechanism of right-shift: it not only stabilizes the deoxy conformation thereby delaying the conformational change to higher oxygen partial pressures, but also adds to the numbers of inter-subunit salt bridges and directly lowers the oxygen affinity of the deoxy form.

3.3 RIGHT-SHIFTERS

A large number of compounds have now been shown to shift the oxygen saturation curve to the right; they include inorganic and multiply-charged organic anions, aldehydes (particularly aromatic) and hydrophobic mono-anions. Very few, however, have been designed with reference to the structure of Hb and attention will be focused largely on these in this section. Binding sites for other compounds have been identified and these will also be discussed since they define alternative binding sites and therefore offer potential sites for future compound design.

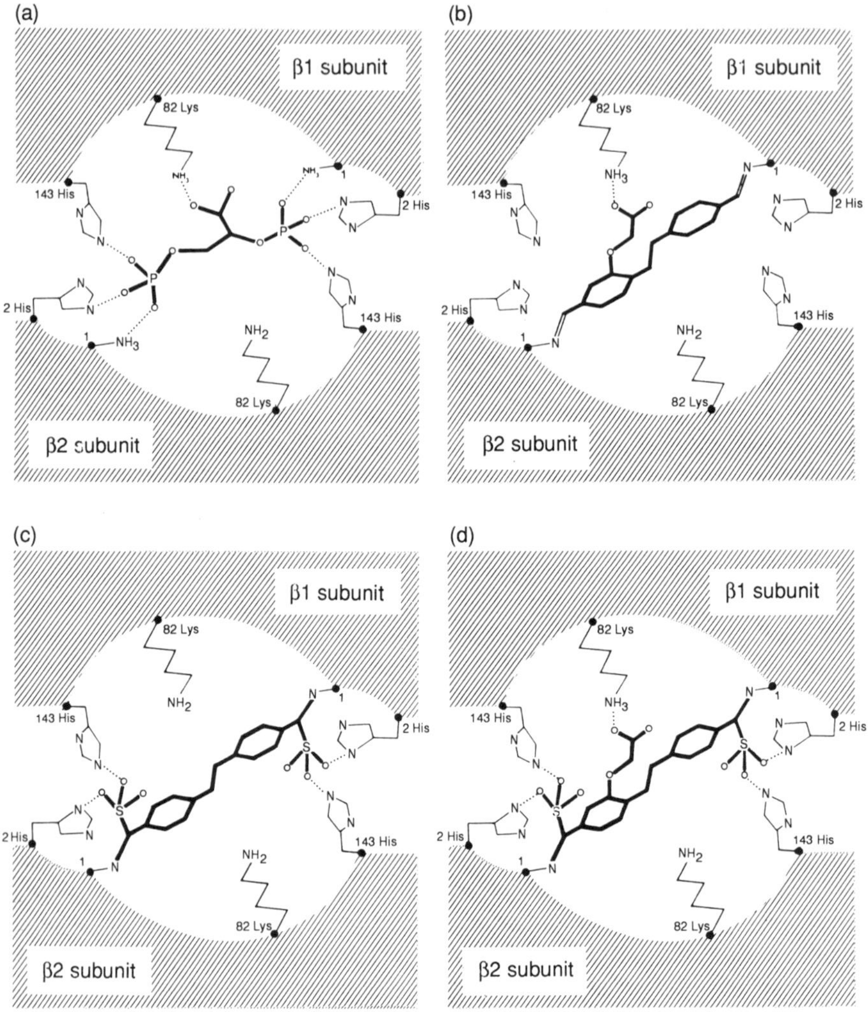

Figure 3.3. Schematic representation of the DPG binding site at the β terminus: (a) with DPG bound as observed by Arnone (1972) and (b)–(d) the postulated binding modes for compounds **1**–**3** respectively. (Reproduced from Beddell *et al.* (1976) by permission of Macmillan Press Ltd)

3.3.1 BIBENZYL DIALDEHYDES

In 1973 a group was set up at the Wellcome Research Laboratories to develop the techniques necessary for the design of compounds to interact at a site of known structure on a protein. The ultimate goal was to use these techniques to design compounds to interact at pharmacological receptor sites in order to identify novel therapeutic agents. At that time, as now, there were no structures of classic pharmacological receptors known, and the choice of a suitable model protein was also limited. Hb was selected largely because of its identified site of interaction with DPG (Arnone, 1972) and the knowledge of the basic mechanism of DPG action, i.e. preferential binding to the deoxy conformation. Possible therapeutic benefits of the ability to shift the position of the oxygen saturation curve were, of course, considered but the main aim was to use Hb as a model system for drug design by the method of receptor fit (Beddell *et al.*, 1976).

The DPG binding site in deoxyHb is a very positively charged region of the protein (Figure 3.3a). In addition to the eight basic nitrogens interacting with DPG, there are also basic nitrogens from Lys 144 and His 146 in this region. There is thus a preference for the binding of multiply charged compounds and a number of inorganic and organic anions have been shown to bind there. An obvious prerequisite for any therapeutic agent is the ability to reach its receptor site *in vivo* and this facet should be considered at an early stage in any drug design exercise. Since the red cell membrane is impermeable to multiply charged anions (DPG is synthesized within the cell) it was necessary at the onset to identify alternative modes of interaction, not involving charged groups, in order to design compounds with any chance of activity in whole blood and *in vivo*. Attention was therefore focused on the amino groups of the lysine and terminal valine residues as reactive functionalities for covalent interactions. Scale models of the β-terminal amino region in the deoxy and oxy conformations were built using the atomic coordinate data for horse deoxy and human met (oxy type) Hbs (Perutz *et al.*, 1968; Bolton and Perutz, 1970) and it became clear that a compound bridging between the two amino groups of the terminal valine residues offered the most favourable opportunity for a 1:1 interaction at the DPG site. Bibenzyl 4,4′-dialdehyde was chosen as a basic structure to give aldehyde groups with an appropriate spacing and geometry to form Schiff base interactions with these groups in the deoxy conformation but not in the oxy. Use of the bibenzyl template gave a degree of rigidity and also offered possibilities for introducing further substitution on the benzene rings to alter the physicochemical properties and/or pick up extra interactions with other groups at the DPG site. Bibenzyl dialdehyde itself was synthesized but proved to be too insoluble to test and modification of the structure to increase aqueous solubility was necessary. This was achieved in two ways and the resulting compounds synthesized are shown as structures **1**–**3**. Their expected modes of binding at the DPG site are shown in Figure 3.3(b–d). Introduction of the carboxyl side chain

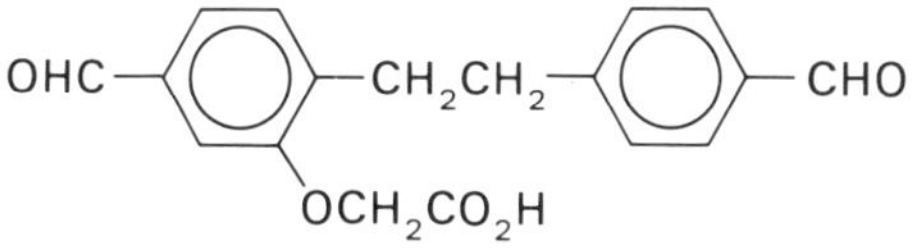

(**1**)

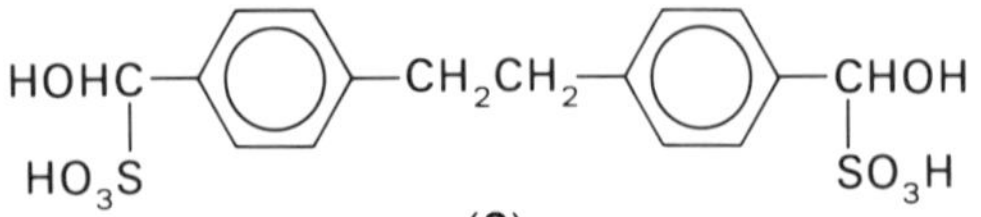

(**2**)

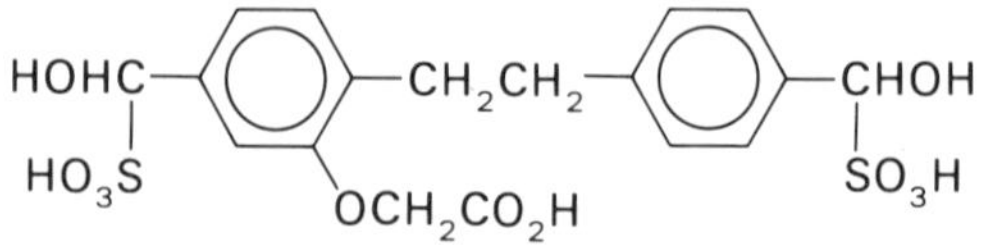

(**3**)

as in **1** produced a sufficient increase in solubility and modelling showed that the carboxyl group should be able to interact ionically with a Lys 82 residue as shown in Figure 3.3(b) and as seen for the carboxylate of DPG. In structure **2** bibenzyl 4,4′-dialdehyde is converted to its bisulphite addition compound. Although this might have acted merely as a solubilizing method with the sulphite group displaced on binding to Hb, it was thought that the sulphite groups might remain bound due to their possible interaction with His 2 and His 143 as shown in Figure 3.3(c). Finally, compound **1** was converted to its bisulphite addition compound to give **3** with the postulated binding shown in Figure 3.3(d). The movement apart of the terminal amino groups and the closing up of the cleft on switching to the oxy conformation would result in a preferential binding to the deoxy form, resulting in a right-shift of the oxygen saturation curve.

The measured effects at a concentration of 2.5 mM for DPG and the compounds on the oxygen saturation curve are shown in Figure 3.4. As expected, all three compounds give significant right-shifts, and it is evident that the magnitude of the shifts parallels the postulated number of interactions at the DPG site. Hb is unique amongst proteins in that the amino acid sequences for Hb from a number of species are known as is the precise point of mutation for hundreds of mutant Hbs. To provide evidence for the validity of the proposed binding modes and to further test the observation that the magnitude of the effect parallels the number of interactions made, the effects of DPG and the compounds were measured for a number of Hbs containing differences in amino acids at the DPG site (Beddell *et al.*, 1979). The Hbs studied are shown in Table 3.1 and involve changes in His 2 (horse), His 143 (HbF_I, HbF_{II}) and Val 1 (HBF_I, HbA_{1c} and Hb Raleigh). The changes at the histidines will reduce the number of possible ionic interactions and the changes at the amino group of Val 1 will prevent both covalent and ionic interactions. The effects of DPG and the compounds at 2.5 mM on human and each of the different Hbs are shown in Figure 3.5; inset in each plot are the calculated numbers of ionic and covalent interactions for each combination of compound and Hb. Although there is some variation in the position of the curves for different Hbs when the postulated

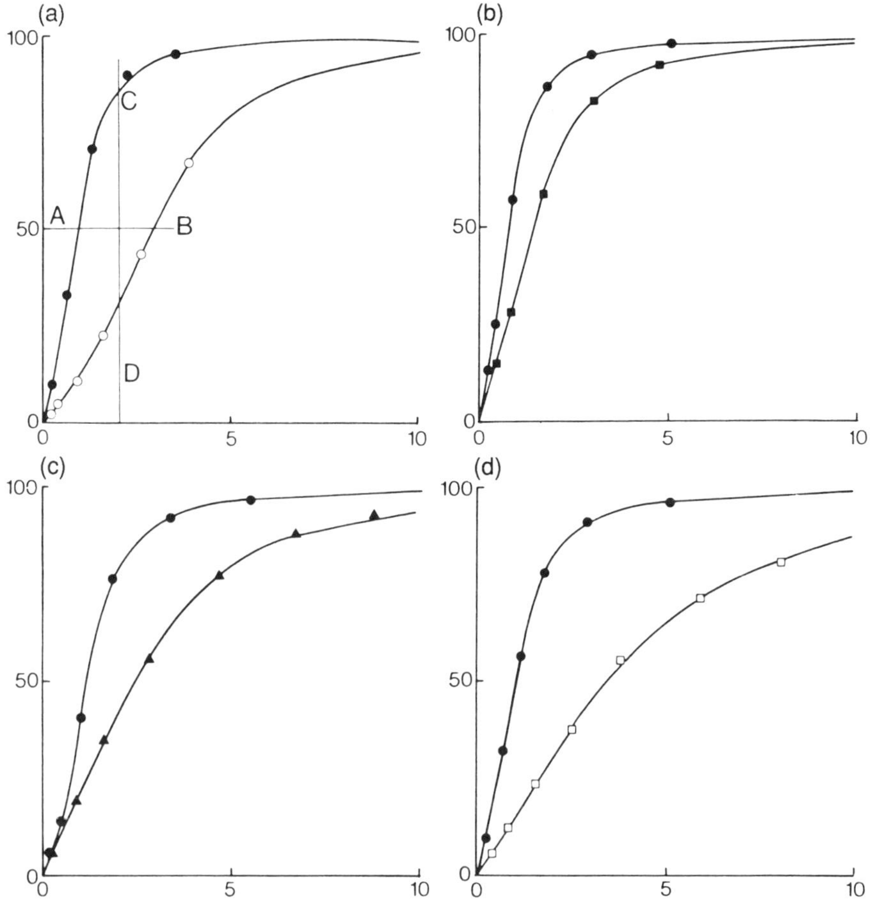

Figure 3.4. Oxygen saturation curves for phosphate free human Hb solutions at a concentration of 20 μM in the presence and absence of: (a) DPG; (b) compound **1**; (c) compound **2**, (d) compound **3**; all at 2.5 mM. The curves were measured in the association mode using an equilibrium specrophotometric method at 37°C in 100 mM Hepes buffer, pH 7.4 containing 35 mM KCl (Reproduced from Beddell *et al.*, 1976 by permission of Macmillan Press Ltd)

interactions are the same, in general the observed shifts appear well correlated with the numbers of postulated interactions.

Association constants for binding to the deoxy form, K_D for each combination of compound and Hb were calculated using equation (1) (Monod *et al.*, 1965; Goodford *et al.*, 1978) which relates saturation values, Y, to the partial pressure represented by α, and compound concentration, d. α is the normalized oxygen pressure, i.e. pressure multiplied by the oxygen affinity of the R state, c is the ratio of oxygen affinities of the R and T states and b is the reduced value of c in

Table 3.1 Modifications at the 2,3-diphosphoglycerate (DPG) receptor site in the haemoglobins studied

Haemoglobin	Modifications at the DPG receptor site
Horse	β2 Histidine → Glutamine
Fetal Major (HbF_{II})	β143 Histidine → Serine
Fetal Minor (HbF_{I})	β143 Histidine → Serine
	β1 Valine acetylated
HbA_{1c}	β1 Valine forms Schiff base with glucose
Raleigh	β1 Valine → acetylalanine

Reproduced by permission of Macmillan Press Ltd from Beddell *et al.* (1979)

the presence of allosteric effector, i.e. the amount by which the effector increases the average strength of the inter-subunit salt bridges for each stage of oxygenation. L is the allosteric constant, the ratio of T to R states in the absence of ligands.

$$Y = \frac{a(1+\alpha)^3(1+dK_O) + Lc\alpha(1+c\alpha)^3 + Lb\alpha(1+b\alpha)^3 dK_D}{(1+\alpha)^4(1+dK_O) + L(1+c\alpha)^4 + L(1+b\alpha)^4 dK_D} \quad (1)$$

In order to obtain reliable estimates it was necessary to fix the values for K_O, the binding constant to the oxy form at the values calculated for the full data set for human Hb. Free energies of interaction (ΔG) were calculated from these K_D values and used to obtain a linear free energy relationship between the ΔG values and the number of ionic, n_I, and covalent, n_C interactions shown as equation (2).

$$\Delta G = -3.14\,n_I - 6.78\,n_C - 8.29 \quad (2)$$
$$(\pm 0.30)\ (\pm 0.68)\ (\pm 1.37)$$

$$n = 29;\ r = 0.928;\ F_{2,26} = 81.2$$

This equation explains 86% of the variation of the ΔG values and shows that to a first approximation the energy of binding of the compounds to the various Hbs may be obtained by summing the postulated numbers of interactions, a covalent bond contributing about twice as much energy (6.8 kJ mol^{-1}) as an ionic interaction (3.1 kJ mol^{-1}). Other estimates of the strength of ionic interactions between ligands and proteins are scarce but tend to be somewhat higher

Figure 3.5. Oxygen saturation curves measured under the same conditions as in the legend of Figure 3.4 for DPG, compound **1** and compound **3** with six different Hb types. Inset in each figure are the postulated numbers of ionic and covalent interactions for each combination of compound and Hb. (Reproduced from Beddell *et al.* (1979) by permission of Macmillan Press Ltd)

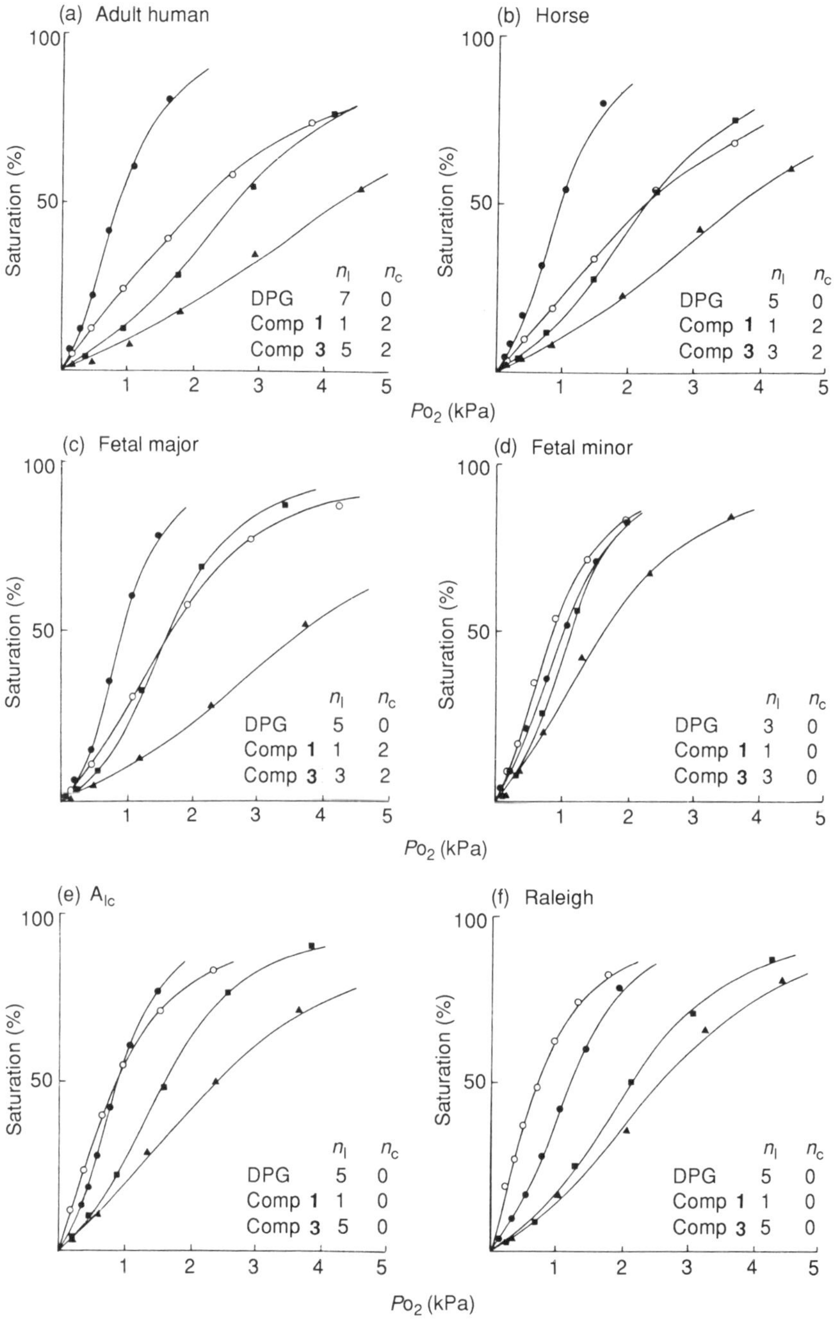
(a) Adult human
100
50
Saturation (%)
n_l n_c
DPG 7 0
Comp **1** 1 2
Comp **3** 5 2
1 2 3 4 5
(b) Horse
100
50
Saturation (%)
n_l n_c
DPG 5 0
Comp **1** 1 2
Comp **3** 3 2
1 2 3 4 5
Po_2 (kPa)
(c) Fetal major
100
50
Saturation (%)
n_l n_c
DPG 5 0
Comp **1** 1 2
Comp **3** 3 2
1 2 3 4 5
(d) Fetal minor
100
50
Saturation (%)
n_l n_c
DPG 3 0
Comp **1** 1 0
Comp **3** 3 0
1 2 3 4 5
Po_2 (kPa)
(e) A_{Ic}
100
50
Saturation (%)
n_l n_c
DPG 5 0
Comp **1** 1 0
Comp **3** 5 0
1 2 3 4 5
(f) Raleigh
100
50
Saturation (%)
n_l n_c
DPG 5 0
Comp **1** 1 0
Comp **3** 5 0
1 2 3 4 5
Po_2 (kPa)

(two to four times) (Wells *et al.*, 1987). However, the energy is critically dependent on the environment, with larger values in buried hydrophobic regions; the DPG site is particularly open and accessible to water and other ions. This treatment is also very simplified, of course, as the ionization of the groups involved is assumed the same as is the energy of the different covalent interactions.

The agreement between the predicted numbers of interactions at the DPG site and the observed shifts provides very good evidence for the interaction of the compounds at this site. This has been further substantiated by proton NMR studies (Brown and Goodford, 1977) which showed that the same histidine resonances shifted on interaction with DPG were also shifted in the presence of the compounds, and by crystallography of crystals soaked in polyethylene glycol solutions containing the compounds (Körber, 1984). Additional electron density in the presence of the compounds was seen only at the DPG cleft. The X-ray data were obtained at only low resolution (5.5 Å) but raised the possibility that the compounds might be binding differently than predicted with the covalent interactions formed between the Lys 82 amino groups and the terminal valine rather than directly between the terminal valines. This interpretation is not necessarily ruled out by the other observations.

A number of other compounds based on the bibenzyl structure were synthesized at the Wellcome Laboratories in an attempt to increase potency but syntheses in this area proved difficult. Also it was found that for those compounds which were able to penetrate the red cell a left-shift was generally produced due to the expected competition for binding with DPG. Work in this series was therefore abandoned.

3.3.2 OTHER RIGHT-SHIFTERS WITH IDENTIFIED BINDING SITES

3.3.2.1 DPG Site

Avian erythrocytes contain inositol pentaphosphate instead of DPG as the naturally occurring allosteric effector and it was observed that an analogue of this, inositol hexaphosphate (IHP, **4**) binds much more tightly to deoxyHb than

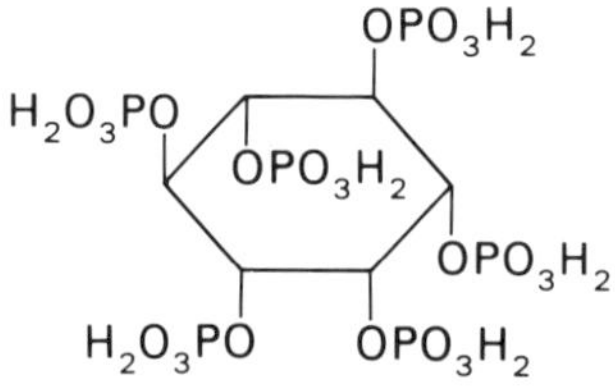

(**4**)

DPG (Edalji *et al.*, 1976; Goodford *et al.*, 1978). IHP was shown to bind at the DPG site using the same eight basic nitrogens and, in addition, forming H-bonds with the amino groups of Asn 139β residues (Arnone and Perutz, 1974). Poly-γ-glutamyl derivatives of folic acid also bind to residues at the DPG site (Benesch *et al.*, 1985) and pteroylheptaglutamate (**5**) binding in a low salt polyethyleneglycol medium has been observed crystallographically (Arnone *et al.*, 1986). The glutamate residues, though ill-defined, were clearly interacting at the DPG site. The pteridine and *p*-aminobenzoic acid portions were clearly visible and extended down the central cavity between the $\alpha\beta$ dimers towards the centre of the tetramer interacting with residues Lys 132 and Ala 135 of the β chains and Phe 36 of the α chains. This is a unique binding mode and may offer possibilities for the design of right-shifters.

Pyridoxal phosphate and some analogues have also been shown to shift the position of the oxygen saturation curve (Benesch *et al.*, 1973, 1975). Three of these have been observed crystallographically (Arnone *et al.*, 1977) after reduction of the reversible Schiff base linkages with borohydride to give permanently modified Hbs. Two of these derivatives, pyridoxal phosphate (PLP, **6**) and nor-formyl pyridoxal phosphate (NFPLP, **7**) produce a right-shifted oxygen saturation curve and bind to the β-terminal amino region and the third pyridoxal sulphate (PLS, **8**) produces a left-shift and binds at the corresponding α-terminal amino region as described in Section 3.4.2. In the PLP derivative each Val 1 amino group is covalently reacted and the phosphates occupy similar positions to the phosphates of DPG interacting with His 2, His 143 and Lys 82. NFPLP reacts to give a derivative cross-linked between the β chains by reaction

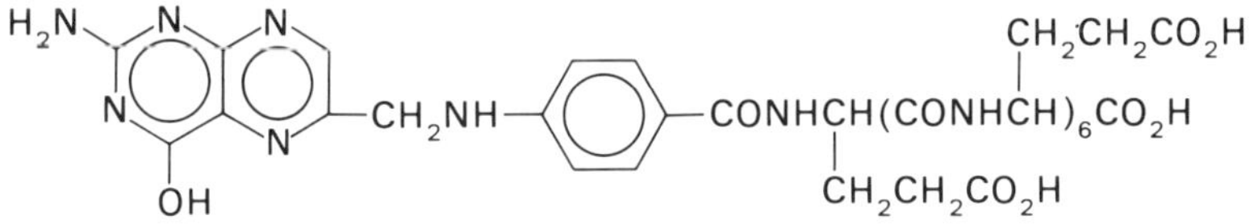

(**5**)

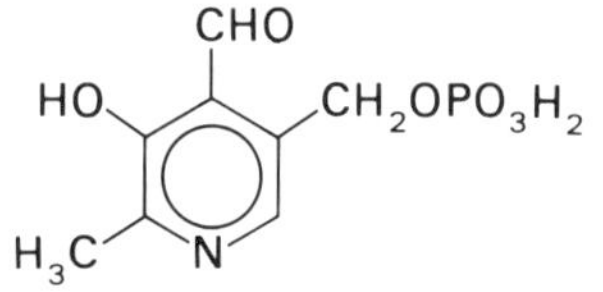

(**6**)

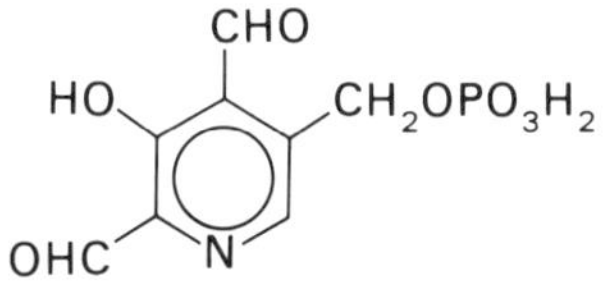

(**7**)

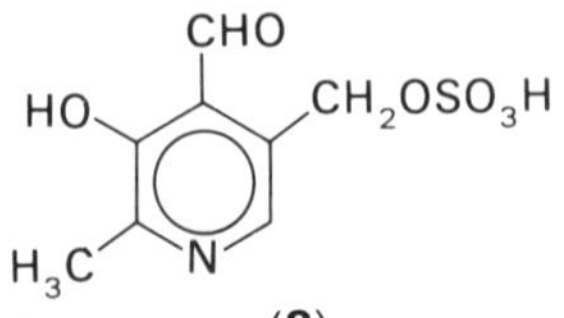

(**8**)

at the Val 1 amino of one chain and the Lys 82 of the opposite chain. The phosphate group again interacts in a similar manner to the phosphates of DPG. Surprisingly, although there is apparently room for two molecules of NFPLP to bind in the DPG cleft, the reaction is 1:1 with Hb. The use of NFPLP reacted Hb as a blood substitute has been considered (Benesch *et al.*, 1984); it has been found to have a much longer half-life than ordinary Hb *in vivo* after infusion and gives a very right-shifted oxygen saturation curve which is better suited to deliver oxygen. A more recent paper (Benesch and Kwong, 1988) describes a series of bis-pyridoxal phosphates in which the phosphate groups are linked onto another phosphate compound such as DPG. These compounds also react covalently with Hb, producing large right-shifts of the oxygen saturation curve. Their binding has not been observed crystallographically but biochemical methods reveal a cross-link between the N-terminal amino group of one β chain and Lys 82 of the other.

3.3.2.2 Right-shifters bound at other sites

In a search for agents to increase the solubility of deoxyHbS two known antihyperlipoproteinaemic drugs clofibrate (CFA, structure **9**) and bezafibrate (BZF, structure **10**) were found to right-shift the oxygen saturation curve (Abraham *et al.*, 1983b; Perutz and Poyart, 1983). BZF in particular gave large right-shifts in Hb solution and also retained some activity in whole blood acting synergistically with DPG, indicating that BZF binds remote from the DPG site. These compounds belong to a class of right-shifting compounds which may be described as hydrophobic mono-anions and a large number of this type have

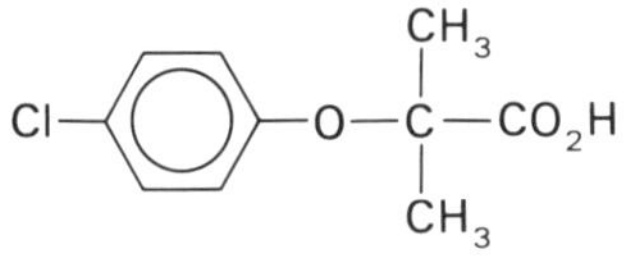

(**9**)

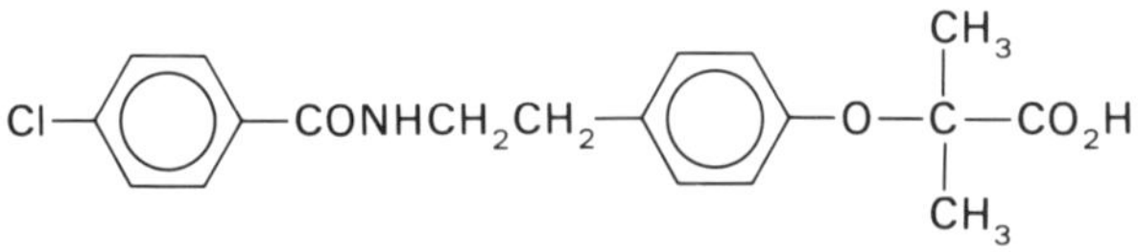

(**10**)

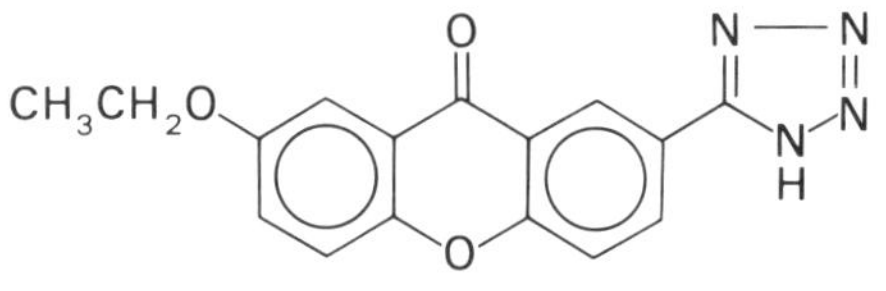

(**11**)

been synthesized and tested at the Wellcome Laboratories (Hyde *et al.*, 1988). Some of these compounds have the additional property of maintaining the DPG level in stored blood and the actions of one typical compound, 2-ethoxy-6-(5-tetrazolyl)xanthone (BWA827C, structure **11**) have been described (Hyde *et al.*, 1984, 1988). It is likely that compounds of this class share common binding sites and, indeed, BWA827C and BZF show competitive behaviour in their right-shifting action (Wootton, unpublished). Binding of BZF has been observed crystallographically (Perutz *et al.*, 1986) and a pair of identical sites was identified in deoxyHb running parallel to the diad axis of symmetry and extending from the α-terminal region along the central cavity towards the centre of the tetramer. Interestingly, the Cls of the BZF molecules interact with the same Phe 36α residues interacting with PteGlu7 which approaches the centre of the tetramer from the other side, i.e. from the β-terminal region (Arnone *et al.*, 1986). The energy of interaction between BZF and deoxyHb derives from a large number of apparently weak polar interactions and van der Waals contacts and the knowledge of the binding sites does not provide any obvious clues towards the design of more potent compounds. However, more traditional medicinal chemical methods have identified analogues of BZF with increased potency (Lalezari *et al.*, 1988; Lalezari and Lalezari, 1989). One compound, LR-16 (structure **12**) binds similarly to BZF but the reasons for its increased potency were not apparent from the crystal structure observations (Lalezari *et al.*, 1988).

Two more compounds of very similar structure to LR-16 have been described more recently (Lalezari *et al.*, 1990) which are yet more potent, but again the crystal structure determinations do not clarify the structure–activity relationships. In fact, the distinction between potency, which relates to the concentration range over which an effect occurs, and efficacy, which is related to the maximum effect given by a particular compound, and for Hb is a measure of the relative affinities for the deoxy and oxy conformations (Goodford *et al.*, 1980), has often been confused in the biochemical literature. For the compounds mentioned above, the steep and variable slopes of the dose/response curves make an assessment of relative potency difficult. Despite increased 'potency'

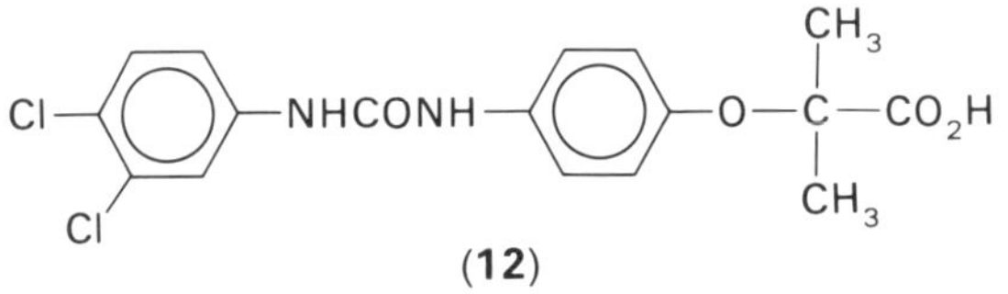

(**12**)

none of these compounds is active in whole blood. Some other close analogues in which the urea linkage is replaced by a carboxamidomethyl (*cf.* BZF, Randad *et al.*, 1991) have been reported to show activity in whole blood but in fact the blood was diluted, and with such steep dose/response curves small changes in relative concentration are critical.

The effects of a range of concentrations of BZF on the oxygen saturation curve of dilute Hb solution have been studied in detail (Wootton, 1984). Figure 3.6 shows two plots, (a) the $\Delta \log P_{50}$ values between control and treated curves as a function of the log of the BZF concentration in the absence and presence of DPG and (b) the effect of increasing concentrations of BZF on the oxygen affinity of the T state with corresponding DPG data for comparison. The first plot shows the very steep log concentration/response curve for BZF indicating multiple oxygen-linked binding sites and four binding sites gave the best fit to the data. Also the shallower curves in the presence of DPG show clearly that there is some competition for binding and in this case the data were well fitted by a model involving two sites competitive with DPG and thus in the vicinity of the β-terminus and two sites non-competitive, resulting in the partial synergy of DPG and BZF effects. Figure 3.6(b) shows that the oxygen affinity of the T state is markedly lowered in the presence of BZF, suggesting that BZF,

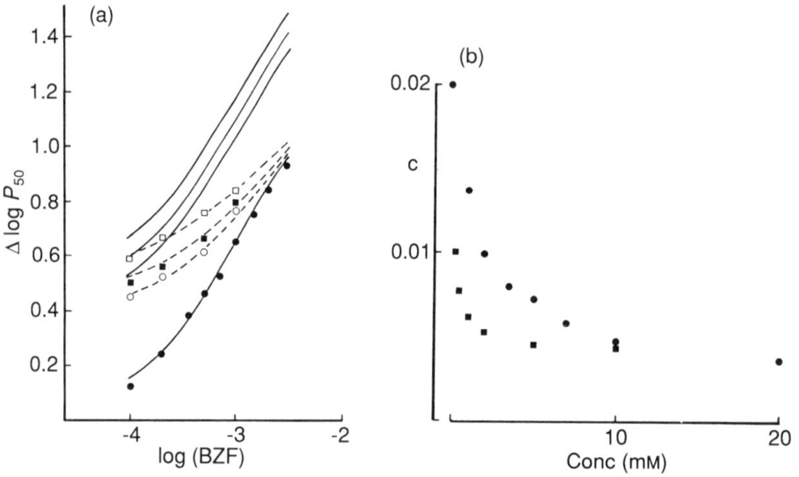

Figure 3.6. (a) Difference in log P_{50} between control and treated curves for a range of BZF concentrations alone (●) or in the presence of DPG at 1 (○), 2 (■) and 5 mM (□). Oxygen saturation curves were measured at 37°C as continuous curves in the dissociation mode using a Hemox Analyzer (TCS Medical Products Division, Southampton, PA, USA) for Hb solutions at 23 μM in 100 mM Hepes buffer, pH 7.4, containing 35 mM KCl. (b) The effect of increasing concentrations of DPG (■) or BZF (●) on the oxygen affinity of the T state as measured by fitting the Monod–Wyman–Changeux (1965) equation to digitized oxygen saturation curves to obtain the parameter c (cf. equation 1)

like DPG, forms additional bonds between subunits across the interface between the $\alpha\beta$ dimers.

These results are cited here to show that the results of crystallographic studies should be treated with some caution, particularly for weakly binding compounds like BZF, as they may not necessarily reflect the binding modes pertaining in solution. No binding at the DPG site was observed crystallographically and the observed binding between the α chains involved very little in the way of inter-subunit bridging interactions. The discrepancy probably arises in this case because the binding was studied under high salt conditions which inhibit electrostatic interactions. Thus, the high salt would be expected to block binding at the DPG site and prevent a likely interaction of the carboxyl group of BZF with the α-terminal amino groups and/or the guanidinium groups of the carboxy terminal arginine residues.

A recent publication (Mehanna and Abraham, 1990) has attempted a correlation of the binding of hydrophobic anions such as bezafibrate, measured under high and low salt conditions, with the binding profiles observed crystallographically (Perutz *et al.*, 1986). However, for deoxyHb the purported low salt solutions contained 50 mM phosphate and 100 mM dithionite, both of which will bind at the α- and β-terminal amino regions and disfavour binding of the compounds involving these loci. Also, examination of the Scatchard plots reveals that the binding data in most cases span only a limited concentration range, and the subsequent resolution into discrete classes and numbers of binding sites represents an over-interpretation of the data. The dangers of such analyses have been demonstrated in a number of publications (see, for example, Klotz, 1982).

3.4 LEFT-SHIFTERS

As mentioned in the introductory section it is only the deoxygenated form of HbS which has a reduced solubility, and a compound which binds preferentially to the oxy form and so maintains a greater proportion of the Hb in the oxy form at a given Po_2 would be expected to show anti-sickling effects. The stabilization of the oxy form with its higher oxygen affinity would also, of course, lead to a left-shift of the oxygen saturation curve. A number of compounds have been identified as left-shifters and will be discussed in Section 3.4.2 but the only agents specifically designed to bind to a defined site on Hb are a series of derivatives of benzaldehyde designed to bind at the α-terminal amino region of oxyHb (Beddell *et al.*, 1984).

3.4.1 SUBSTITUTED BENZALDEHYDES

Although the interface between the $\alpha\beta$ dimers in deoxyHb is relatively open and accessible to compounds as large as PteGlu7 and BZF for instance, the α- and β-terminal regions are particularly open. In oxyHb the β-terminal cleft closes up as the terminal amino groups move further apart, but at the α terminus there is still a sizeable cleft in the oxy form. In designing compounds to stabilize the oxy form, attention was therefore focused on this region; it seemed unlikely that compounds sufficiently large to achieve tight binding could be accommodated at the β terminus in the oxy form and act as competitive antagonists for DPG. This would be the classical pharmacological method of inhibiting the agonistic effect, i.e. the right-shift, due to DPG.

The basic structure for binding at the α-terminal region was selected as compound **13** and its postulated binding mode is shown schematically in Figure 3.7(a). The basis of the design, as for right-shifters, was to link the terminal amino groups but in this case, to ensure adequate solubility, this was achieved by a Schiff base interaction with one amino group and an ionic interaction at the other. Also, since the α-terminal region is less polar than the β terminus, the biphenyl template was expected to form favourable hydrophobic interactions with residues at the sides of the cleft, particularly the α77 proline residues. *Ortho*-hydroxyaldehydes (salicylaldehydes) form more stable Schiff base linkages than corresponding benzaldehydes (Zaugg *et al.*, 1977) and the

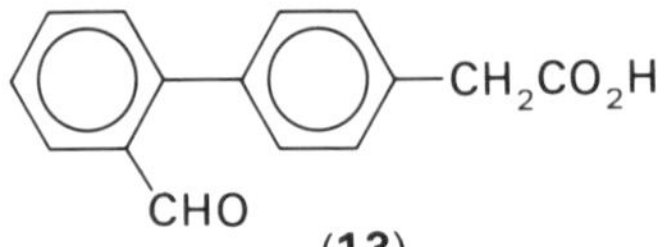

(**13**)

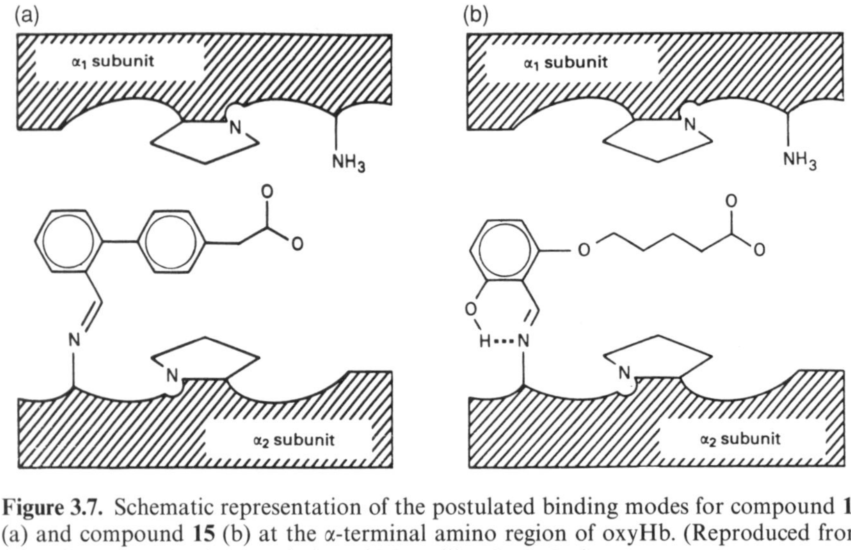

Figure 3.7. Schematic representation of the postulated binding modes for compound **13** (a) and compound **15** (b) at the α-terminal amino region of oxyHb. (Reproduced from Beddell *et al.* (1984) by permission of Macmillan Press Ltd)

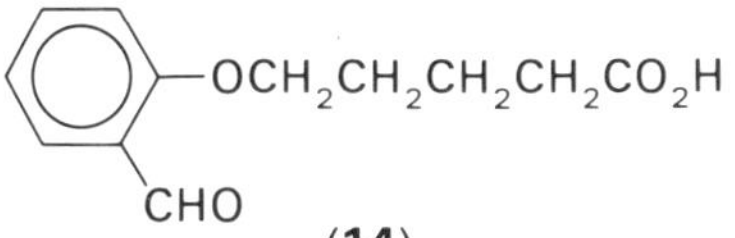

(**14**)

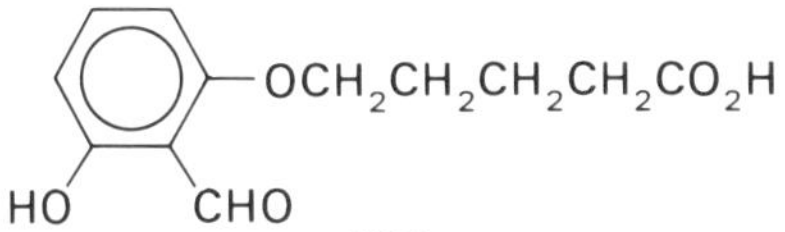

(**15**)

introduction of an *ortho*-hydroxy group into compound **13** was attempted but proved to be synthetically difficult. An alternative structure linking the aldehyde and acid groups was therefore sought and compounds **14** and **15** were made. The postulated binding mode for compound **15** is shown in Figure 3.7(b) and illustrates the proposed H-bond at the Schiff base linkage which is thought to lead to the increased stability. The designed complementarity of compound structure and design site is lost in the switch to the deoxy conformation and the compounds should therefore stabilize the oxy form. Schiff base interactions are still possible at both α- and β-terminal amino groups in the deoxy form and the binding would not be completely specific for the oxy form.

All three compounds gave the expected left-shift of the oxygen saturation curve in both Hb solution and whole blood but compound **15** (BW12C) was much more potent than the others. The effects of a range of BW12C concentrations are shown in Figure 3.8(a) for dilute Hb solution and in Figure 3.8(b) for

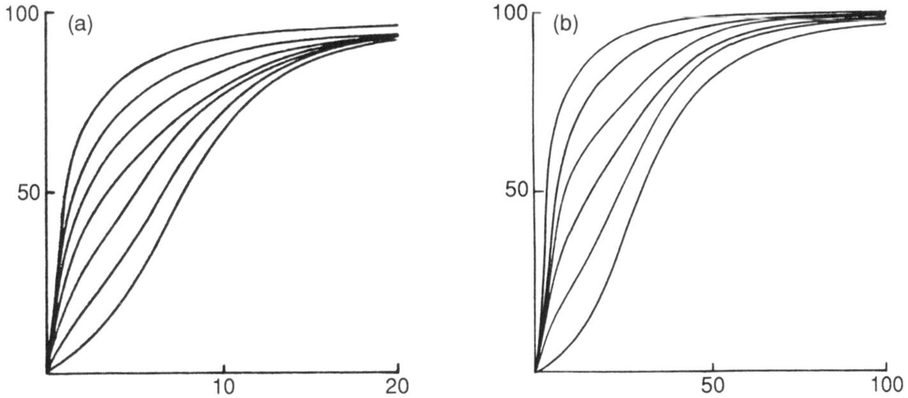

Figure 3.8. Oxygen saturation curves for human Hb with a range of concentrations of BW12C. (a) Solution measurements determined as described in the legend to Figure 3.6 for BW12C concentrations (right to left): 0, 10, 20, 30, 40, 50 and 100 μM. (b) Fresh whole blood measurements (Hem-O-Scan) for BW12C concentrations (right to left): 0, 2, 4, 6, 8 and 10 mM

whole blood. About 4 moles of BW12C per mole of Hb are required to fully shift the curve under both conditions but, due to the tight binding of the compound, this is achieved at concentrations some 100 times lower in solution at a Hb concentration of 23 μM. The anti-sickling action of BW12C at low oxygen partial pressures has also been established *in vitro* (Beddell *et al.*, 1984). Figure 3.9 shows sickle cell erythrocytes incubated at a $P\text{O}_2$ of 28 mmHg (3.72 kPa) in the absence and presence of BW12C. Again a molar ratio of BW12C to Hb of 4:1 completely inhibits the sickling process under these conditions. Similar molar ratios of compound to Hb have also been shown to restore the filterability of sickle cell suspensions to normal under low $P\text{O}_2$ conditions (Kenny and Stuart, 1983).

The binding of BW12C to oxyHb has not been observed crystallographically; no crystals suitable for X-ray analysis could be grown in the presence of compound and preformed crystals of carbonmonoxyHb (oxy form) disintegrated when soaked in solutions containing BW12C (Merrett *et al.*, 1986; Perutz *et al.*, 1986). However, a number of other techniques have been used to show that, although the overall binding pattern is complex, the compound does show a preferential binding at the α-terminal amino region in oxyHb (Merrett *et al.*, 1986). Reaction of oxyHb with ^{14}C-labelled BW12C at a 1:2 molar ratio followed by $NaBH_4$ reduction resulted in almost exclusive labelling of the α-chains and tryptic digestion of the isolated chains confirmed the terminal amino group as the site of reaction. Higher BW12C ratios gave considerable labelling of the β chains but still only half of the α chains were labelled as expected from the predicted binding between the α chain terminal amino groups. Under deoxy conditions the specificity for α chain binding was lost and α and β chains were labelled equally. At higher ratios additional reaction at both α and β chains was seen and much more than half the α chains were singly labelled. The isoelectric focusing pattern for oxyHb reacted with BW12C/Hb ratios of 0.5–4:1 was complex and gave four modified bands with the strongest band corresponding to two BW12C molecules bound; equilibrium dialysis measurements also indicated two strong binding sites with affinities $>10^6$ M^{-1}. The existence of two strong binding sites in oxyHb is not in conflict with the results of the chain labelling experiments since it was noted that about half of the BW12C was lost on disruption of the tetramer by urea, suggesting binding which does not involve Schiff base formation or perhaps is inaccessible to $NaBH_4$ when the cavity between the α chains is blocked. The location of this additional binding is not known but the lack of effect of BW12C on Hb carbamylated at α and β termini or just at the α termini, as shown by isoelectric focusing studies and oxygen dissociation curve measurements, may indicate that this second site only exists when the BW12C occupies the α-terminal site.

The binding of BW12C is thus complex, involving binding to α- and β-terminal amino groups in both oxy and deoxy conformations and probably to lysine residues at higher molar ratios. This also means that the left-shifting effect

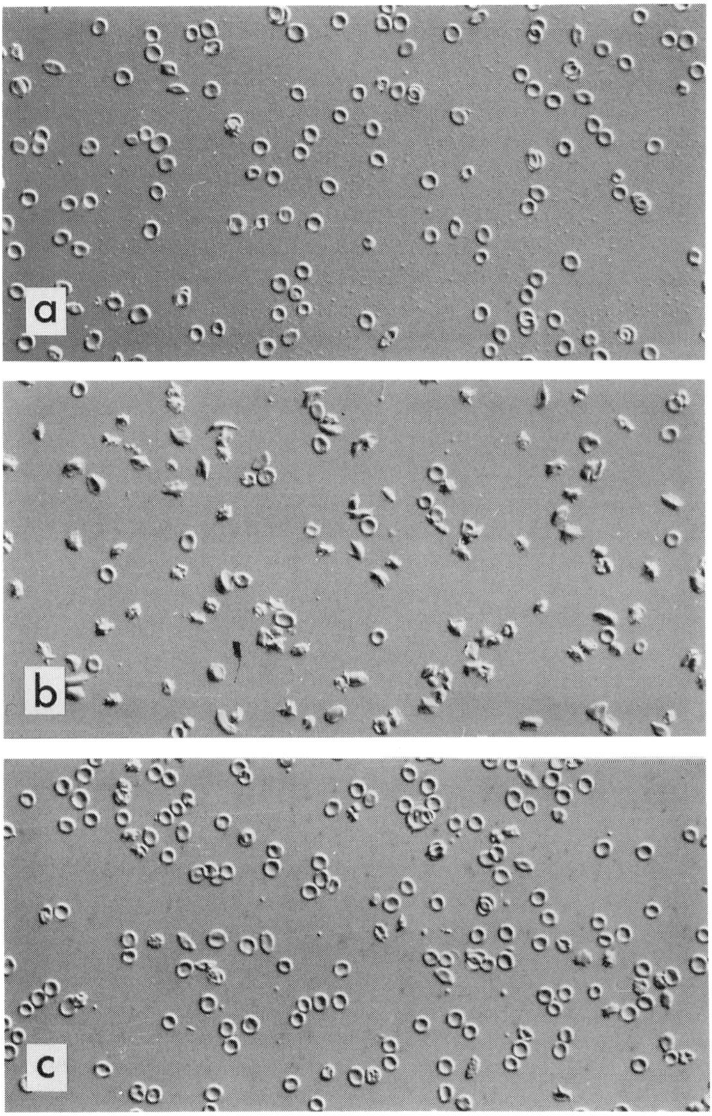

Figure 3.9. Homozygous sickle cell blood suspensions: (a) in normal air; (b) at a reduced partial pressure of oxygen of *ca.* 28 mmHg (3.72 kPa); (c) at the same reduced Po_2 in the presence of a saturating concentration of BW12C

of BW12C in whole blood is more complex than a simple preferential binding to the oxy form. This probably represents the main effect of BW12C but the binding at the α terminus in the deoxy form will also be left-shifting. This is because the α-terminal amino groups in deoxyHb are involved in the salt-bridge network responsible for its low oxygen affinity as discussed in Section 3.2 and formation of the Schiff base adducts will lead to the loss of two important salt bridges and thus an increase in the oxygen affinity of the deoxy form. Also in the whole blood case the binding of BW12C to the β-terminal amino groups will be competitive with DPG binding and may thus antagonize the action of DPG and again lead to a left-shift.

Over 100 analogues of BW12C have been synthesized at the Wellcome Laboratories and the structure–activity relationships which emerge (G. Kneen, R. D. White and R. Wootton, unpublished) largely support the postulated binding mode for BW12C shown in Figure 3.7(b). The main points are:

(a) an aldehyde group is essential;
(b) a hydroxyl group *ortho* to the aldehyde leads to a large increase in affinity;
(c) introduction of polar groups into the alkyl side chain or shortening this chain reduce affinity;
(d) an acidic function such as the carboxyl in BW12C is not essential for potency.

This latter finding was surprising but was confirmed for a number of compounds in which the carboxylate was converted to a neutral derivative. However, for the weaker compounds which lack the *ortho*-hydroxyl then the charged carboxyl does contribute towards the binding. The main binding energy for BW12C therefore comes from the Schiff base interaction and the hydrophobic component. This latter contribution must be large since BW12C binds much more tightly than pyridoxal compounds.

As for the right-shifting compounds discussed in Section 3.3.1, the effect of changes in the postulated binding sites on the potency of BW12C binding was explored (Wootton *et al.*, 1986). In particular, Hbs with changes in the α77 proline residues were sought to assess the importance of the hydrophobic contribution. There are no mutant Hbs with changes in this position but chicken Hbs provide a useful comparison. Chicken blood contains two main components, chicken major (AII, 75%) and chicken minor (AI, 20%). The major component has the proline residues replaced by serines and the minor component, in addition, has the α-terminal valine residues replaced by methionines. The effect of BW12C on the isolated chicken Hbs are shown in Figure 3.10(a) and (b). The replacement of the hydrophobic prolines by hydrophilic serines would be expected to reduce the binding of BW12C at the α terminus and in fact concentrations more than ten times higher are required to produce the same magnitude of left-shift as produced with human Hb. At lower concentrations BW12C actually right-shifts the curve at the P_{50} and this was explained by

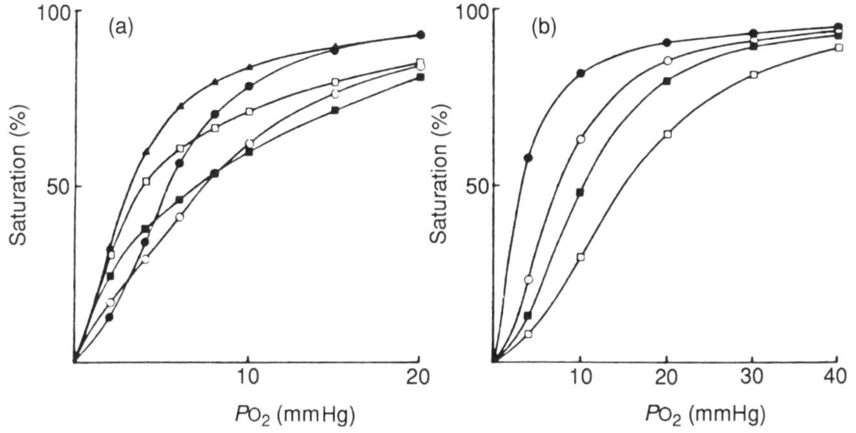

Figure 3.10. Effect of increasing concentrations of BW12C on the oxygen saturation curves of chicken Hbs measured using a Hemox as described in the legend to Figure 3.6. (a) Chicken major Hb with BW12C concentrations (mM) (●) 0, (○) 0.1, (■) 0.2, (□) 0.35, (▲) 1.0. (b) Chicken minor Hb with BW12C concentrations (mM) (●) 0, (○) 0.1, (■) 0.2, (□) 0.5. The data were recorded as continuous curves; the symbols are added merely as an aid to identification

postulating that the strongest binding sites are at the β terminus in the deoxy form but at higher concentrations binding at the α terminus becomes dominant with possible binding to both the oxy form and to the deoxy form causing an increase in its oxygen affinity as discussed earlier.

Figure 3.10(b) shows the remarkable observation that BW12C is a right-shifter with the chicken minor component with a potency comparable to that of DPG for human Hb. Two possible reasons were advanced for this effect. The replacement of the α-terminal valines by methionine would disfavour Schiff base formation at this locus since studies of the interaction of pyridoxal with amino acids have shown that higher affinities are seen for amino acids branched at the β carbon (Metzler, 1957). However, chicken minor has an additional modification at the α-terminal region; residue 138α which is serine in human Hb is replaced by glutamate. This residue is positioned such that it could form a salt bridge with the α-terminal amino groups in the deoxy form and this would also inhibit the Schiff base interaction which requires a free lone pair on the nitrogen atom. In chicken major, residue 138α is also changed from human but in this case to alanine which would not produce this effect.

Crystals of deoxyHb with BW12C bound have recently been analysed using X-ray crystallography (Wireko and Abraham, 1991). BW12C was reacted with oxyHb in a 2:1 molar ratio and cyanoborohydride used to reduce Schiff base linkages. The reaction solution was then crystallized under deoxy conditions after evacuation and flushing with nitrogen. The resulting electron density maps were interpreted as showing two molecules of BW12C bound, one on each α-terminal amino group, and it was suggested that this result was evidence that the original design hypothesis of a single molecule of BW12C spanning between the amino groups was incorrect. In fact, other work mentioned previously (Merrett *et al.*, 1986) showed that, under deoxy conditions, both α-terminal amino groups are indeed reacted with BW12C and, in this crystal structure determination, BW12C binding could only be detected if the reacted Hb solution was crystallized without removing excess compound. This suggests that additional BW12C is bound during the conformational change accompanying deoxygenation. This work could therefore be interpreted as providing support for a lower stoichiometry of binding in the oxy conformation at the α-terminal amino region in accordance with the design hypothesis.

3.4.2 OTHER LEFT-SHIFTERS

PLP and NFPLP which are dianions at neutral pH have been shown to right-shift the oxygen saturation curve and bind at the β-terminal amino region as discussed in Section 3.3.2.1. In contrast, pyridoxal derivatives which are only singly charged produce small left-shifts and show preferential binding to the α terminus in oxyHb (Benesch *et al.*, 1973, 1975). Pyridoxal sulphate (PLS, **8**) binding has been observed crystallographically after borohydride reduction of the Schiff base linkages and crystallization in the deoxy form (Arnone *et al.*, 1977). Two molecules are bound per tetramer with the covalent linkage at the α-terminal amino groups as expected and the sulphonate groupings forming salt bridges between the secondary amine of the reacted terminal amino and the guanidinium group of the carboxy terminal arginine of the opposite α chain. This salt bridge replaces the normal salt bridge formed by an inorganic anion at this locus (O'Donnell *et al.*, 1979) which explains the relatively small effects of PLS on the oxygen saturation curve compared with BW12C and its analogues which are unable to form such a replacement salt bridge in deoxyHb. Similar correlations showing the importance of this salt bridge have been observed. Carbamylation of Hb or reaction with glyceraldeyde (Kilmartin and Rossi-Bernardi, 1969; Nigen and Manning, 1977) produces a left-shift but reaction with glyoxalate (DiDonato *et al.*, 1983) which produces a carboxyl group results in a right-shift. Covalent reaction with isothiocyanato-benzene derivatives with anionic or neutral substituents show a similar trend but it is less clear cut in this case, presumably because the anionic group in the *para* position is not well positioned to form a salt bridge (Currell *et al.*, 1982; Bellelli *et al.*, 1986).

A diuretic drug, ethacrynic acid (ECA, **16**), which has structural similarities with CFA and BZF discussed in Section 3.3.2.2, has been shown to produce a marked left-shift of the oxygen saturation curve and to be a good anti-gelling agent. This compound binds covalently through its acrylophenone grouping

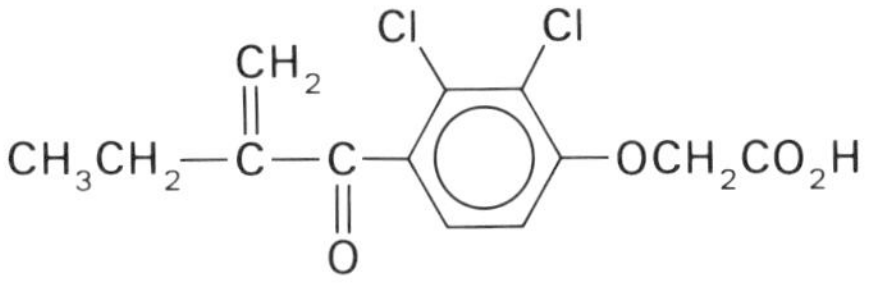

(**16**)

and its binding under oxy and deoxy conditions has been observed crystallographically (Perutz *et al.*, 1986). Two pairs of binding sites were revealed under deoxy conditions but one of these was formed by the interface between two adjacent Hb molecules. This may be of relevance to the anti-gelling action but is not relevant to solution conditions. The other sites involved a covalent reaction with Cys 93β residues which resulted in a displacement of the carboxy terminal histidine residues of the β chains and the breaking of the salt-bridge interactions with Lys 40α (and Asp 94β) mentioned in Section 3.2.3. Rupture of these salt bridges would lead to an increase in the oxygen affinity of the deoxy form as observed at the α terminus and studies of the loss of cooperativity observed for ECA reacted Hb supports this mechanism (Perutz *et al.*, 1986). The crystal structure determination did not explain the direct anti-gelling action of ECA which occurs in addition to the anti-sickling effect of the left-shift. The diuretic action of ECA would preclude its use as an anti-sickling agent, but the possibility of using other analogues devoid of diuretic activity was suggested as was the possible use of ECA itself extracorporeally, i.e. removing a quantity of blood, reacting it *in vitro* and transfusing it back into the patient (Kennedy *et al.*, 1984). This would obviously require hospitalization and be very costly and is a poor alternative to an orally active compound. Some analogues of ECA have now been shown to have potent anti-gelling activity without any diuretic effect (Abraham *et al.*, 1989). However, these compounds also produce adverse effects on red cell membranes, reducing the erythrocyte volume and increasing their rigidity which would again prevent their use as therapeutic anti-sickling agents.

3.5 ANTI-GELLING AGENTS

In theory a compound increasing the solubility of deoxyHbS but having no other effects on Hb function would represent the best approach to an anti-sickling agent based on direct interaction with Hb. The search for compounds with such a direct anti-gelling action has been extensive but has met with mixed success. The best example of a series of designed compounds is for a series of compounds known as diaspirins based on the initial finding of a weak interaction of Hb with aspirin (**17**) itself (Klotz and Tam, 1973) and this is described in detail.

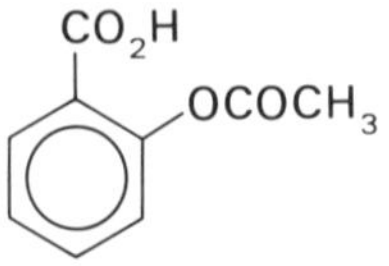

(**17**)

3.5.1 DIASPIRINS

The observation that cyanate reacts covalently with the terminal amino groups of the α and β chains of HbS with a resultant increase in deoxyHbS solubility (Gillette *et al.*, 1971) led Klotz and Tam (1973) to examine the reactivity of HbS towards acylation by aspirin, a more pharmacologically acceptable reactant. Aspirin itself reacted only weakly with either HbS in solution or HbS containing erythrocytes, but 3,5-dibromo aspirin (**18**) gave much improved reactivity with *ca.* 60% of the HbS inside erythrocytes modified after a 30 min incubation with 5 mM compound at 37°C (Walder *et al.*, 1977). The presence of the hydrophobic and electron-withdrawing bromine substituents increased cell penetration and reactivity of the compound towards nucleophilic attack. The treated erythrocytes had a left-shifted oxygen saturation curve and showed a pronounced decrease in the amount of sicked cells at a $P\text{O}_2$ of 3 mmHg (0.40 kPa). An increase in solubility of treated deoxyHbS solutions was also observed indicating a direct anti-gelling action independent of the left-shift.

Reaction with aspirin occurs at a number of sites but the major reaction is not at the α- and β-terminal amino groups but at lysine residues 144β, in the vicinity of the β-terminal cleft (Shamsuddin *et al.*, 1974). The existence of other amino groups close by (Lys 82β and Val 1β) and the known propensity of this region to bind multiply charged anions led to the synthesis and testing of alternative agents termed diaspirins such as bis(3,5-dibromosalicyl)succinate (**19**) and the corresponding fumarate (**20**) with the aim of forming cross-links between two

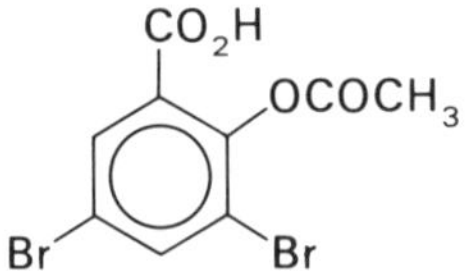

(**18**)

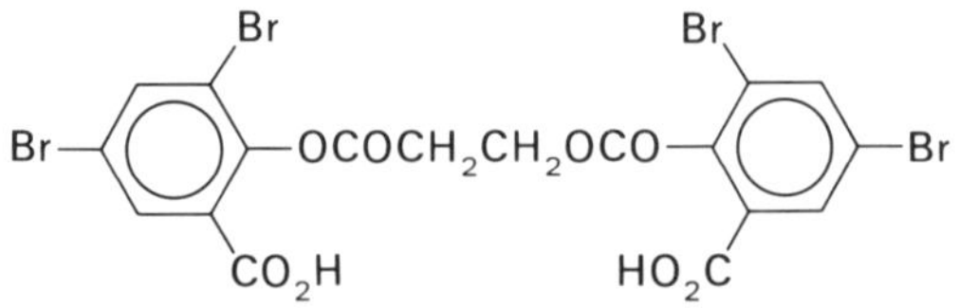

(**19**)

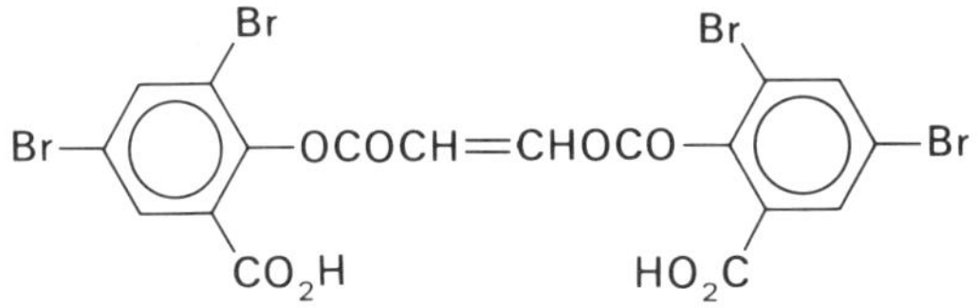

(**20**)

amino groups (Walder *et al.*, 1979; Zaugg *et al.*, 1980). These compounds again produce extensive modification of HbS *in vitro* at low doses and do indeed form cross-links between the β subunits as revealed by the presence of dimers when the Hb is denatured. Comparison of the length of the cross-linking bridge and the distances between amino groups at the β terminus suggested that the cross-link would be formed between the Lys 82 amino groups (Walder *et al.*, 1979) and this was verified by a crystal structure determination of Hb reacted under oxy conditions and crystallized under deoxy conditions (Walder *et al.*, 1980). This same study also confirmed that formation of the cross-linked fumarate derivative has little effect on oxygen affinity and that the anti-sickling action is thus a direct anti-gelling effect. Although formation of the cross-link is only specific for the β terminus under oxy conditions, the crystal structure analysis showed that it is also present under deoxy conditions with no major distortion of structure and should not therefore lead to any large change in the relative stability of the two forms. The presence of the cross-link does, however, lead to a movement of the Lys 82 residues closer together and this also moves other residues on the same section of chain (Phe 85 and Leu 88) which have been shown to be involved in the gelation process. This movement explains the anti-gelling effect since it would result in a non-alignment of important intermolecular contact regions in the deoxyHbS gel.

A large number of disaspirin analogues have been prepared in order to define structure–activity relationships. These have involved both changes in the linkage between the ester groups and in the substituents on the benzene rings (Wood *et al.*, 1981; Delaney *et al.*, 1984). The 4-carbon linkage provides the greatest reactivity with the rigid link of fumarate overall best but 5- and 6-carbon links are also tolerated. Electron-withdrawing substituents on the benzene rings are required for high reactivity and hydrophobic groups facilitate transport. Unfortunately no real improvement in properties over bis(dibromosalicyl)fumarate has apparently been found and one of the main problems with compounds of this type is a lack of selectivity for binding to Hb compared with other proteins; binding at the cell membrane represents a particular problem which would probably preclude even extracorporeal treatment of blood with such compounds (Chatterjee *et al.*, 1984). Following the lead of the work with the above compounds, mono-(3,5)-dibromosalicylfumarate has been shown to react at the Lys 82β residues to give a modified Hb with a right-shifted oxygen saturation curve (Bucci *et al.*, 1989). This modified Hb also had a prolonged half-life *in vivo* in the rat and its use as a blood substitute, as suggested for pyridoxylated Hb, was considered.

3.5.2 COMPOUNDS DESIGNED TO OTHER SITES

The contact sites between neighbouring molecules in the HbS polymer have been identified (Benesch *et al.*, 1982) and include the $\beta 6$ mutation site itself. In the polymer, the Val 6β of one HbS molecule nestles in a small hydrophobic pocket incorporating residues Phe 85β and Leu 88β of the neighbouring molecule. In addition, a hydrogen bond is formed between Thr 4β of one molecule and Asp 83β of the other. This information has been used in an attempt to find small molecules to interact at the 'donor' site, i.e. around Val 6β, or at the 'acceptor' site, i.e. around Phe 85β, with varying degrees of success. Linear peptides mimicking the amino-terminal end of the β chain showed moderate anti-gelling activity (Kubota and Yang, 1977; Votano *et al.*, 1977); peptides containing aromatic amino acids are most active (Gorecki *et al.*, 1980; Noguchi *et al.*, 1985; Burley *et al.*, 1987). Cyclic tetrapeptides mimicking the donor site were essentially inactive (Sheh *et al.*, 1987) as were proline analogues designed to the donor site (Abraham *et al.*, 1983a) or the acceptor site (Abraham *et al.*, 1984). The actual sites of interaction with Hb have not been identified for any of these compounds and with generally weak binding and possible multiple binding sites (Perutz *et al.*, 1986) it may be difficult to determine their precise mechanism of action and improve potency. Attempting to design small molecules to interact at surface sites on proteins rather than clefts may well prove to be a rather difficult task, generally.

3.6 LESSONS FROM HAEMOGLOBIN AND FUTURE WORK

Although the structure of Hb has been known for over 20 years now and numerous attempts have been made to design compounds to bind to Hb with therapeutic applications in mind, there are as yet no drugs in the clinic arising from this approach. This is perhaps partly due to a lack of interest in this area by pharmaceutical companies since the main perceived therapeutic target, sickle cell anaemia, represents a small market with poor prospects for recouping the enormous cost of modern drug development. It is also a reflection of the fact that the design of the drug to the receptor is only one factor in the search for acceptable drugs. *In vivo* activity after oral dosing and minimal toxic effects are also required and most designed compounds do not meet these requirements. The left-shifting compounds described in Section 3.4.1 are perhaps closest to the market. BW12C itself is absorbed very rapidly after oral dosing as measured by oxygen saturation curves of *ex vivo* samples but produces gastric mucosal

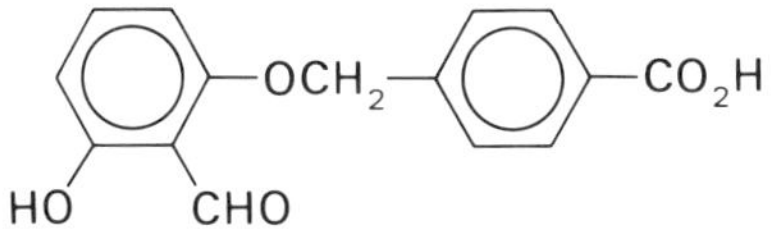

(**21**)

damage similar to that caused by aspirin but not as severe. This precludes the use of oral dosage for this compound but another analogue, BW589C (**21**), is absorbed more slowly and does not present any problems in oral dosing (White and Wootton, 1989). BW589C also appears to be more tightly bound than BW12C and has a very long half-life in man (Rolan *et al.*, 1989). No other toxicities have been encountered with BW12C in animal studies with prolonged dosages which were not directly attributable to the hypoxia of left-shift at high doses. It has been estimated that a maintenance of some 20–30% of the HbS *in vivo* in a non-deoxy form would be necessary to prevent sickling (Sunshine *et al.*, 1978) and modifications of the Hb of normal volunteers and sickle cell patients in the steady state of around 20% have been obtained (Fitzharris *et al.*, 1985; Keidan *et al.*, 1986). In further studies, a modification of around 40% has been achieved for about 2 hours by i.v. infusion in a number of healthy volunteers with no adverse effects (Nicholls *et al.*, 1989). The efficacy of BW12C in the palliation of sickle cell disease has not yet been assessed but clinical trials are underway. In addition, the use of BW12C to induce an oxygen deficit *in vivo* is also being explored in connection with cancer therapy since tumours are particularly sensitive to alterations in oxygen supply (Adams *et al.*, 1986, 1989; Honess *et al.*, 1989).

The advantage that BW12C and BW589C have over other compounds designed to bind to Hb in showing *in vivo* activity with no serious toxic effects stems largely from their potency and specificity for binding to Hb compared with other proteins. Generally, in control systems involving hormone receptor interactions the dissociation constant for the hormone lies in the range of hormone concentrations which occur since this allows maximum control of receptor occupancy and thus response levels. Hb is a rather special case, however, since its whole blood concentration is about 2.5 mM, very much higher than most receptors. This has perhaps led to the notion that tight binding is not necessary in order to modulate oxygen affinity and some spectacularly large right-shifts have been observed *in vitro* for weakly binding compounds, particularly of the hydrophobic acid type discussed in Section 3.3.2.2. These effects are usually very much reduced in whole blood, however, although bezafibrate still gives good shifts at higher doses. This is simply due to binding to plasma proteins which, for most compounds, is stronger than the binding to Hb, and, with bezafibrate as an exception, appears unsaturable with large numbers of binding sites (Hyde, unpublished). BW12C shows similar binding to plasma proteins but its binding to Hb is much tighter and plasma protein binding has little effect on whole blood or *in vivo* activity. Large doses of BW12C are still required *in vivo*, of course, in order to react a significant amount of Hb, particularly since more than one binding site is involved, but tight binding minimizes the necessary dose for therapeutic effect. It also reduces the level of unbound compound *in vivo* and is therefore important for minimizing toxic side effects; although free Hb and bound Hb are present at millimolar levels, the amount of free compound is determined by its dissociation constant and if this is submicromolar, as for BW12C, then the levels of free compound will also be less than micromolar.

The relatively tight binding of BW12C to Hb must reflect, in part, an important contribution of hydrophobic/van der Waals contact interactions; pyridoxal analogues discussed in Section 3.3.2 which make similar Schiff base and ionic interactions do not bind so tightly (Arnone *et al.*, 1977). Although it has not been possible to visualize such interactions using crystallography, the study of the effects of changes in the Hb structure at the postulated binding site using different species (chicken) and chemical modification (carbamylation) has provided valuable information regarding the relative importance of the different interactions, as did the classical SAR studies in which the structure of BW12C was chemically modified. Studies of mutant and different species of Hb were similarly valuable in providing evidence for the predicted binding modes for the right-shifting bibenzyls discussed in Section 3.3.1. It is now much easier to modify the structures of proteins at selected residues by site-directed mutagenesis (Fersht *et al.*, 1986; Nagai *et al.*, 1986; Wells *et al.*, 1987) and this is much more satisfactory than using Hbs from other species as it involves only changes at a single residue. In a particularly elegant study, Nagai *et al.* (1987) modified residues in the vicinity of the iron atoms to show that the basic mechanism of cooperativity as described in brief in Section 3.2 is valid. Although such studies, and also NMR studies in certain instances, are of great value in providing binding information under solution conditions, there is really no substitute for direct visualization of binding sites using crystallography, particularly in providing information on how to modify the structures of compounds to increase the number, and possibly quality, of interactions in order to achieve tighter binding. If only weakly binding compounds are available or inappropriate crystallization conditions are used, however, then the crystal structure information should be used with caution as simple modifications of the structure of compounds could result in completely different binding modes with different functional effects.

Hb research is apparently undergoing a fairly quiet period at the moment; now that the fundamentals of the mechanism of its homotropic and heterotropic interactions with ligands are apparently well understood, it may seem that the excitement is over. However, the physiology of the delivery of oxygen, an absolutely essential and fundamental process for almost all forms of life, is by no means fully understood and the means of controlling this delivery *in vivo* is not yet available. Reducing oxygen supply can be achieved by dosing with BW12C or, more simply, by inducing anaemia, but increasing oxygen delivery, which would be very desirable in many disease states, is not yet possible. Many compounds which right-shift the oxygen saturation curve of Hb in solution are now known but none of these is sufficiently potent and specific for Hb to right-shift, and thus increase oxygen delivery, at feasible doses *in vivo*. It has been argued that compensatory mechanisms will come into play should such a right-shift be achieved to restore normal oxygen delivery but until a compound is available the consequences of a right-shift will remain unknown. Drug design to the Hb molecule is still, therefore, a very interesting endeavour.

REFERENCES

Abraham, D. J., Mokotoff, M., Sheh, L. and Simmons, J. E. (1983a). *J. Med. Chem.*, **26**, 549–54.

Abraham, D. J., Perutz, M. F. and Phillips, S. E. V. (1983b). *Proc. Natl. Acad. Sci. USA*, **80**, 324–8.

Abraham, D. J., Gazze, D. M., Kennedy, P. E. and Mokotoff, M. (1984). *J. Med. Chem.*, **27**, 1549–59.

Abraham, D. J., Mehanna, A. S., Williams, F. S., Cragoe, E. J. and Woltersdorf, O. W. (1989). *J Med. Chem.*, **32**, 2460–7.

Adams, G. E., Barnes, D. W. H., du Boulay, C., Loutit, J. F., Cole, S., Sheldon, P. W., Stratford, I. J., van den Aardweg, G. J. M. J., Hopewell, J. W., White, R. D., Kneen, G., Nethersell, A. B. W. and Edwards, J. C. (1986). *Int. J. Radiat. Oncol. Biol. Phys.*, **12**, 1299–302.

Adams, G. E., Stratford, I. J., Nethersell, A. B. W. and White, R. D. (1989). *Int. J. Radiat. Oncol. Biol. Phys.*, **16**, 1179–82.

Arnone, A. (1972). *Nature*, **237**, 146–9.

Arnone, A. and Perutz, M. F. (1974). *Nature*, **249**, 34–6.

Arnone, A., Benesch, R. E. and Benesch, R. (1977). *J. Mol. Biol.*, **115**, 627–42.

Arnone, A., Rogers, P. H., Benesch, R. E., Benesch, R. and Kwong, S. (1986). *J. Biol. Chem.*, **261**, 5853–7.

Baldwin, J. and Chothia, C. (1979). *J. Mol. Biol.*, **129**, 175–220.

Beddell, C. R., Goodford, P. J., Norrington, F. E., Wilkinson, S. and Wootton, R. (1976). *Br. J. Pharmac.*, **57**, 201–9.

Beddell, C. R., Goodford, P. J., Stammers, D. K. and Wootton, R. (1979). *Br. J. Pharmac.*, **65**, 535–43.

Beddell, C. R., Goodford, P. J., Kneen, G., White, R. D., Wilkinson, S. and Wootton, R. (1984). *Br. J. Pharmac.*, **82**, 397–407.

Bellelli, A., Ippoliti, R., Currell, D., Condo, S. G., Giardina, B. and Brunori, M. (1986). *Eur. J. Biochem.*, **161**, 329–33.

Benesch, R. and Benesch, R. E. (1967). *Biochem. Biophys. Res. Commun.*, **26**, 162–7.

Benesch, R. and Benesch, R. E. (1969). *Nature*, **221**, 618–22.

Benesch, R., Benesch, R. E., Yung, S. and Edalji, R. (1975). *Biochem. Biophys. Res. Commun.*, **63**, 1123–9.

Benesch, R., Triner, L., Benesch, R. E., Kwong, S. and Verosky, M. (1984). *Proc. Natl. Acad. Sci. USA*, **81**, 2941–3.

Benesch, R. E. and Kwong, S. (1988). *Biochem. Biophys. Res. Commun.*, **156**, 9–14.

Benesch, R. E., Yung, S., Suzuki, T., Bauer, C. and Benesch, R. (1973). *Proc. Natl. Acad. Sci. USA*, **70**, 2595–9.

Benesch, R. E., Kwong, S. and Benesch, R. (1982). *Nature*, **299**, 231–4.

Benesch, R. E., Kwong, S., Benesch, R. and Baugh, C. M. (1985). *J. Biol. Chem.*, **260**, 14653–8.

Bolton, W. and Perutz, M. F. (1970). *Nature*, **228**, 551–2.

Brown, F. F. and Goodford, P. J. (1977). *Br. J. Pharmac.*, **60**, 337–41.

Bucci, E., Razynska, A., Urbaitis, B. and Fronticelli, C. (1989). *J. Biol. Chem.*, **264**, 6191–5.

Burley, S. K., Wang, A. H.-J., Votano, J. R. and Rich, A. (1987). *Biochemistry*, **26**, 5091–9.

Chanutin, A. and Curnish, R. R. (1967). *Arch. Biochem. Biophys.*, **121**, 96–102.

Chatterjee, R., Iwai, Y., Walder, R. Y. and Walder, J. A. (1984). *J. Biol. Chem.*, **259**, 14863–73.

Currell, D. L., Nguyen, D. M., Ng. S. and Hom. M. (1982). *Biochem. Biophys. Res. Commun.*, **106**, 1325–30.

Dean, J. and Schechter, A. N. (1978). *New Engl. J. Med.*, **299**, 752–63, 804–11, 863–70.

Delaney, E. J., Massil, S. E., Shi, G-Y. and Klotz, I. M. (1984). *Arch. Biochem. Biophys.*, **228**, 627–38.

DiDonato, A., Fantl, W. J., Acharya, A. S. and Manning, J. M. (1983). *J. Biol. Chem.*, **258**, 11890–5.

Edalji, R., Benesch, R. E. and Benesch, R. (1976). *J. Biol. Chem.*, **251**, 7720–1.

Fersht, A. R., Leatherbarrow, R. J. and Wells, T. N. C. (1986). *Nature*, **322**, 284–6.

Fitzharris, P. McLean, A. E. M., Sparks, R. G., Weatherley, B. C., White, R. D. and Wootton, R. (1985). *Br. J. Clin. Pharmac.*, **19**, 471–81.

Gillette, P. N., Manning, J. M. and Cerami, A. (1971). *Proc. Natl. Acad. Sci. USA*, **68**, 2791–3.

Goodford, P. J., St-Louis, J. and Wootton, R. (1978). *J. Physiol.*, **283**, 397–407.

Goodford, P. J., St-Louis, J. and Wootton, R. (1980). *Br. J. Phamac.*, **68**, 741–8.

Gorecki, M., Votano, J. R. and Rich. A. (1980). *Biochemistry.*, **19**, 1564–8.

Honess, D. J., White, R. D., Nethersell, A. B. W. and Bleehen, N. M. (1989). *Int. J. Radiat. Oncol. Biol. Phys.*, **16**, 1187–90.

Hyde, R. M., Livingstone, D. J., Paterson, R. A., Batchelor, J. F. and King, W. R. (1984). *Lancet*, *ii*, 15–16.

Hyde, R. M., Livingstone, D. J., Hersey, A., Dawson, J., Rahr, E. and Rose, V. S. (1988). *J. Drug. Dev.*, **1**, 109–17.

Imai, K. (1982). *Allosteric Effects in Haemoglobin*. Cambridge University Press, Cambridge.

Keidan, A. J., Franklin, I. M., White, R. D., Joy, M., Huehns, E. R. and Stuart, J. (1986). *Lancet*, *i*, 831–4.

Kennedy, P. E., Williams, F. L. and Abraham, D. J. (1984). *J. Med. Chem.*, **27**, 103–5.

Kenny, M. W. and Stuart, J. (1983). *Br. J. Haematol.*, **55**, 465–71.

Kilmartin, J. V. and Rossi-Bernardi, L. (1969). *Nature*, **222**, 1243–6.

Klotz, I. M. (1982). *Science*, **217**, 1247–9.

Klotz, I. M. and Tam, J. W. O. (1973). *Proc. Natl. Acad. Sci. USA*, **70**, 1313–5.

Körber. F. (1984). PhD Thesis, University of Leeds.

Kubota, S. and Yang, J. T. (1977). *Proc. Natl. Acad. Sci. USA*, **74**, 5431–4.

Lalezari, I. and Lalezari, P. (1989). *J. Med. Chem.*, **32**, 2352–7.

Lalezari, I, Rahbar, S., Lalezari, P., Fermi, G. and Perutz, M. F. (1988). *Proc. Natl. Acad. Sci. USA*, **85**, 6117–21.

Lalezari, I., Lalezari, P., Poyart, C., Marden, M., Kister, J., Bohn, B., Fermi, G. and Perutz, M. F. (1990). *Biochemistry*, **29**, 1515–23.

Mehanna, A. S. and Abraham, D. J. (1990). *Biochemistry*, **29**, 3944–52.
Merrett, M., Stammers, D. K., White, R. D. Wootton, R. and Kneen, G. (1986). *Biochem. J.*, **239**, 387–92.
Metzler, D. E. (1957). *J. Am. Chem. Soc.*, **79**, 485–90.
Monod, J., Wyman, J. and Changeux, J.-P. (1965). *J. Mol. Biol.*, **12**, 88–118.
Nagai, K., Perutz, M. F. and Poyart, C. (1986). *Phil. Trans. R. Soc. Lond. A*, **317**, 443–7.
Nagai, K., Luisi, B., Shih, D., Miyazaki, G., Imai, K., Poyart, C. De Young, A., Kwiatkowsky, L., Noble, R. W., Lin, S.-H. and Yu, N.-T. (1987). *Nature*, **329**, 858–60.
Nicholls, A. J., Parker, J. E. and Posner J. (1989). *Br. J. Clin. Pharmac.*, **28**, 215P.
Nigen, A. M. and Manning, J. M. (1977). *Proc. Natl. Acad. Sci. USA*, **74**, 367–71.
Noguchi, C. T., Luskey, K. L. and Pavone, V. (1985). *Mol. Pharmac.*, **28**, 40–4.
O'Donnell, S., Mandara, R., Schuster, T. M. and Arnone, A. (1979). *J. Biol. Chem.*, **254**, 12204–8.
Perutz, M. F. (1976). *Br. Med. Bull.*, **32**, 195–208.
Perutz, M. F. and Poyart, C. (1983). *Lancet*, *ii*, 881–2.
Perutz, M. F., Muirhead, H., Cox, J. M. and Goaman, L. C. G. (1968). *Nature*, **219**, 131–9.
Perutz, M. F., Fermi, G., Abraham, D. J., Poyart, C. and Bursaux, E. (1986). *J. Am. Chem. Soc.*, **108**, 1064–78.
Randad, R. S., Mahran, M. A., Mehanna, A. S. and Abraham, D. J. (1991). *J. Med. Chem.*, **34**, 752–7.
Rolan, P., Ingram, J., Leavens, W., Parker, J. E., Wootton, R., Gray, S. J. and Posner, J. (1989). *Eur. J. Clin. Pharmac.*, **36**, A110.
Shamsuddin, M., Mason, R. G., Ritchey, J. M., Honig, G. R. and Klotz, I. M. (1974). *Proc. Natl. Acad. Sci. USA*, **71**, 4693–7.
Sheh, L., Mokotoff, M. and Abraham, D. J. (1987). *Int. J. Pept. Prot. Res.*, **29**, 509–20.
Sunshine, H. R., Hofrichter, J. and Eaton, W. A. (1978). *Nature*, **275**, 238–40.
Votano, J. R., Gorecki, M. and Rich, A. (1977). *Science*, **196**, 1216–19.
Walder, J. A., Zaugg, R. H., Iwaoka, R. S., Watkin, W. G. and Klotz, I. M. (1977). *Proc. Natl. Acad. Sci. USA*, **74**, 5499–503.
Walder, J. A., Zaugg, R. H., Walder, R. Y., Steele, J. M. and Klotz, I. M. (1979). *Biochemistry*, **18**, 4265–70.
Walder, J. A., Walder, R. Y. and Arnone, A. (1980). *J. Mol. Biol.* **141**, 195–216.
Wells, J. A., Powers, D. B., Bott, R. R., Graycar, T. P. and Estell, D. A. (1987). *Proc. Natl. Acad. Sci. USA*, **84**, 1219–23.
White, R. D. and Wootton, R. (1989). *Br. J. Pharmac.*, **98**, 772P.
Wireko, F. C. and Abraham, D. J. (1991). *Proc. Natl. Acad. Sci. USA*, **88**, 2209–11.
Wood, L. E., Haney, D. N., Patel, J. R., Clare, S. E., Shi, G.-Y., King, L. C. and Klotz, I. M. (1981). *J. Biol. Chem.*, **256**, 7046–52.
Wootton, R. (1984). *FEBS Lett.*, **171**, 187–91.
Wootton, R., Stammers, D. K. and White, R. D. (1986). *FEBS Lett.*, **209**, 129–33.
Zaugg, R. H., Walder, J. A. and Klotz, I. M. (1977). *J. Biol. Chem.*, **252**, 8542–8.
Zaugg, R. H., Walder, J. A., Walder, R. Y., Steele, J. M. and Klotz, I. M. (1980). *J. Biol. Chem.*, **255**, 2816–21.

4 Drug Interactions with Target Enzymes of Known Structure

BARBARA ROTH
University of North Carolina at Chapel Hill, NC 27599, USA
AND DAVID K. STAMMERS
Wellcome Research Laboratories, Beckenham, Kent, UK

4.1. INTRODUCTION

In what a medicinal chemist might consider an ideal world, the three-dimensional structures of all molecular receptors would be known, and design of a drug targeted to interfere with or augment a reaction at that receptor would merely involve manipulation of small molecules to fit into the known ligand pocket. In actual fact, the number of known proteins of established structure which represent the target of action for drugs on the world market can be easily counted on the fingers of one hand. Furthermore, even with a molecular receptor which is a well defined enzyme, design of a ligand which fits into the

The Design of Drugs to Macromolecular Targets. Edited by C. R. Beddell
© 1992 John Wiley & Sons Ltd

active site cleft does not necessarily create a potent inhibitor or alternative substrate. The problem is much more complex than was originally realized.

This chapter discusses three enzymes which represent therapeutic targets, for which 3-D structural information is available. By far the greatest amount of information published to date concerns dihydrofolate reductase (DHFR, EC 1.3.1.5), an important enzyme which is widely distributed in nature. Several drugs have specificity for one or more of the forms of this enzyme which occur in different species and several 3-D structures for DHFR containing various ligands have been elucidated and refined to better than 2 Å by X-ray crystallographic methods. Spectroscopic techniques, particularly NMR, have also provided important structural information. Therefore this enzyme occupies the major topic in this chapter. We will not be limited to discussion of drugs on the market, but will, where appropriate, include closely related compounds whose interactions are known with the enzyme. The two other enzymes discussed are carbonic anhydrase and D-Ala-D-Ala-peptidase. Other enzymes or proteins are known which interact with drugs, but which are not the target enzyme. For example, the structure of the complex of penicillin and lysozyme (Johnson, 1967) has been determined but since this is not the target of action of this drug, we have elected to omit such cases.

4.2. DIHYDROFOLATE REDUCTASE (DHFR)

4.2.1 FUNCTION OF THE ENZYME

DHFR catalyses the NADPH dependent reduction of 7,8-dihydrofolate (**1a**) to 5,6,7,8-tetrahydrofolate (**1b**) thereby generating a chiral centre at C-6. In at least 17 catalytic reactions, this reduced folate is able to accept single carbon moieties in various states of oxidation to form 5- or 10- or 5,10-substituted derivatives,

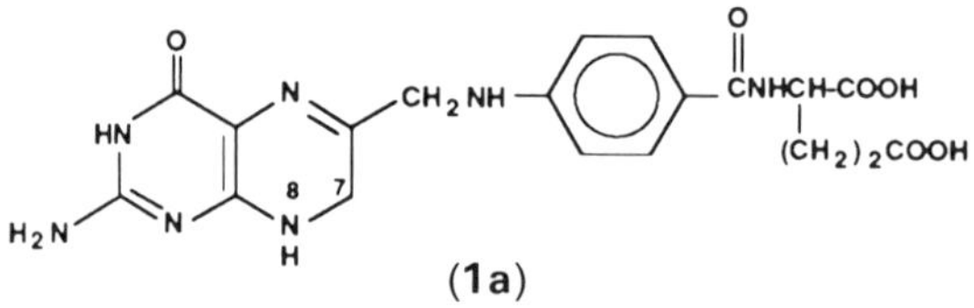

(**1a**)

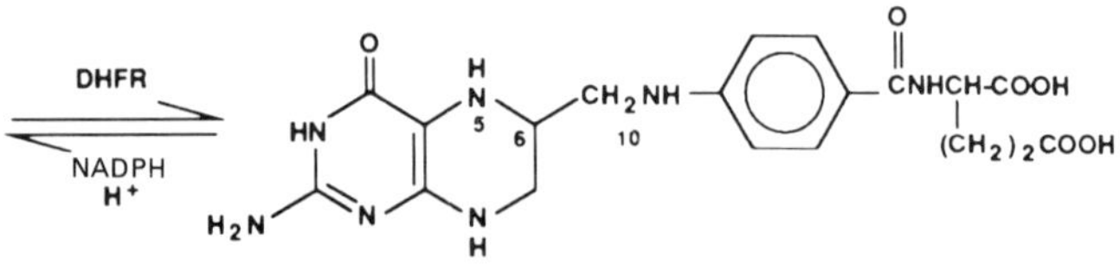

(**1b**)

such as 5,10-methylenetetrahydrofolate. These single carbon functions are then transferred enzymatically to form certain amino acids and building blocks for the nucleic acids. For example, the above-mentioned 5,10-methylene derivative presents its methylene group to the 5-position of deoxyuridylate in a reaction catalysed by thymidylate synthetase, thus forming thymidylate, and in this single case, also forming dihydrofolate as the other product of the reaction. The fact that this oxidation occurs may represent the most important function for DHFR, in cyclic reduction of dihydrofolate. Blockade of the DHFR reaction leads to cell death due to depletion of nucleic acid components. The phenomenon has been referred to as 'thymineless death' (Cohen, 1971).

4.2.2 SELECTIVE AND NON-SELECTIVE INHIBITORS AS DRUGS

Methotrexate (MTX, **2**), a very potent inhibitor of almost all DHFRs, was developed shortly after the elucidation of the folic acid structure, and before DHFR was a known entity (Seeger *et al.*, 1949). It soon became an important anticancer agent, and maintains this status today. It differs from folic acid only in the replacement of the 4-oxo function by an amino group, and in the addition of a methyl group to the N-10 nitrogen. Pyrimethamine (**3**), which was developed shortly thereafter, was found to be very active as an antimalarial, with reasonably low mammalian toxicity (Russell and Hitchings, 1951). Trimethoprim (TMP, **4**) also a product of the 1950s, has potent broad spectrum antibacterial activity but is well tolerated in the mammalian host (Hitchings and

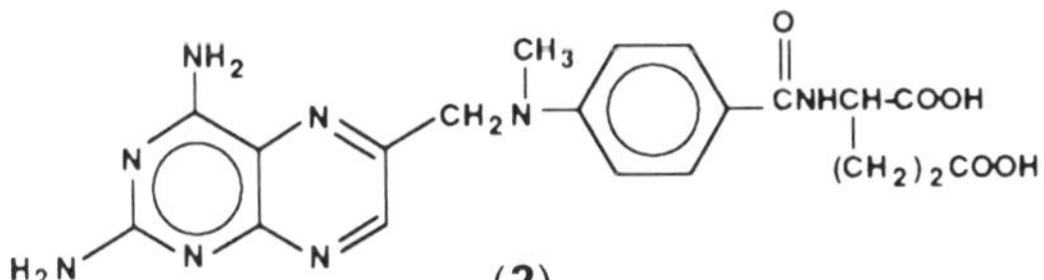

(**2**)

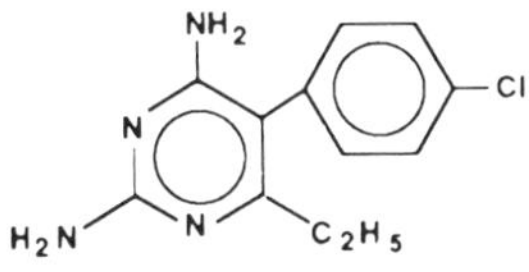

(**3**)

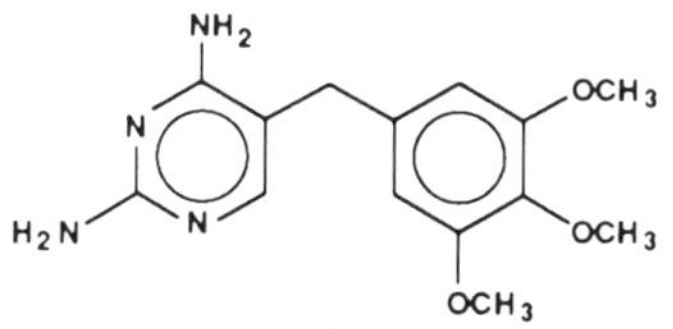

(**4**)

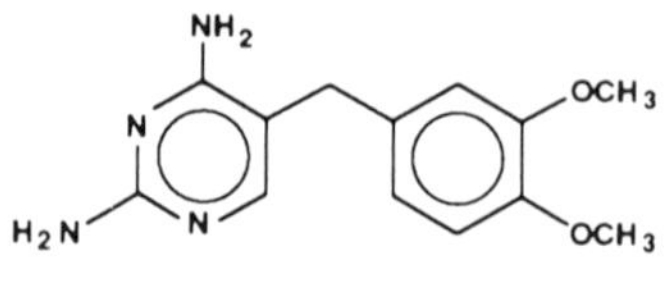

(5)

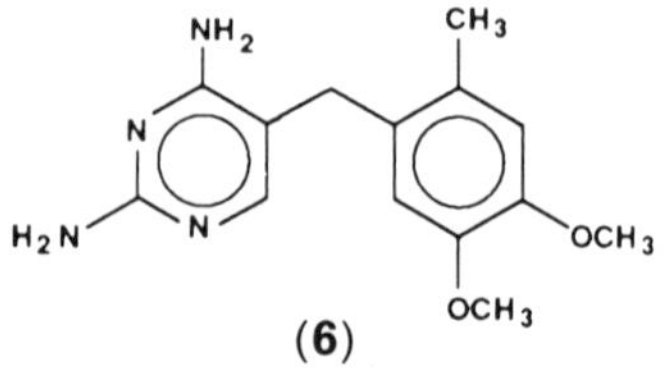

(6)

Roth, 1959; Roth, *et al.*, 1962). Its close relatives diaveridine (**5**) and ormetoprim (**6**) were found to be useful in combination with a sulpha drug against coccidiosis in chicks (Orton and Hambly, 1971; Versenyi, 1974). A dihydrotriazine (**7**) was discovered during that period to be the active metabolite of a biguanide antimalarial drug (Carrington *et al.*, 1951), and owes its activity to the inhibition of DHFR. Various direct analogues of the above compounds have been tested clinically, and some have found special applications. Two compounds currently in clinical trial as lipophilic anticancer agents, trimetrexate (**8**) and piritrexim (**9**) (Bertino *et al.*, 1979; Grivsky *et al.*, 1980), are also being tested against pneumocystis infections, a complication of AIDS (Queener *et al.*, 1987).

4.2.3 AMINO ACID SEQUENCES OF DIHYDROFOLATE REDUCTASES

By the 1960s it had become evident that this enzyme, although performing the same function in many species, must differ structurally from species to species (Burchall and Hitchings, 1965). This has led to the purification and sequencing

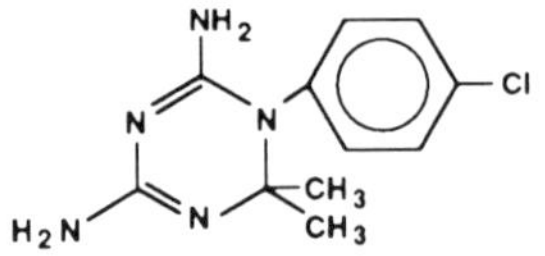

(7)

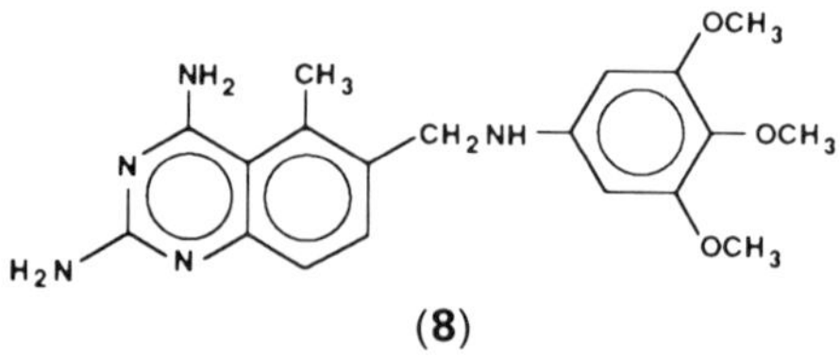

(8)

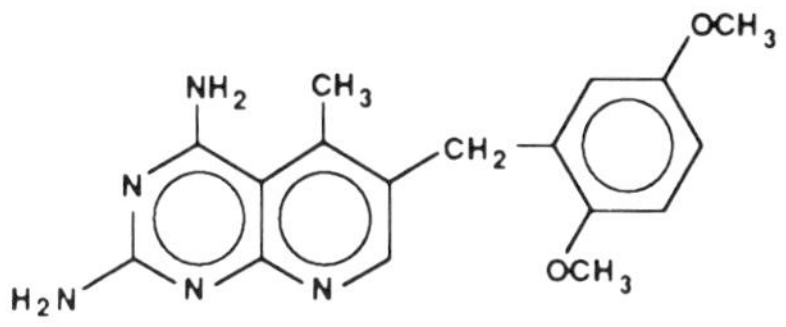

(9)

of more than 14 DHFRs (see Champness *et al.*, 1986a, for 14 comparisons, see also Roth, 1989). Sequences for the DHFRs discussed in this chapter are shown in Figure 4.1.

The bacterial DHFRs, which contain 159–162 residues, were found to have low homology (approximately 25–40%). The vertebrate enzymes contain about 30 more residues, and have a much higher homology to each other (about 75–90%), and low homology to bacterial species (of 20–30%). The R plasmid R67 enzyme from *E. coli* is an entirely different enzyme, with two identical subunits, each of 78 residues. It has little in common with the other DHFRs besides reducing dihydrofolate, whilst MTX and TMP inhibit only weakly. It is evident from the above that it was essential to determine the 3-D structures of several DHFRs from different species to determine their common features, and also to try to learn the basis for inhibitor selectivity. The following sections deal with structural results to date on the DHFRs of Figure 4.1.

4.2.4 STRUCTURE AND LIGAND INTERACTIONS

4.2.4.1 *Escherichia coli* DHFR

The first three-dimensional structure announced for a DHFR enzyme was that of a binary complex of *E. coli* (Ec) DHFR complexed with MTX (Matthews *et al.*, 1977). This was later refined to 1.7 Å (Bolin *et al.*, 1982). The main feature of the protein is an eight-stranded β sheet beginning at the amino terminus and ending as an antiparallel strand at the carboxyl end. Four helical regions are inserted into the loops connecting successive β strands. A *cis* Gly–Gly peptide bond is present at residues 95 and 96, which appears to be invariant with this enzyme. A 15 Å cavity cuts across one face of the enzyme; here MTX is bound. This is shown in Figure 4.2.

The pteridine ring of MTX is bonded through its N-1 and 2 amino group to Asp 27, which is invariant in bacterial species and replaced by Glu in the vertebrate enzymes sequenced to date, as well as in *Saccharomyces cerevisiae* DHFR. NMR experiments with the MTX and TMP complexes indicate that the pK_a values at N-1 of both bases are raised to 10 or above when complexed with the enzyme, so that a protonated ligand is interacting with Asp 27 anion (Roberts *et al.*, 1981; Cocco *et al.*, 1983). The 2-amino group is further H-bonded

```
Ec                M I S L . A . L A V D R V . . M E N A M . . – N . P A D L A W . K . N . L – – – – – –
Lc                  T A F L W . Q N R D G L . . . . . H . . . – H . P D D L H . . R A Q . V – – – – – –
Ng            M L K I T I . A . C A E . L C . . A G N A M . . – H I P E D F A F . K V Y . L – – – – – –

Cl            V R S L N S I V A V C Q N M G I G K D G N L P W P P L R N E Y K Y F Q R M T S T S H V E G
Human         . G . . . C . . . . S . . . . . . . N . D . . . . . . . . . F R . . . . . . T . . S . . .
L1210         . . P . . C . . . . S . . . . . . . N . D . . . . . . . . . F . . . . . . . T . . S . . .

R67       M E R S S N E V S N P V A G N F V F P S N A T F G M G D R V R K K S G A A W Q G Q I V G W Y C T

Ec      N K P . . . . R H . . E . . G – – – . . . . P G . K . . I L S S Q P –     G T D D R – V T W V . . V
Lc      G K I M V V . R R . Y E . F . – – K . . . P E . T . V . . T H Q E D       Y Q A Q . – A V V V H D V
Ng      G K P . . . . R . . . E . L . – – V K . . P G . R . . . I . . Q A D       Y C A A . – A E T V A . .

Cl      Q N A V I M G K K T W F S I P E K N R P L K D R I N I V L S R E L K       E A P K G A H Y L S K S L
Human   . . L . . . . . . . . . . . . . . . . . . . G . . . L . . . . . . .       . P . Q . . . F . . R . .
L1210   . . L . . . . R . . . . . . . . . . . . . . . . . . . . . . . . . .       . P . R . . . F . A . . .

R67     L T P E G Y A V E S E A H P G S V Q I Y P V A A L E R I N

Ec      E . I . A C G D V P – – – – – – – E I M V I . . G R . . E Q F L – – . K A Q K . Y L . H . D A Q V .
Lc      A V F . Y A K Q H L D Q – – – – – E L V . A . . A Q I F T . F K – – D D V D T . L . . . L A G S . .
Ng      V . . . . C A G A . – – – – – – – E A V . M . . A Q I . G Q . – – – . L A T D . R I . E V D L S V .

Cl      D A L A L L D S P E L K S K V D M V W I V G G T A V Y K A A M E K P I N H R L F V T R I L H E F E
Human   . . . K . T E Q . . . A N . . . . . . . . . . S S . . . E . . N H . G H L K . . . . . . . M Q D . .
L1210   . . . R . I E Q . . . A . . . . . . . . . . . S S . . E Q . . N E . G H L . . . . . . . . M Q . . .

Ec      . . H . . D Y E P D . W E S V F S – – – – – – – E F H D A . A Q N   S H S . C . . I L E R R
Lc      . . K M I P L N W D . . T K V S S – – – – – – – R T V . D T N – P   A L T H T Y . . W . . K A
Ng      . A . . . . . . R T H W R E A E R T E – – – – – R R V S S K . – –   – V A . T . V H . L G K

Cl      D T F F P E I D Y K D F K L L T E Y P G V P A D I Q E E D G – –   – I Q Y K F E V Y Q K S V L A Q
Human   . . . . . . . . L E K Y . . . P . . . . . L S . V . . . K . – –   – . K . . . . . . . E . N D
L1210   . . . . . . . . L G K Y . . . P . . . . . L S E V . . . . . . – –   – . K . . . . . . . E . K D
```

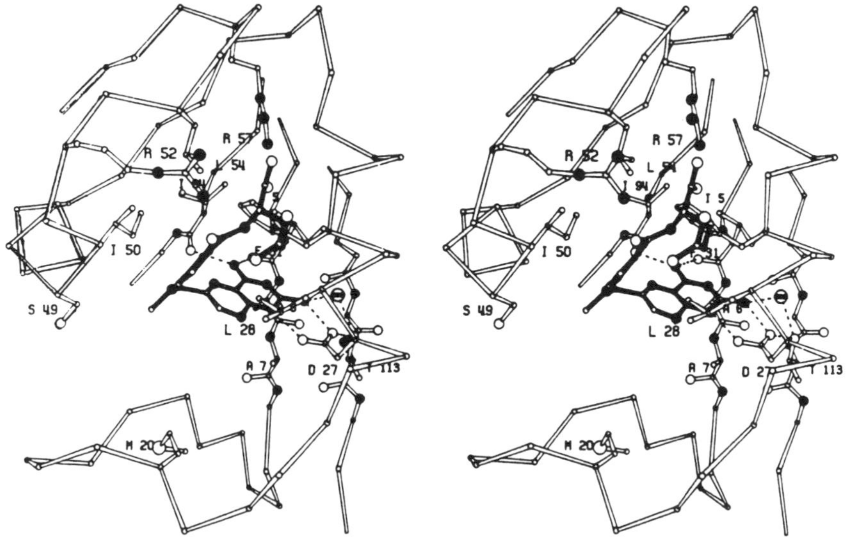

Figure 4.2. Stereo view of the binding of methotrexate to *E. coli* dihydrofolate reductase. Methotrexate is indicated by solid bonds and the protein by open bonds. Carbon atoms are represented by smaller open circles, oxygen atoms by larger open circles and nitrogen by blackened circles. Hydrogen bonds between the pyrimidine portion of methotrexate and the protein are indicated by dashed lines. (Reproduced from Matthews *et al.* (1985a) by permission of the American Society for Biochemistry and Molecular Biology)

through a water molecule to Thr 113. The 4-amino group is H-bonded to the backbone carbonyl functions of Ile 5 and Ile 94. Several van der Waals contacts are present between the pteridine and enzyme side chains. The *p*-aminobenzoyl moiety lies in a hydrophobic pocket comprising the side chains of Leu 28, Phe 31, Ile 50, Leu 54, and Ile 94. The glutamate portion interacts through its α-carboxyl group with Arg 57. There is no direct hydrogen bond from the γ-carboxyl function which interacts only with several water molecules that are in turn hydrogen bonded to Lys 32 and the α-carboxyl and amide groups of MTX.

The structure of the binary complex of TMP with Ec DHFR was reported in 1982 (Baker *et al.*,), and later refined to 2.3 Å (Matthews *et al.*, 1985a). The

---

Figure 4.1. Comparison of dihydrofolate reductase amino acid sequences. Alignments are based on those presented by Roth (1989). The one letter code for amino acids is used, (.) indicates a sequence identity with Cl DHFR, (–) indicates a deletion. The sources of the amino sequences are as follows: Ec, *Escherichia coli* (Stone *et al.*, 1977); Lc, *Lactobacillus casei* (Bitar *et al.*, 1977); Ng, *Neisseria gonorrhoeae* (Baccanari *et al.*, 1984); Cl, Chicken liver (Kumar *et al.*, 1980); Human, Human (Masters and Attardi, 1983); L1210, Mouse L1210 (Stone *et al.*, 1979); R67, R Plasmid R67 (Stone and Smith, 1979). (Reproduced in modified form from Roth (1989) by permission of J. R. Prous SA) pyrimidine ring occupies the same locus as the corresponding ring of MTX, and has the same H-bonding interactions. The geometry of the TMP molecule and its remaining interactions are best grasped by reference to Figure 4.3. The aromatic ring is found in the hydrophobic pocket described for the benzene ring of MTX. One edge of this ring and part of one face lies exposed to solvent. The torsional angles describing the rotations around C-5–C-7 and C-7–C-1′ designated as θ_1 and θ_2, are 177° and 76°, respectively, which positions the pyrimidine and benzene rings nearly perpendicular to each other.

The structure of Ec DHFR was used by Kuyper *et al.*, (1982, 1985) to design more potent analogues of trimethoprim. These contained carboxyalkoxy substituents in place of one *meta*-methoxy group, with the aim of forming an electrostatic interaction with the guanidinium group of Arg 57. Modelling studies suggested that compounds with three to five methylenes would form optimal interactions. Such analogues were synthesized and the relative affinities for Ec DHFR were measured (Table 4.1). The compound with R = $O(CH_2)_5COOH$ (**10a**), was found to have optimal affinity in both binary and ternary complexes. The corresponding esters showed no such enhancement. Crystallographic studies confirmed the proposed binding mode (Figure 4.4).

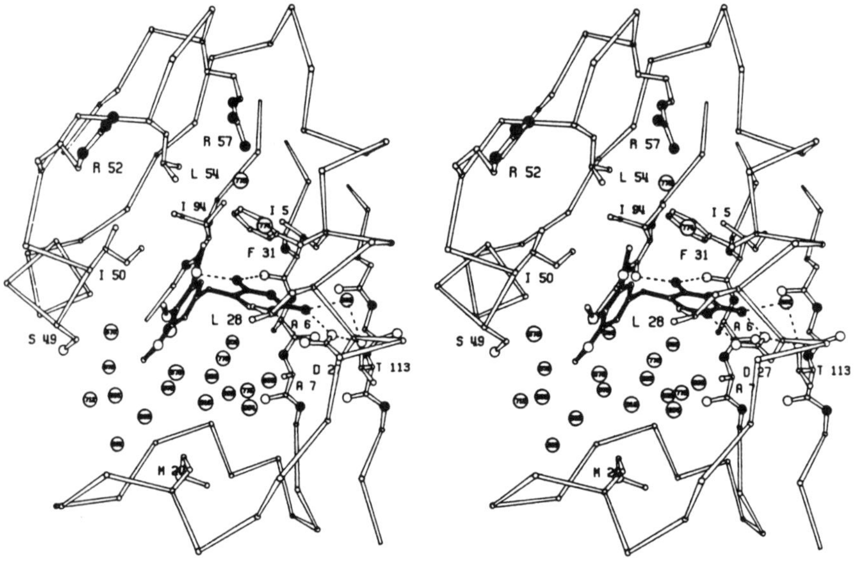

Figure 4.3. Stereo view of the binding of trimethoprim to *E. coli* dihydrofolate reductase. Trimethoprim is indicated by solid bonds and protein by open bonds. Carbon atoms are represented by smaller open circles, oxygen atoms by larger open circles, and nitrogen atoms by blackened circles. Large numbered circles represent fixed solvent molecules. (Reproduced from Matthews *et al.* (1985a) by permission of the American Society for Biochemistry and Molecular Biology)

Table 4.1. Relative affinities of some analogues of trimethoprim for *E. coli* DHFR

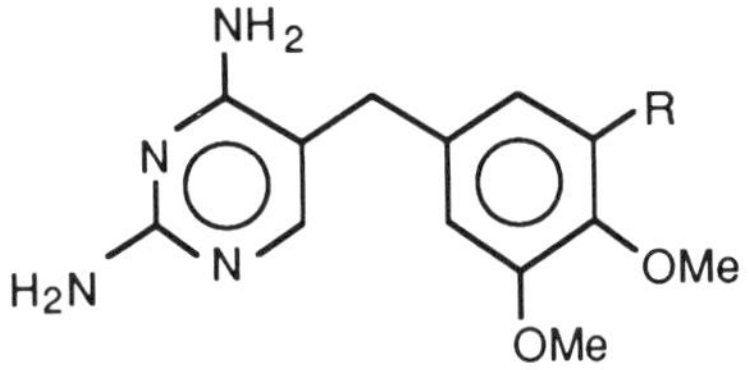

R	Affinity relative to trimethoprim	
	Binary complex	Ternary complex
OCH_3 (trimethoprim) (**4**)	1.0	1.0
OCH_2CO_2H	0.8	0.5
$O(CH_2)_2CO_2H$ (**10b**)	3.4	3.5
$O(CH_2)_3CO_2H$	6.7	37.0
$O(CH_2)_4CO_2H$	7.7	20.0
$O(CH_2)_5CO_2H$ (**10a**)	16.0	54.0
$O(CH_2)_6CO_2H$	7.7	26.0
$OCH_2CO_2CH_3$		0.1
$O(CH_2)_3CO_2CH_2CH_3$		2.8
$O(CH_2)_4CO_2CH_3$		1.7
$O(CH_2)_5CO_2CH_3$		1.5
$O(CH_2)_6CO_2CH_3$		0.7

Reproduced in modified form from Beddell (1984)

The analogue with R = $O(CH_2)_2COOH$ (**10b**) was not optimally positioned, there being only a single hydrogen bond to the arginine (Kuyper *et al.*, 1985).

This concept was further developed such that the carboxyl group interaction of **10a** with Arg 57 was replaced by a sulphonamide group (Hyde *et al* ., 1983). The compound **11** had enhanced potency over TMP for Ec DHFR and also inhibited dihydropteroate synthetase, although this did not contribute to the antibacterial activity. The proposed binding mode to DHFR was confirmed by X-ray crystallography (Hyde *et al.*, 1983).

A number of additional ligands have been crystallized with Ec DHFR. A

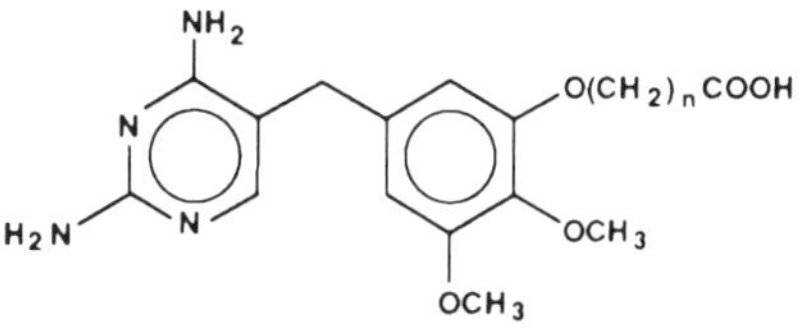

(**10a** *n* = 5)
(**10b** *n* =2)

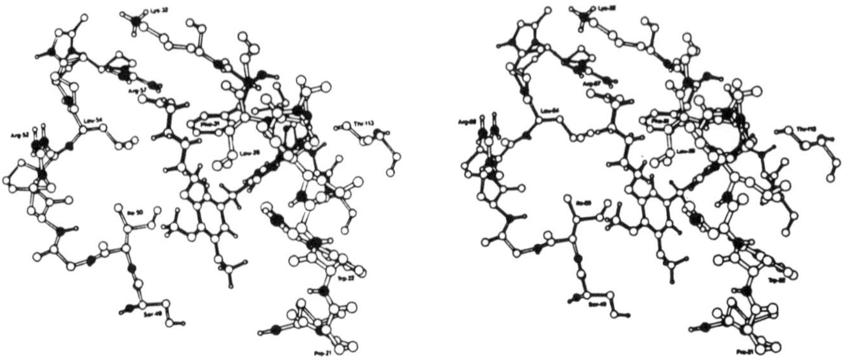

Figure 4.4. Stereo view of the binding of compound **10a** to *E. coli* DHFR. Compound **10a** is indicated by solid bonds and the protein by open bonds. (Reprinted with permission from Kuyper, L. F., Roth, B., Baccanari, D. P., Ferone, R., Beddell, C. R., Champness, J. N., Stammers, D. K., Dann, J. G., Norrington, F. E. A., Baker, D. J. and Goodford, P. J. (1985), *J. Med. Chem.*, **28**, 303–11. Copyright (1985) American Chemical Society)

pyridine analogue of TMP (Rauckman and Roth, 1980) in which N-3 was replaced by —CH═, and which was 300-fold less potent than TMP as an enzyme inhibitor despite the fact that it was a stronger base than TMP, was bound in the same manner as TMP to Ec DHFR. However, the electron density difference map compared to TMP showed a feature of negative density centred on C-3 and extending over the adjacent C-2 and C-4, which is explained by an increase in thermal vibration of the heterocyclic ring. The 3-position is near the main chain of Ile 5 and Ala 6, and the additional hydrogen atom at position 3 in the pyridine would seem to be unfavourably close to these residues (Baker *et al.*, 1983; Poulos and Finzel, 1985). Indeed bulkier substituents than H at C-3 progressively reduce potency (Rauckman and Roth, 1980).

Several studies have been reported recently on single-residue mutations constructed by oligonucleotide-directed mutagenesis of the cloned wild-type Ec DHFR gene; one very thorough report included X-ray examination of the mutants (Howell *et al.*, 1986). The authors found that the Asp 27 → Asn and Asp 27 → Ser mutants had K_D values, respectively, of 1.9 nM and 210 nM at pH 7.0, versus 0.07 nM for the wild type. The X-ray structures were determined for both

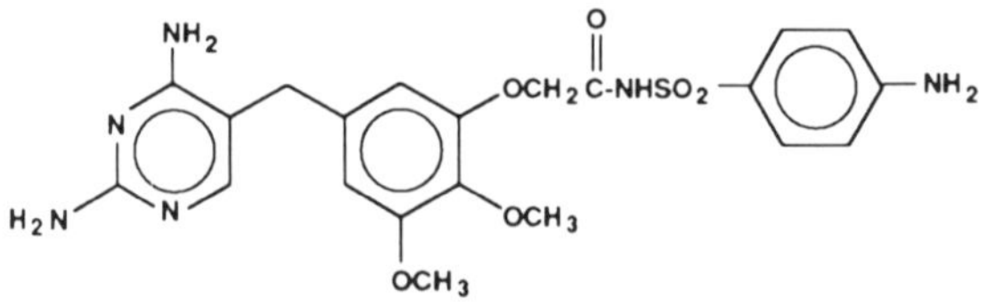

(11)

mutants and refined at 1.9 Å. Difference maps showed that with the Asn 27 structure the largest changes involved shifts of two fixed water molecules. With Ser 27, the changes were more complex due to the smaller size of Ser. A new water molecule was introduced such that Asp 27 was replaced by Ser hydrate. No other coordinate changes greater than 0.2 Å occurred. MTX was not protonated in either case, as determined by difference spectroscopy. The authors deduced new H-bonding patterns and calculated an apparent binding energy loss of only 1.8 kcal mol $^{-1}$ upon eliminating the ionic interaction between MTX and Asp 27.

Kinetic studies have shown that there is cooperativity in the binding of inhibitors and cofactor to DHFR. The magnitude of this effect varies between DHFRs from different species. TMP shows a 40-fold tighter binding in the ternary complex with NADPH and Ec DHFR than in the binary complex, whilst only three- to four-fold tighter binding for comparable complexes for mouse L1210 DHFR occurs (Baccanari *et al.*, 1982). Thus an overall factor of ten in selectivity of TMP results from this. Possible structural explanations of this have been investigated crystallographically. The ternary complex of Ec DHFR with TMP and NADPH has been crystallized and its structure determined by molecular replacement (Champness *et al.*, 1986b). A stereo view is shown in Figure 4.5. TMP was found to have torsion angles similar to those in the binary complex, with new hydrophobic contacts due both to the addition of the cofactor and the fact that the protein folding was altered in the region of residues 15–21 so that the chain moved up to close the cleft. This involved rotation of the side chain of Met 20 so that it now was in contact with the

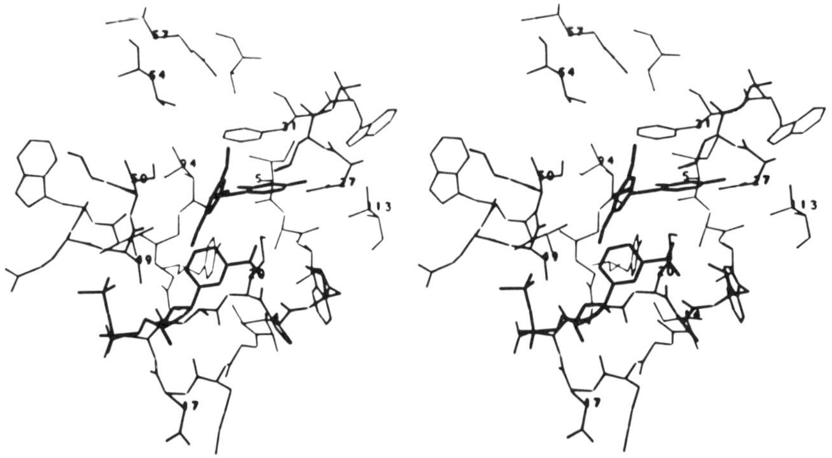

Figure 4.5. Stereo view of the binding of TMP and NADPH to Ec DHFR. TMP and part of the cofactor are drawn with bold lines. (Reproduced with permission from Champness *et al.*, 1986b)

aromatic region of the ligand, and in particular with a *meta*-methoxy group which was surrounded by solvent in the binary complex.

4.2.4.2 *Lactobacillus casei* DHFR

The structure of the ternary complex of *L. casei* (Lc) DHFR with MTX and NADPH was reported in 1978 (Matthews *et al.*) and refined to 1.7 Å in 1982 (Bolin *et al.*). Its backbone structure was compared with that from Ec DHFR in binary complex with MTX. Despite the many sequence differences the backbone structures were very similar except for the 'teen loop' region, where this segment of the chain was found to move by about 3 Å in Lc, compared to Ec. The overall binding pattern of MTX was practically identical with that from Ec DHFR, despite the many differences in side chains. Predictably, however, the residues involved in the principal contacts were mostly invariant. In the binding area for the Glu moiety of MTX, the α-carboxy group again interacted with Arg 57, but the γ-carboxy group now interacted with His 28, a residue not present in Ec DHFR.

The NADPH cofactor is bound to the enzyme in an extended conformation, and occupies a long cleft in the enzyme surface, stretching across five strands of β sheet, with a distance of about 17 Å between the nicotinamide- and adenine-binding pockets. As might be expected from its function, the nicotinamide ring lies close to the pyrazine ring of MTX. Another feature of interest is that three protein oxygen atoms lie very close to ring carbons C-2, C-4 and C-6 of the nicotinamide and in plane with the ring. The authors suggest a possible involvement of these carbonyls in the stabilization of the transition state. It is well established that overall hydride transfer takes place from the A side of the nicotinamide ring (Charlton *et al.*, 1979; Fonticella-Camps *et al.*, 1979).

One startling observation that could be quickly made upon observing the relation between the nicotinamide and the pyrazine ring of MTX was that the reduced pyridine was in a position to deliver its hydrogen atom to the *wrong side* of the pyrazine, since the absolute stereochemistry of 5,10-methylenetetrahydrofolate derived from the biologically active diastereoisomer of folinic acid is known. This geometry is the opposite of what is observed in the MTX complex with Lc DHFR and NADPH, however. One could only conclude that the inhibitor has its pteridine ring flipped so that the pteridine ring presents its opposite face to the cofactor.

A large number of sophisticated NMR experiments have been carried out by Roberts and co-workers on Lc DHFR. The TMP binding geometry was predicted and found to correspond closely with X-ray data on Ec DHFR (Cayley *et al.*, 1979; Birdsall *et al.*, 1983).

When MTX was converted to its α, γ, and α,γ amides, the binding to Lc DHFR was decreased 100-fold with the α amide, 10-fold with the γ amide, and a little more than 100-fold with the di-amide. However, NMR studies of the histidine residues plus other NMR evidence indicated that the glutamate moiety must undergo a substantial conformational change when the α-carboxy group is modified (Antonjuk *et al.*, 1984).

All of the 60 methyl resonances of the valine and leucine residues in Lc DHFR have been identified by 2-D NMR (Searle *et al.*, 1986). In addition, all the histidine and tyrosine resonances and several isoleucine and phenylalanine resonances have been assigned (Hammond *et al.*, 1986) which means about 30% of the protein has been assigned. Since such studies are carried out in solution, very valuable data on conformational movements can be obtained by such techniques. It should be noted that X-ray data for the Lc ternary complex largely matches data found by NMR in solution.

Conformational changes produced by cofactor binding have been thoroughly investigated (Roberts, 1987). The complex of Lc DHFR with TMP and $NADP^+$ exists in solution as a mixture of two approximately equal amounts of slowly interconverting conformational states (Gronenborn *et al.*, 1981; Birdsall *et al.*, 1984a). In one form the nicotinamide ring has swung away from the enzyme into solution.

In 3′,5′-difluoro-MTX bound to Lc DHFR, the benzoyl group was found to flip about its symmetry axis at a rate increased by addition of coenzyme (Clore *et al.*, 1984). Studies on the design of more potent analogues of trimethoprim using the X-ray structure of Lc DHFR have been described by Birdsall *et al.* (1984b). The analogues contained *m*-carboxyalkoxy substituents which were intended to bind to Arg 56. These are thus related to the compounds described by Kuyper *et al.* (1982). Compound **12** was 50-times more potent than brodimoprim, itself an analogue of trimethoprim. A compound containing a second carboxyl group designed to interact with His 28 was about 10^3 times more potent on Lc DHFR than brodimoprim. The predicted binding site was confirmed by NMR studies.

An irreversible inhibitor of Lc DHFR was designed to interact with His 28, by converting MTX to a derivative with alkylating potential (**13**); the irreversible reaction occurred as predicted, and the structure was proven (Rosowsky *et al.*, 1982). Another irreversible inhibitor (**14**) was prepared before the DHFR structures were known, in a search for irreversibility (Roth *et al.*, 1981). Although **14** did not interact irreversibly with Ec, rat, human or chicken DHFRs, it did with Lc DHFR, as well as with the *Neisseria gonorrhoeae* (Ng) DHFR (see section below) (Tansik *et al.*, 1984). Molecular modelling of **14** with

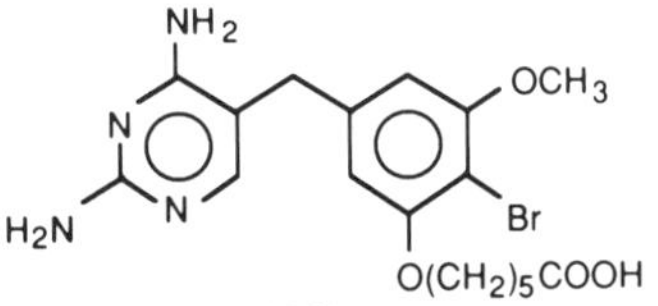

(**12**)

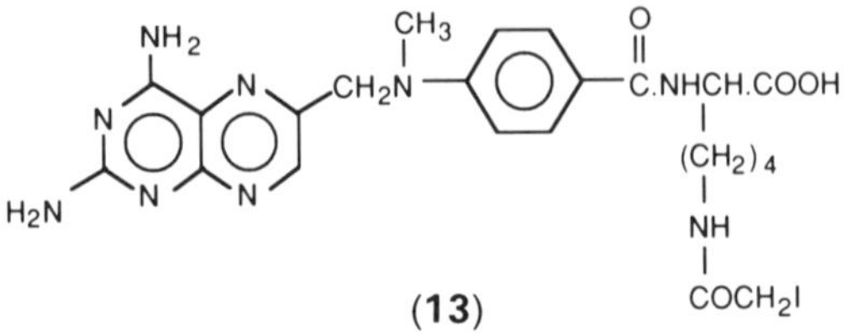

(**13**)

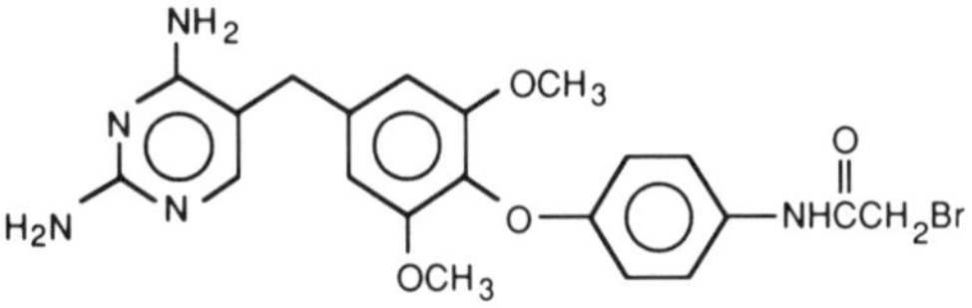

(**14**)

Lc DHFR suggested potential interactions at His 18, 22, or 28. The problem was investigated by molecular mechanics (Chan *et al.*, 1985), and the lowest energy complex was found to be that at the ring nitrogen of His 22. This interaction was confirmed to be with residue His 22 by 2-D NMR (Roberts, Birdsall, private communication, 1986). It will be noted by reference to Figure 4.1 that Ec chicken and human DHFRs have no suitably reactive side chains in the region between residues 18 and 28.

4.2.4.3 *Neisseria gonorrhoeae* DHFR

Although this enzyme has been sequenced (Figure 4.1), no X-ray structure is available. However, a model has been constructed from the Ec structure (Kuyper, personal communication). The chief difference from Ec DHFR in the active site region is the exchange of Leu 28 by Phe, which, due to its larger size, creates a more hydrophobic cleft.

TMP has been used successfully in the clinic for treating gonorrhoeal infections but high doses are required (Austin *et al.*, 1973). It is a relatively weak inhibitor of Ng DHFR ($I_{50} = 45 \times 10^{-8}$ M) (Roth *et al.*, 1987). However, a number of more lipophilic TMP derivatives are considerably more active, such as **15** ($I_{50} = 2.9 \times 10^{-8}$ M) (Roth *et al.*, 1987, 1988). We have noted previously that in Ec DHFR the edge of the trimethoxyphenyl ring lies partially exposed to

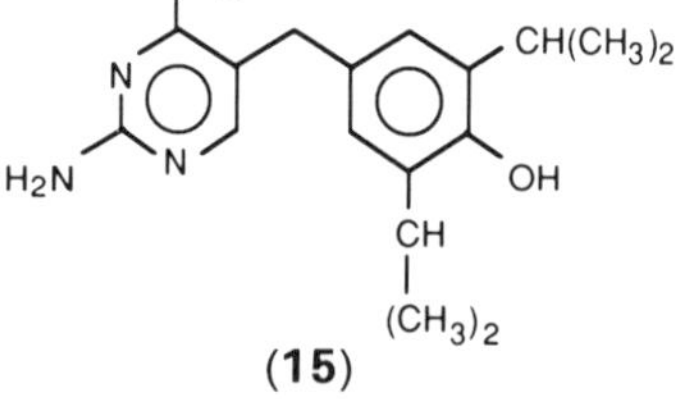

(**15**)

solvent. In the Ng DHFR enzyme the methoxy groups would presumably be shielded from solvent by the Phe residue which corresponds to Leu 28, but desolvation energy would be required on forming the complex. Thus, a more lipophilic derivative would be expected to be more active. Compound **14** is an irreversible inhibitor of Ng DHFR. Its single site of action was proven to be His 25 (equivalent to His 22 of Lc DHFR) by sequence analysis (Tansik *et al.*, 1984). (Note that in contrast to Lc DHFR, the Ng enzyme has only one histidine in the 18–28 region.) This compound represents the first case of a 'site specific' irreversible inhibitor of DHFR with specificity for a DHFR of therapeutic interest. (Figure 4.1 should be consulted again to note that human DHFR has no appropriate side chains for interaction with **14**.) Since a suggestion of reactivity with other proteins has been observed, **14** has not been developed as a potential drug.

4.2.4.4 Chicken Liver DHFR

The 3-D structure of chicken liver (Cl) DHFR has been solved in ternary complex with a number of ligands plus NADPH at 2.2 Å resolution (Volz *et al.*, 1982; Matthews *et al.*, 1985a, b). The holoenzyme (binary complex with NADPH) has also been structured at 2.0 Å (Matthews *et al.*, 1985a). A dihydrotriazine ligand was investigated initially (Volz *et al.*, 1982), since such compounds bind to the enzyme reasonably well. Of greater interest, however, was the structure of the ternary complex with TMP, since this was expected to provide the long sought answer to the specificity question.

The Cl protein has 189 residues, compared with 159 for Ec DHFR, so the first question concerned the backbone structure. It was found that many of the extra residues were accommodated in loops near the surface of the protein that interconnect helices and β strands. In addition, a large insertion in β strand G, termed a β blowout (Volz *et al.*, 1982) was present. These extra protein segments are accommodated without undue interference in the overall fold although the width of the active site cleft of Cl DHFR is about 1.5 Å wider than in Ec DHFR. A superposition of the α carbon atoms of the two proteins is shown in Figure 4.6.

A stereo view of the active site region of the TMP ternary complex with NADPH is shown in Figure 4.7. Two hydrophobic cavities are found, whereas the bacterial enzymes have room for only one. A very obvious difference in conformation of TMP from that in the Ec DHFR complex is observed. The two rings now lie almost as an open book, rather than being nearly perpendicular to each other. Torsional angles (θ_1, θ_2) between the two rings are $-85°$ and $102°$, respectively. The pyrimidine ring of TMP is inserted more deeply into the active-site cleft than is a phenyldihydrotriazine ligand, or than TMP or MTX bound to Ec DHFR. The result is that the 4-amino group of TMP cannot now form two hydrogen bonds to backbone carbonyl atoms as it did in Ec DHFR. There is left only one hydrogen bond, to Ile 7. However, the dihydrotriazine is

Figure 4.6. Stereo view of a comparison of the α-carbon backbones of Cl DHFR (dashed line) and Ec DHFR (solid line). (Courtesy of Dr L. Kuyper)

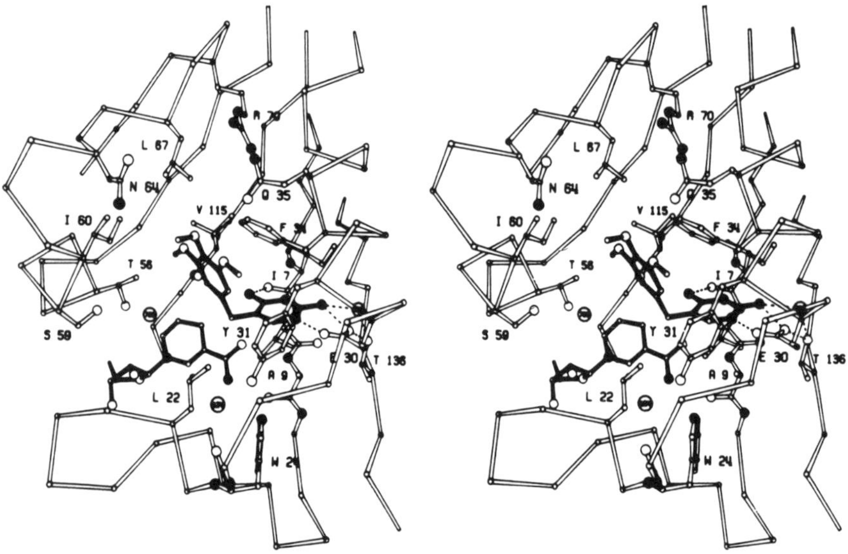

Figure 4.7. Stereo view of the active site of Cl DHFR showing the binding of TMP and NADPH. TMP is indicated by solid bonds, the enzyme by open bonds and a portion of NADPH is also shown. Carbon atoms are shown by small open circles, oxygen atoms by larger open circles and nitrogen atoms by solid black circles. Fixed solvent molecules are shown by large numbered circles. (Reproduced from Matthews *et al.* (1985a) by permission of the American Society for Biochemistry and Molecular Biology)

hydrogen bonded to both Ile 7 and Val 115 respectively (equivalent to Ile 5 and 94 in Ec DHFR). TMP lies with its trimethoxybenzyl portion in the upper hydrophobic cleft, in a more hydrophobic environment than was the case in the bacterial enzyme. One *meta*-methoxy group is completely buried. On comparing the positions of the enzyme side chains in the TMP complex with those in the holoenzyme it was found that Tyr 31 underwent a very large conformational shift (about 5 Å) upon binding TMP, and the phenolic group now formed a new hydrogen bond to a carbonyl of Trp 24.

It has been argued (Matthews *et al.*, 1985b), that the large tyrosine movement probably requires little energy, since in a complex where TMP is replaced by its *para*-isopropenyl analogue the tyrosine shift does not occur. The conclusion is that the main source of the 3000-fold weaker binding energy for TMP to Cl DHFR than to the Ec enzyme is the loss of the hydrogen bond at the 4-amino group. The remaining bonding loci have been carefully compared. Protonated TMP interacts with Glu 30 in the same manner as it does with Asp 27 in Ec DHFR. The trimethoxybenzyl moiety is surrounded by a hydrophobic pocket consisting of the side chains of Phe 34, Met 52, Ile 50, Leu 67, and Val 115, as well as the methyl of Thr 56. The nicotinamide ring of the cofactor, however, is at no point less than 4 Å from the inhibitor. The large cooperative effect of adding the coenzyme is not observed here, although there is a minor additivity in binding.

A further contribution to species selective inhibition of DHFR by TMP is attributable to the methoxy groups. The triethyl isostere, **16**, has nearly the same potency as TMP against Ec DHFR, and it is 25-fold more active against Cl DHFR (Roth and Aig, 1987). When the compound was soaked into crystals of Cl DHFR holoenzyme and a difference map calculated it was observed that a very similar binding position as TMP was present (Matthews *et al.*, 1985b). It seems possible that differences in desolvation of the inhibitors might account for the different selectivity, methoxy being more hydrophilic than ethyl. In Ec DHFR the methoxy groups of TMP (and presumably the methylene groups of the triethyl isostere) are in contact with the solvent, and perhaps little desolvation energy is required for binding. In Cl DHFR these functions (particularly one *meta* substituent) are largely buried and hence more desolvation energy will be required for TMP than for the triethyl compound.

To analyse properly all of the causes for the specificity of TMP for bacterial DHFR is an exceedingly difficult problem, which will require study of the energetics of each atomic interaction, not neglecting solvent effects. Phenyldi-

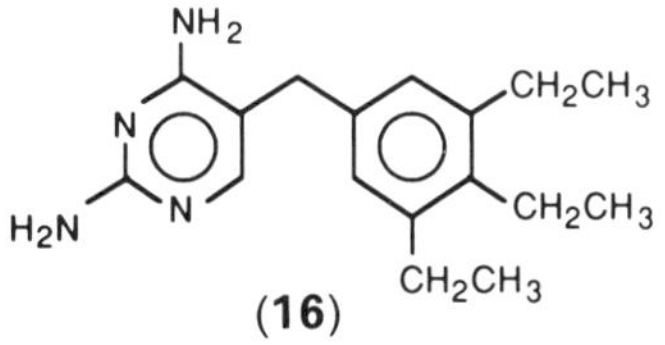

(**16**)

hydrotriazines interact with Cl DHFR in approximately the same manner as pyrimidines (Matthews *et al.*, 1985b). Thus analogues of **7** behave like puckered pyrimidines, where the 5-carbon has been replaced by a substituted nitrogen. The other nitrogen attached to the side of the gem-dimethyl groups is the protonated centre which interacts with Glu 30, along with the adjacent amino group. For steric reasons the aromatic and dihydrotriazine rings lie perpendicular to each other; when the *p*-methoxy analogue of **7** was superimposed on TMP in Cl DHFR the two aromatic rings were found to be nearly perpendicular to each other but were centred in the same space. A 5-adamantylpyrimidine was found to bind to Cl DHFR in almost the same manner as the phenyldihydrotriazines (Cody and Zakrzewski, 1982; Matthews *et al.*, 1985b). Two analogues of **7** with long chains in the *meta*-position (Hansch *et al.*, 1984) were crystallized with Cl DHFR. These chains were bound in a narrow channel leading from the extreme upper portion of the active site cleft to the protein surface (Matthews *et al.*, 1985b). This channel lies near Arg 70 (equivalent to Arg 57 in Ec DHFR). A series of 5-phenylpyrimidines containing *meta*-carboxyalkoxy chains in the benzene ring was investigated to see whether binding might be enhanced by interaction with Arg 70 (Roth, 1986). The expected increase in binding was, however, not observed.

4.2.4.5 Mouse L1210 DHFR

The X-ray structure of a ternary complex of L1210 DHFR with TMP and NADPH was solved at 2.5 Å (Stammers *et al.*, 1983) and refined to 2.0 Å (Stammers *et al.*, 1987). A corresponding MTX complex was similarly refined to 2.5 Å (Stammers *et al.*, 1987). These are the first structures reported for a mammalian DHFR.

The binding of TMP to the L1210 enzyme is essentially the same as that reported for Cl DHFR. Once more a single H-bond exists between the 4-amino group and Ile 7; the amino function is 0.6 Å farther from Val 115 than in Cl DHFR. TMP has inter-ring torsion angles of $-96°$ and $85°$, with the valence angle at the methylene group opened out to 119°, where it is normally about 115° (Figure 4.8). The position of Phe 31, which replaces Tyr 31 of Cl DHFR, is similar to that observed in the Cl DHFR holoenzyme, suggesting that it has not undergone much of a conformational change upon binding TMP. The opened out valence angle for the methylene group appears to optimize van der Waals contacts of the drug with the side chain of Phe 31. Once more the cleft is somewhat wider than it is in the bacterial enzymes, but with less of a difference than with Cl DHFR. MTX binding is of interest, since this complex was not obtained with Cl DHFR. The pteridine binding is similar to that found in bacterial DHFRs, despite the differences in the active site region. The 4-amino group of the pteridine is now H-bonded to the main chain carbonyl oxygens of both Ile 7 and Val 115. Among the non-polar interactions the relative positions

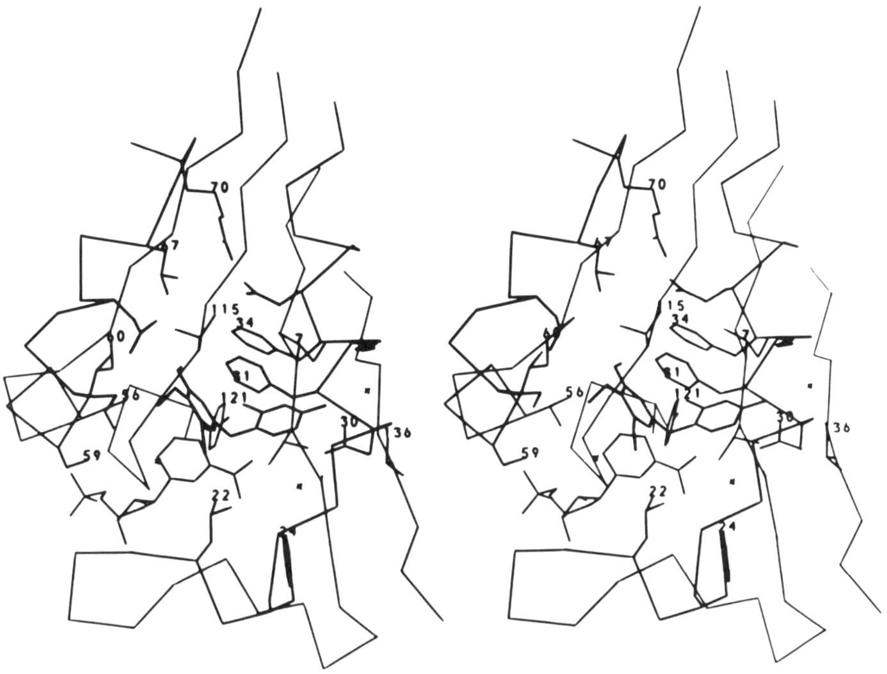

Figure 4.8. Stereo view of the binding of trimethoprim to L1210 DHFR

of the side chains of Ec DHFR Leu 28 and Phe 31, which are perpendicular to each other, are approximately parallel in this DHFR (L1210 Phe 31 and Phe 34). Apparently as a result of this the benzoyl group of MTX is rotated by about 60° compared to its orientation in bacterial DHFR. This in turn alters other contacts.

The binding of the glutamate moiety is less clearly defined in this structure than are the pteridine or benzoyl parts. The best interpretation has the γ-carboxyl group making no H-bond interaction with the protein, while the α-carboxyl group is hydrogen bonded via a water molecule to Gln 35; no interaction with Arg 70 is observed. This result is surprising, since MTX is as tightly bound to L1210 DHFR as in other DHFRs. It should be noted that the MTX complex was prepared by soaking crystals of the enzyme/NADPH/TMP complex in the inhibitor, which may conceivably have prevented significant conformational changes.

4.2.4.6 Human DHFR

The crystal structure of recombinant human DHFR with folate bound in the active site has recently been announced, at 2 Å resolution (Oefner *et al.*, 1988).

Binary complexes with MTX and with TMP have also been investigated at 3.5 Å. The results show clearly that the pteridine rings of folate and MTX have inverse orientations in the enzyme relative to the *p*-aminobenzoylglutamate portions of the molecule, as postulated earlier by Bolin *et al.* (1982). Figure 4.9 and 4.10 compare the binding of folate and MTX to this enzyme. The 2-amino and 3-NH functions of folate are hydrogen bonded to Glu 30, which in turn is hydrogen bonded to Thr 136. The 4-oxo group is hydrogen bonded by way of a water molecule to invariant Trp 24. Interestingly, the authors state that neither N-5 nor N-8 have hydrogen bonds either to protein or solvent, although a proton is required to initiate hydride transfer from the NADPH. They suggest that proton transfer may arise by intermediary protonation of the 4-oxo group. They also suggest that the 8-NH of dihydrofolate might be hydrogen bonded to the backbone carbonyl of Ile 7, thus positioning the substrate for hydride transfer to C-6, rather than C-7 of the pteridine. Although NADPH is not present in the figures, it would be situated with the dihydropyridine ring behind the pteridine, near the 6- and 7-positions of the latter.

The binding of the pteridine ring of MTX is as described for the bacterial and L1210 enzymes. However, the *p*-aminobenzoylglutamate moiety is not as found for the L1210 enzyme, but rather places the α-carboxyl group in appropriate alignment for hydrogen bonding to Arg 70, as was the case with the Lc and Ec enzymes. Phe 31, at the front of the cleft, now has its aromatic function

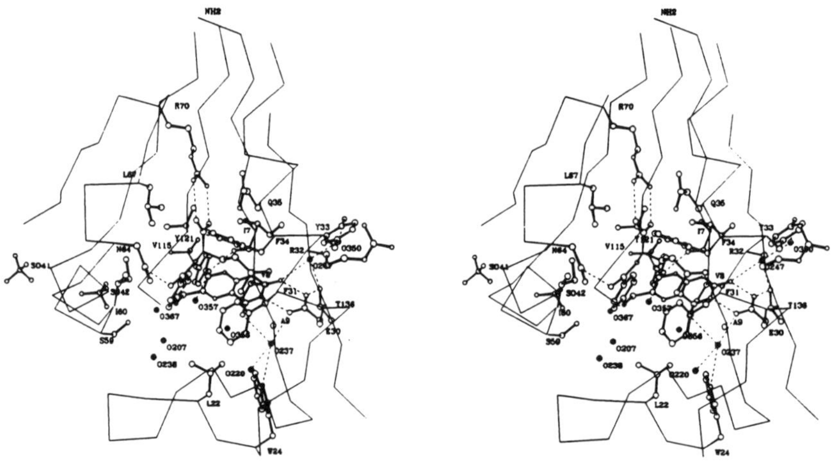

Figure 4.9. Stereo view of human DHFR complexed with folic acid. Side and main chains of selected residues are shown superimposed on the α-carbon backbone. Hydrogen bonds are indicated by broken lines and water molecules by blackened circles. Two bound sulphate ions are shown as S041 and S042. (Reproduced from Oefner *et al.* (1988) by permission of Springer-Verlag)

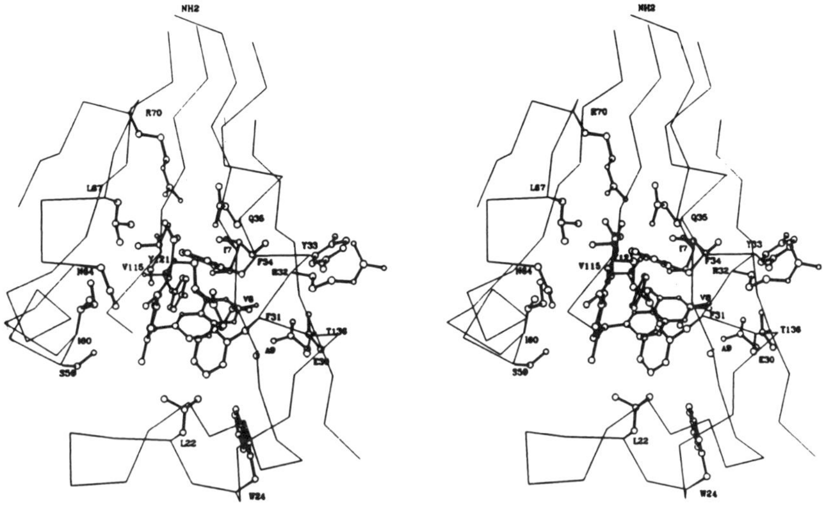

Figure 4.10. Stereo view of the binding of MTX to human DHFR represented as in Figure 4.9. (Reproduced from Oefner *et al.* (1988) by permission of Springer-Verlag)

approximately perpendicular to invariant Phe 34, rather than parallel, as in the L1210 DHFR.

Trimethoprim is bound in the orientation described for Cl DHFR. However, one important difference between the human and Cl enzymes is that the cleft is not significantly wider for human DHFR than for the Ec bacterial enzyme, a factor which Matthews *et al.* (1985b) considered important with Cl DHFR relative to specificity. The reason for the change in conformation and consequent poor binding of TMP to vertebrate DHFR does not seem to have been adequately explained to date.

Modelling studies using the human DHFR sequence compared to the L1210 structure have been carried out (Stammers *et al.*, 1987). Of 21 residue differences, 19 occur near the surface, in solvent accessible regions. The single difference at a ligand-binding site is the substitution of Arg 54 a 2′-phosphate binding site for NADPH by Lys in human DHFR.

4.2.4.7 R-Plasmid Trimethoprim-resistant DHFR (R67 DHFR)

An R-plasmid DHFR enzyme has been isolated and purified from *E. coli* which is almost totally resistant to TMP and MTX, yet reduces dihydrofolate in the presence of the coenzyme NADPH (Smith *et al.*, 1979). The X-ray structure has now been solved at 2.8 Å (Matthews *et al.*, 1986) as the apoenzyme, which is a dimeric molecule with two identical 78 amino acid subunits, each folded into a barrel conformation. A stereo view of its backbone conformation is shown in

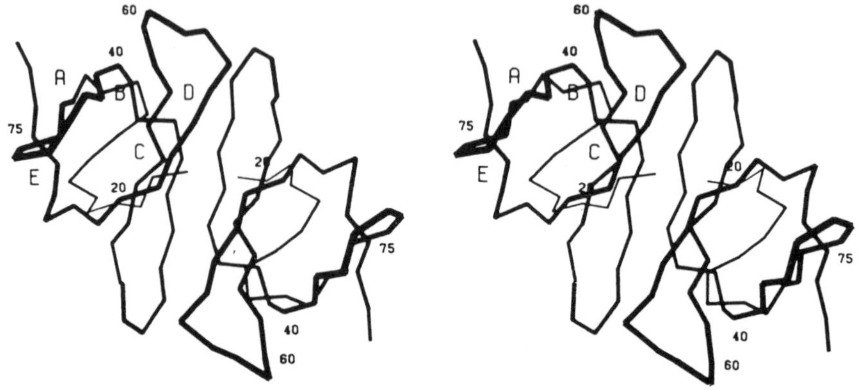

Figure 4.11. Stereo view of the α-carbon backbone conformation of the R67 dihydrofolate reductase dimer viewed approximately along the molecular twofold. (Reprinted with permission from Matthews, D. A., Smith, S. L., Baccanari D. P., Burchall, J. J., Oatley, S. J. and Kraut, J. (1986). *Biochemistry*, **25**, 4194–204. Copyright (1986) American Chemical Society)

Figure 4.11. The enzyme structure bears no resemblance to chromosomal DHFR. In fact it more closely resembles the R domain of papain. A hypothetical model has been build (Figure 4.12) to suggest that the coenzyme binds between and along the length of the two subunits (Matthews *et al.*, 1986). A pteridine binding site is suggested, which does not include the rest of the molecule. It is observed that Glu 58 and Thr 48 of this enzyme resemble the conserved

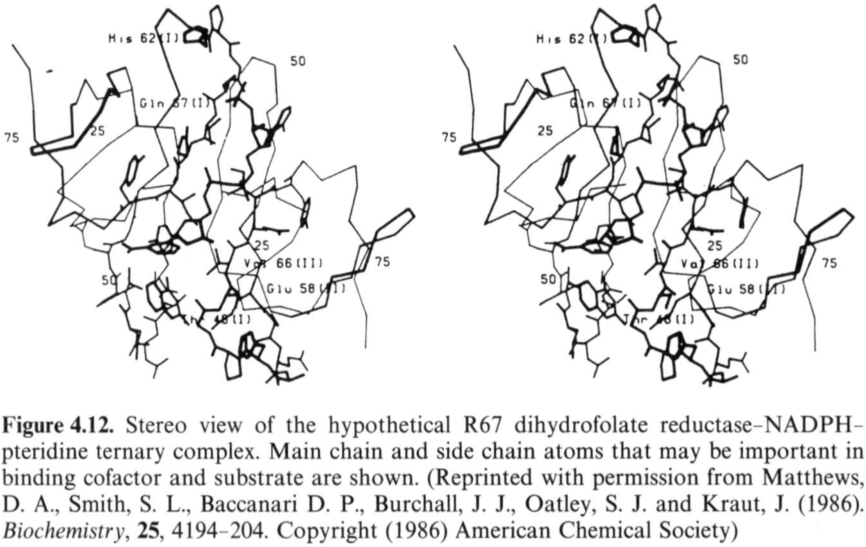

Figure 4.12. Stereo view of the hypothetical R67 dihydrofolate reductase–NADPH–pteridine ternary complex. Main chain and side chain atoms that may be important in binding cofactor and substrate are shown. (Reprinted with permission from Matthews, D. A., Smith, S. L., Baccanari D. P., Burchall, J. J., Oatley, S. J. and Kraut, J. (1986). *Biochemistry*, **25**, 4194–204. Copyright (1986) American Chemical Society)

carboxyl and threonine active site residues (Glu 30 and Thr 136, Cl DHFR) of the chromosomal enzymes in their loci, except that their positions are reversed in the two enzymes. A similar double mutation in Ec chromosomal DHFR produced an enzyme which still functioned although at greatly reduced efficiency (Howell *et al.*, 1987).

Matthews *et al.*, argue that 2,4-diaminopyridines could not bind in the same manner as the substrate in the plasmid enzyme, but the argument is not convincing, since it would seem that the 4-amino group could serve as a hydrogen donor to Tyr 46 and Thr 51 (or 48), just as a 4-oxo group accepts protons from these functions. However, such a compound might prefer to bind to Glu 58 through its 1-N and 2-amino group, rather than through the two protons of the 2-amino group, which would completely change the conformation relative to the enzyme and provide no hydrophobic pocket—a necessity for tight binding. In any event, one can see no possible hydrophobic pocket for TMP in any conformation, although one can visualize a pocket for an extended pteroylglutamate chain extending up towards a Tyr-containing region. A crystal structure with ligands would be of extreme interest.

4.3 CARBONIC ANHYDRASE

Primary aromatic and heterocyclic sulphonamides are inhibitors of carbonic anhydrase. One such compound, acetazolamide (**17**), has a number of pharmacological actions including diuretic properties and the inhibition of aqueous humour secretion (Friedland and Maren, 1984). Such properties have largely been ascribed to the inhibition of carbonic anhydrase. The clinical use of acetazolamide has been mainly as a treatment for glaucoma, to reduce intraocular pressure.

Carbonic anhydrase (CA) catalyses the reversible hydration of carbon dioxide.

$$CO_2 + H_2O \rightleftharpoons HCO_3^- + H^+ \qquad (1)$$

It is monomeric of molecular weight 30 kDa and contains one zinc ion that is essential for activity (Lindskog *et al.*, 1971). A number of different isozyme forms have been identified that show extensive amino acid sequence homology. CAI and CAII are approximately 60% homologous but show differences in catalytic activity and inhibition by sulphonamides (Hewett-Emmett *et al.*, 1984; Sanyal, 1984). A further isoenzyme (CAIII) has been identified in muscle tissue (Holmes, 1977).

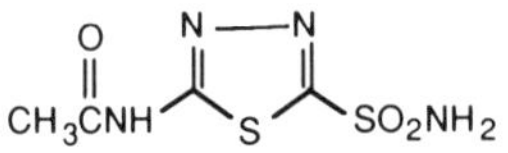

(**17**)

The turnover number for CO_2 hydration by human CAI at pH 9 and 25°C is $2 \times 10^5\ s^{-1}$ whilst CAII is even more active with a turnover number of $1 \times 10^6\ s^{-1}$ under similar conditions (Khalifah, 1971). The K_i for acetazolamide is 0.2 μM for CAI and 0.01 μM for CAII (Sanyal *et al.*, 1982). For cat muscle CAIII a turnover number of $4 \times 10^3\ s^{-1}$ at pH 7.5 and 25°C has been measured together with a K_i for acetazolamide of 300 μM (Sanyal *et al.*, 1982).

The three-dimensional structure of native human carbonic anhydrases I and II together with complexes with acetazolamide have been determined by Kannan and co-workers (Kannan *et al.*, 1977). The initial isomorphous replacement studies have been extended by refinement of the CAI structure (Kannan *et al.*, 1984). The protein part of the CAI–acetazolamide complex has been refined and an improved difference map obtained for acetazolamide from the refined phases. Some preliminary details of the drug–enzyme interactions have been presented (Chakravarty *et al.*, 1985). More recently the structure of CAII has been refined at 2 Å resolution (Eriksson *et al.*, 1988a) as have complexes of CAII with acetazolamide, thiocyanate and 3-acetoxymercuri-4-aminobenzenesulphonamide (AMS) (Eriksson *et al.*, 1988b).

The enzyme has extensive secondary structure consisting of ten β strands and seven helices. The active site is contained in a cleft region where the zinc ion is coordinated tetrahedrally by the imidazoles of three histidines, 94, 96 and 119. The fourth coordination position of the zinc is occupied by a solvent molecule presumed to be water. The binding of acetazolamide from the refined CAI and CAII structures shows that the drug interacts with the enzyme in two ways. First, there are mainly hydrophilic interactions at the sulphonamide end whilst the heterocyclic ring is placed in a mainly hydrophobic pocket (Figure 4.13). There appears to be minimal conformational rearrangement in the protein on drug binding. The sulphonamide group interacts with the zinc ion, displacing the previously bound water. The amino group thus occupies the fourth coordination position. In the CAI–acetazolamide complex the oxygen of the sulphonamide group is within 2.9 Å of the zinc, giving a fifth coordination site which had previously been proposed as the CO_2 binding site in the dehydration reaction (Chakravarty *et al.*, 1985). In the CAII complex with AMS (and by implication with acetazolamide) one oxygen of the sulphonamide is 3.1 Å from the zinc and is hence not considered by Eriksson *et al.*, (1988b) to be coordinated to the zinc. Further refinement of both complexes at higher resolution is required to resolve this difference. Additional interactions with the sulphonamide group occur with the side chain of Thr 199 which in turn is hydrogen bonded to Glu 106. The close approach of the sulphonamide group with the zinc thus explains why substitution on this group leads to the loss of binding activity. The heterocyclic ring and amide group make van der Waals contacts with residues Leu 198, Phe 91 and Leu 141 (CAI) and with Leu 198, Glu 92, Val 143 and Val 121 (CAII).

A network of hydrogen bonds connecting the zinc water molecule in CAI

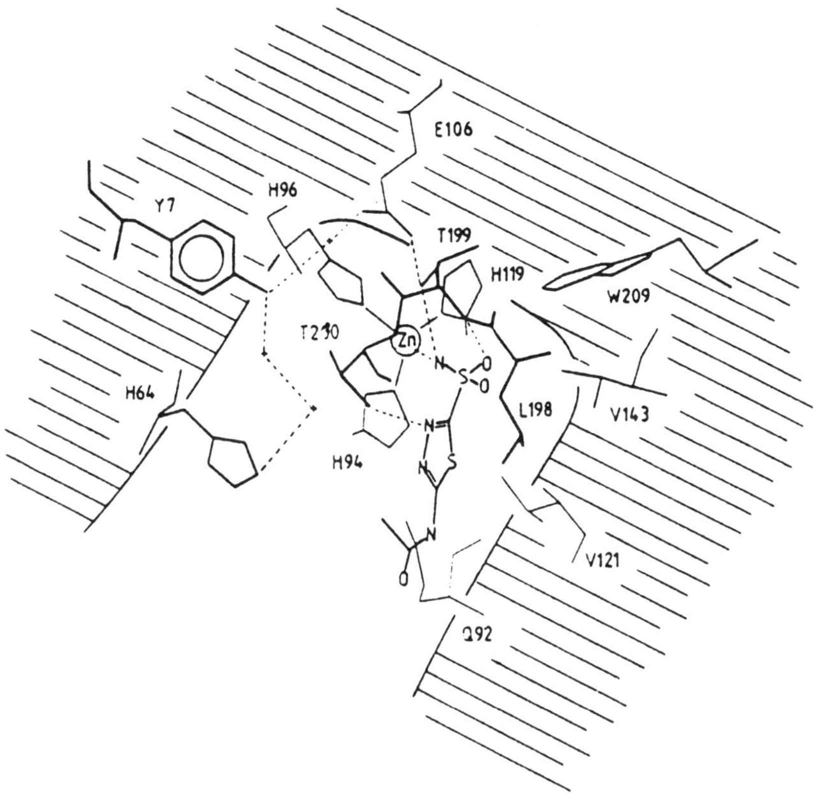

Figure 4.13. Schematic diagram of the binding of acetazolamide to CAII. Side chains are labelled with the one-letter amino acid code. (Reproduced from Eriksson *et al.* (1988b) by permission of John Wiley & Sons, Inc.)

consisting of Thr 199–Glu 106–water–Tyr 7–water–His 64–water–His 200 is also observed in the CAI–acetazolamide complex. This network has been postulated as being involved in a proton transfer step both in catalysis and also in the binding of sulphonamide as the anionic form (Chakravarty *et al.*, 1985). However, site-directed mutagenesis of His 64 indicates it is not essential for proton transfer (Forsman *et al.*, 1988). Eriksson *et al.* (1988b) have observed pentacoordination of the zinc in a complex with the SCN^- ion. This is rationalized on the basis that Thr 199 cannot hydrogen bond to the SCN^- as it can only act as an acceptor since its hydrogen is already bonded to the fully ionized carboxyl of Glu 106. Thus the SCN^- is located in a different position, allowing a water molecule to also bind to the zinc and in turn hydrogen bond to Thr 199.

Model building studies by Eriksson *et al.* (1988b) show that the bicarbonate ion can bind to the zinc utilizing these two coordination positions.

Further CA–sulphonamide complexes are being refined by Kannan and colleagues who intend to calculate binding enthalpies from the enzyme-inhibitor models in an attempt to correlate these with observed K_i data (Chakravarty *et al.*, 1985). Kinetic studies, however, indicate that such a simple relationship might not be observed (King and Burgen, 1976). The binding of sulphonamides to CA does not conform to a simple bimolecular diffusion reaction as indicated by the variation in the overall association rate constant observed in one sulphonamide series. This was best explained by a two-step binding mechanism:

$$E + I \longrightarrow [EI]_A \longrightarrow [EI]_B \qquad (2)$$

Evidence for this was obtained from the binding of sulphonamides to the zinc free (apo) enzyme. The affinity constants were shown to be pH independent and correlated with Log P, typical of a hydrophobic complex. It was proposed that a noncoordinate complex $[E1]_A$ would initially form via hydrophobic interactions followed by a slower rearrangement so that the sulphonamide group enters the zinc coordination sphere to form $[E1]_B$. The overall observed K_i is thus a product of the two equilibria. The X-ray structure of the CA–sulphonamide would be of the low energy $[E1]_B$ complex. Although this complex makes a much greater contribution to the K_i than does the initial complex, the latter should also be taken into account in attempts to use the X-ray structure to optimize sulphonamide–protein interactions.

Acetazolamide shows selectivity in its inhibition of different CA isozymes. Thus the K_i for binding to CAI is 20-fold lower than for CAII (Sanyal *et al.*, 1982). In order to investigate possible structural explanations of the selectivity Kannan *et al.* (1977) have compared the X-ray structure of acetazolamide bound to CAI and CAII. From unrefined data, altered contacts in the two isozyme complexes, including the possible additional interaction with Val 121 in CAII compared with Ala 121 in CAI, were observed. It appears from the data of Kannan *et al.* (1984) and Eriksson *et al.* (1988b) that this additional interaction in the CAII–acetazolamide complex is confirmed in the refined structures.

Using these refined structures it may be possible to attempt to design more isozyme selective inhibitors of CA using the structural data although the kinetic arguments mentioned earlier would need to be borne in mind.

Although it is unknown whether such isozyme selective inhibition of CA could be of clinical use, some of the principles derived from this might be of use in other systems where selective inhibition of tissue specific isozymes of high homology could have therapeutic applications.

An early attempt at using the CAII structure to design more potent inhibitors has been described by Gill *et al.* (1983). They initially modelled in acetazolamide and postulated the thiadiazole nitrogen interacted with main chain carbonyls via water molecules. An isobutyl thiophene analogue of acetazolamide was designed such that the alkyl chain could bond in the hydrophobic pocket. This was shown to be twice as potent as acetazolamide in inhibiting CAII. It would be of interest to measure the binding of the compound to the zinc free enzyme to determine whether the initial hydrophobic complex is contributing to this lower K_i value. An independent study of the docking of acetazolamide and of 27 other sulphonamides to CAII has been carried out by Vedani and Meyer (1984) using computer graphics. They proposed two binding modes for acetazolamide; one is essentially the same as Gill *et al.* (1983) and the other has the thiadiazole ring flipped. A crystallographic study of the binding of some novel thienothiopyran-2-sulphonamides (**18**) to CAII has been described (Baldwin *et al.*, 1989). These compounds have been shown to reduce intraocular blood pressure by topical administration, thus avoiding side effects due to systemic administration. The two optical isomers of **18** show different potency against CAII with K_i values of 0.7 and 50 nM for *R* and *S* respectively. The compounds were crystallized with CAII and refined at 1.5 Å resolution (Figure 4.14, see colour section). This shows that, unlike the case of acetazolamide binding to CAII, there is a conformational change in the protein involving the movement of the side chain of His 64. The centre of the imidazole ring moves a total of 3.1 Å. The sulphonamide portions of both inhibitors are located in a spatially similar position binding to the zinc. The dihedral angles about the N—S—C—S bonds differ by 14°. The C—N—C—C dihedral adopts a less favourable *gauche* conformation to avoiding a clash with the enzyme surface. The terminal methyl groups of both isomers bind in essentially similar positions. The conformational differences in the compounds give rise to different interaction distances with the protein that help to explain the different K_i values. Thus the amide nitrogen of residue Gln 92 and the closer sulphone oxygen have different distances apart for the two isomers as do a sulphone oxygen and the *para* carbon in the benzene ring of Phe 131.

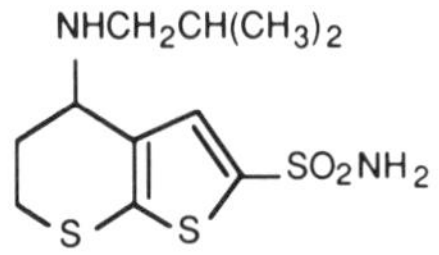

(**18**)

4.4 D-ALA-D-ALA PEPTIDASE

Enzymes catalysing the final stages of the synthesis of bacterial cell walls, involving the cross-linking of peptidoglycan units, are thought to be the lethal targets for penicillin and cephalosporin (β-lactam) antibiotics (Waxman and Strominger, 1983). Cross-linking is achieved by carboxypeptidation of the R-D-Ala-D-Ala substrate followed by transpeptidation. These reactions can be catalysed by separate enzymes or in some cases both activities are associated with a single protein. These enzymes are membrane bound and are not easily amenable to X-ray structural studies. An exocellular D-Ala–D-Ala peptidase (D–D peptidase) which has both carboxypeptidase and transpeptidase activities has been isolated from streptomyces R61 and has been studied as a model of the membrane bound enzymes (Ghuysen *et al.*, 1979). The enzyme is of 37 kDa molecular weight and contains an active site serine that can be labelled by diisopropyl fluorophosphate (DFP). Enzymes such as R61 D–D peptidase catalyse transpeptidation by first releasing D-Ala and forming an acyl enzyme intermediate which is subsequently attacked by the amino acid acceptor to form a cross-link. β-Lactams can inhibit this reaction by acylating the enzyme but in this case the rate of deacylation is slow. According to the substrate analogue hypothesis the β-lactam mimics the D-Ala–D-Ala part of the substrate (Tipper and Strominger, 1965). A general scheme for R-D-Ala–D-Ala and β-lactam processing is

$$E + L \xrightarrow{k_1} E.L \xrightarrow{k_2} E.L' \xrightarrow{k_3} E + P \qquad (3)$$

where E.L is the non-covalent enzyme–ligand complex, E.L′ is the acyl–enzyme intermediate. Cephalosporin C (**19**) forms a stable complex with R61 D–D-peptidase with a half-life of 12 000 min. Several related semisynthetic derivatives of cephalosporin C have clinical uses for treatment of various bacterial infections.

The structure of R61 D–D-peptidase was initially determined to 2.8 Å resolution by Kelly, Knox and coworkers (Kelly *et al.*, 1985). The resolution has now been extended to 2.3 Å (Kelly *et al.*, 1989). The binding of cephalosporin C and other β-lactams have been observed by difference maps. The enzyme has a region consisting of α helices as well as a domain containing a five stranded β sheet and α-helical motif. The active site, which is generally a positively charged area, is located between these domains. The N-terminus of the α helix (62–78) contains the reactive Ser 62, this being the only serine labelled by DFP. This residue is part of the conserved sequence Ser–X–X–Lys found in many β-lactamases and D–D-peptidases. As this sequence is at the beginning of a helix the Lys side chain is also positioned in the active site (Figure 4.15). The presence of the helix dipole which would confer a partial positive charge might activate the serine. The basic Thr–Ser–Gly motif found 50–60 residues from the C-termini of penicillin binding enzymes is positioned on the β strand on the inner edge of

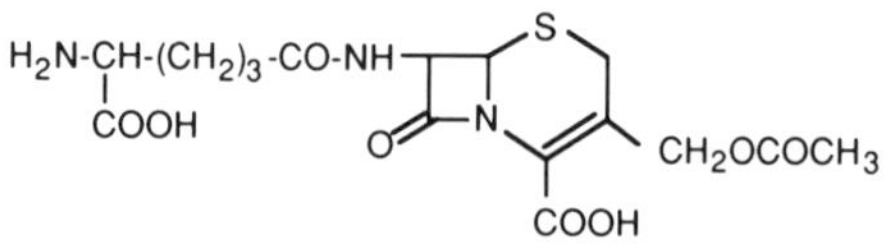

(**19**)

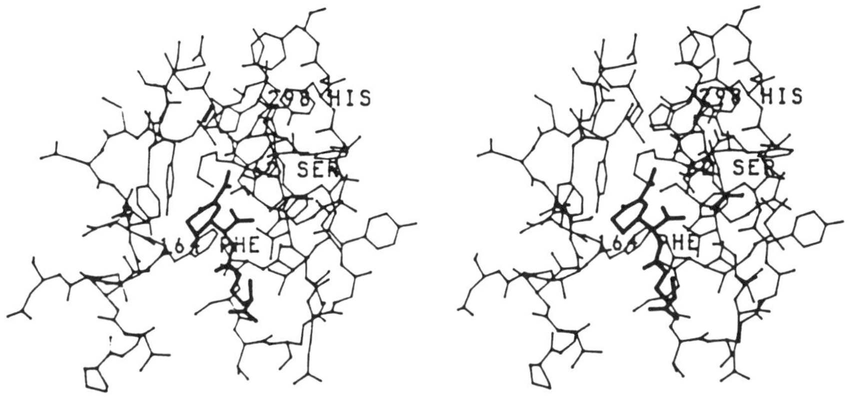

Figure 4.15. Stereo view of the active site of R61 D–D peptidase with cephalosporin C covalently bound (bold lines). Innermost 2 strands of β sheet are on the right side and the helix beginning with the reactive serine 62 is in upper centre. Serine 62 has been left in its native position. The second oxygen on the β-lactam carbon represents the serine 62 γ-oxygen in covalent acyl–enzyme complex. The covalent bond has been omitted for clarity. (Reproduced with permission from Kelly *et al.*, 1989)

the β sheet. The Gly is the closest residue of this triad to Ser 62. Cephalosporin C binds at Ser 62 but also abutts the β sheet. Fitting of the small molecule X-ray structure of cephalosporin C indicated that a number of changes had occurred on binding to the enzyme (Figure 4.15). Thus the C3′ side chain acetate group is absent, the β-lactam ring is open since the C_6 and C_7 atoms have tetrahedral geometry and the C_6–C_7 bond is rotated by 50°. These changes are consistent with the formation of the acyl enzyme. The drug molecule therefore appears covalently bound to Ser 62. The C-7 side chain is in a position to hydrogen bond to the innermost strand of β sheet. Modelling studies of the binding of the drug before ring opening indicate that the dihydrothiazene ring and its carboxylate would be positioned close to the side chain of Lys 69 (Figure 4.16). *O*-Iodobenzamido-penicillanic acid and the monobactam SQ-26,324 have been observed to bind in the same region as cephalosporin C (Kelly *et al.*, 1985, 1989).

The target enzymes for β-lactams are unique to bacteria. Hence selectivity in relation to a host enzyme is not a consideration, in the way that it is for the design of inhibitors of DHFR. Nevertheless, selectivity versus a group of bacterial enzymes that destroy β-lactams is important. These β-lactamases are related to cell wall synthesizing enzymes and indeed might have evolved from them. X-ray studies of an A type β-lactamase by Kelly and co-workers shows a very similar arrangement of secondary structural elements to that observed in R61 D–D-peptidase (Kelly *et al.*, 1986).

This similarity in protein folding has also been observed for a class C β-lactamase refined at 2.0 Å resolution by Oefner *et al.* (1990), who also studied

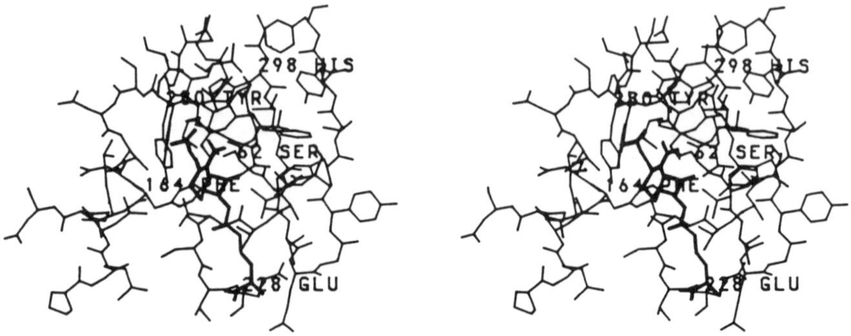

Figure 4.16. Stereo view of the active site of R61 D–D peptidase with modelled cephalosporin C (non-acylated). The β-lactam carbonyl carbon is fixed to the position observed in the covalent complex. (Reproduced with permission from Kelly *et al.*, 1989)

the binding of the monobactam aztreonam (**20**) to the enzyme. The drug appeared covalently bound to the enzyme via Ser 64 with the β-lactam ring opened in an analogous manner to that observed for β-lactam binding to R61 D–D-peptidase (Figure 4.17).

With the availability of complete amino acid sequence data together with further refinement, detailed information of β-lactam D–D-peptidase interactions will become available. This could be of use in designing new analogues of β-lactams. Existing β-lactams generally bind comparatively weakly to D–D-peptidases in the initial non-covalent complex with K_Ds of 0.1–10 mM (Ghuysen *et al.*, 1979) comparable with the K_m for R-D-Ala-D-Ala substrates (Rasmussen and Strominger, 1978). In view of the proposed transition-state-like geometry of

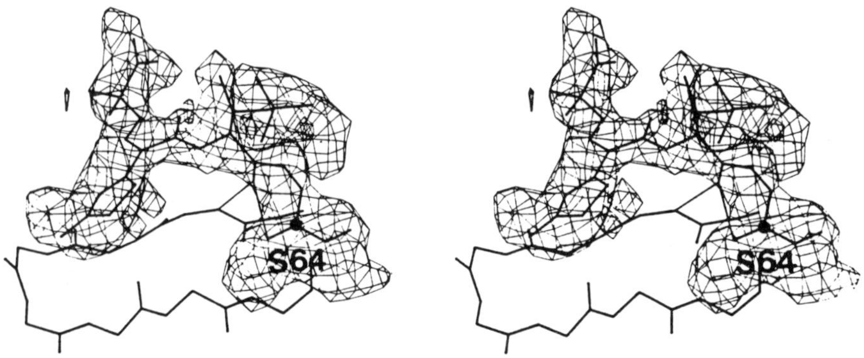

Figure 4.17. Stereo view of the binding of (**20**) as an ester-linked acyl enzyme complex with *C. freundii* β-lactamase. The complexed structure, refined at 2.5 Å resolution, is superimposed on the final (2 Fobs-Fcalc) map. (Reprinted by permission from *Nature* vol. **343**, pp. 284–8. Copyright © 1990 Macmillan Magazines Ltd)

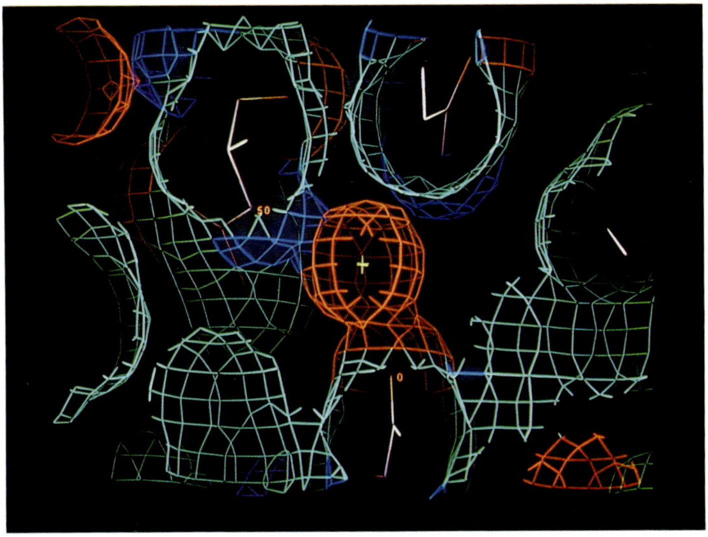

Figure 2.1 (a) Part of inhibitor binding site in HIV-1 protease (Brookhaven Data Bank coordinates 4HVP, Miller *et al.*, 1989) with contoured van der Waals surface illustrating water-mediated contact between N–H of residue 50 and a ligand carbonyl group.

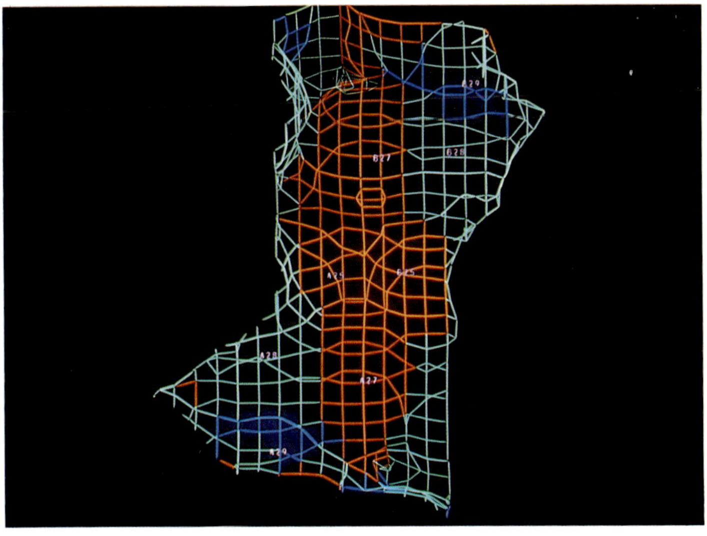

Figure 2.1 (b) Part of double radius surface computed from the inhibitor binding region of the inhibitor-bound HIV-1 protease (coordinate set 4HVP); colour-coding is as described in text, red and blue representing H–bond acceptor and donor regions respectively.

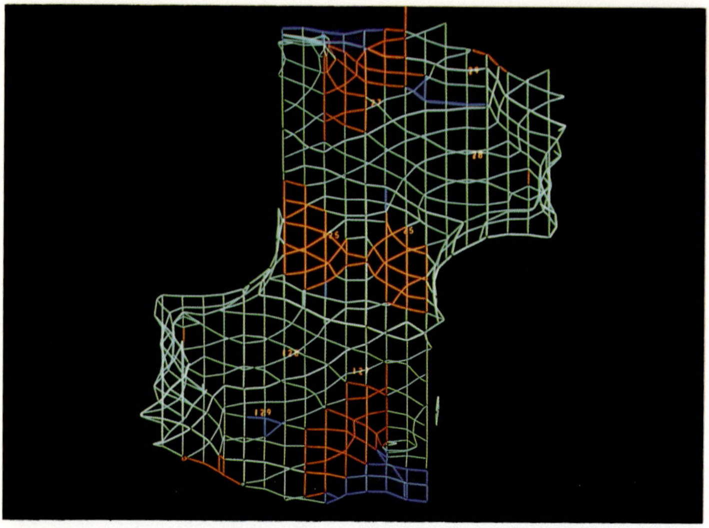

Figure 2.1 (c) Part of double radius surface computed from the inhibitor binding region of the apoprotein HIV-1 protease (coordinate set 3HVP); the apparent decrease in H–bond acceptor capacity results from a re-orientation away from the surface of the carbonyl group of residue 27

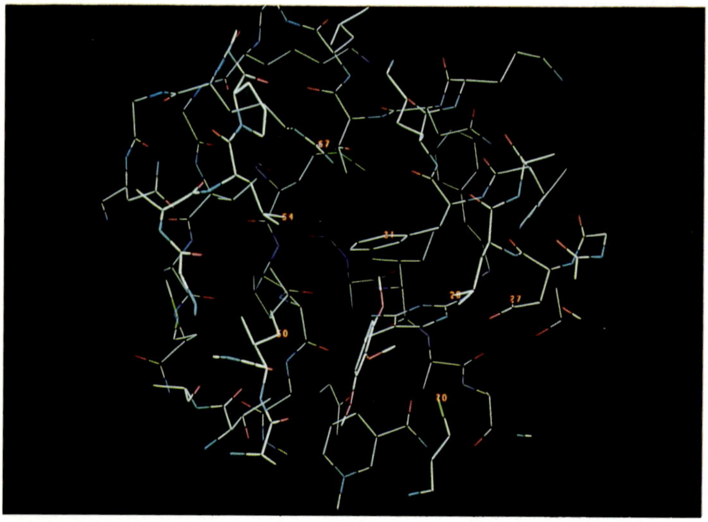

Figure 2.2 (a) Active-site region of dihydrofolate reductase in ternary complex with trimethoprim and NADPH (Champness *et al.*, 1986b). The diaminopyrimidine moiety of the drug molecule interacts via H–bonds with the side chain of Asp 27 (right) and through its 4-amino group with carbonyls of Ile 5 and Ile 94 (behind). The trimethoxybenzyl moiety (coming forward in picture) is in van der Waals contact with several surrounding residues; Met 20, Leu 28, Phe 31, Ile 50 and the nicotinamide of NADPH (behind).

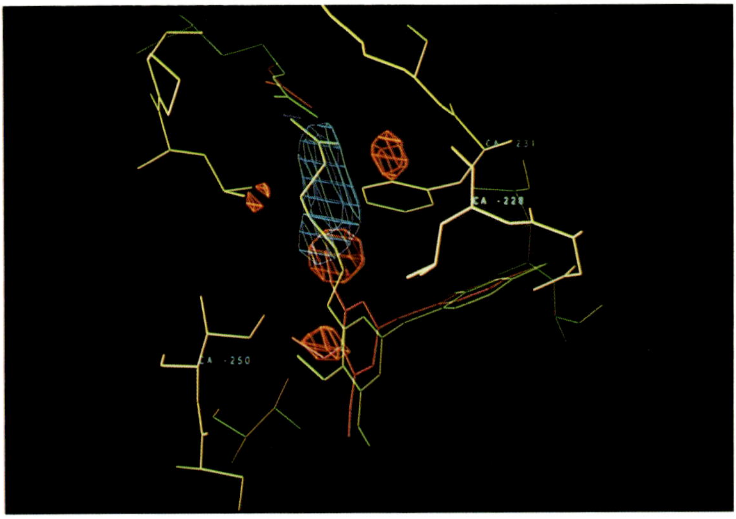

Figure 2.2 (b) Difference-Fourier map for dihydrofolate reductase with bound carboxalkoxy (*n*=5) derivative of TMP versus TMP-bound enzyme. Difference density contours are superimposed on a skeletal model of the complex. The magenta-coloured skeleton represents the structure of the TMP-bound form for comparison. The blue contours represent the new electron density of the extended chain, orange contours indicate regions from which electron density has moved slightly. (Reproduced from Champness *et al.* (1986a) by permission of Elsevier Science Publishers (Biomedical Division))

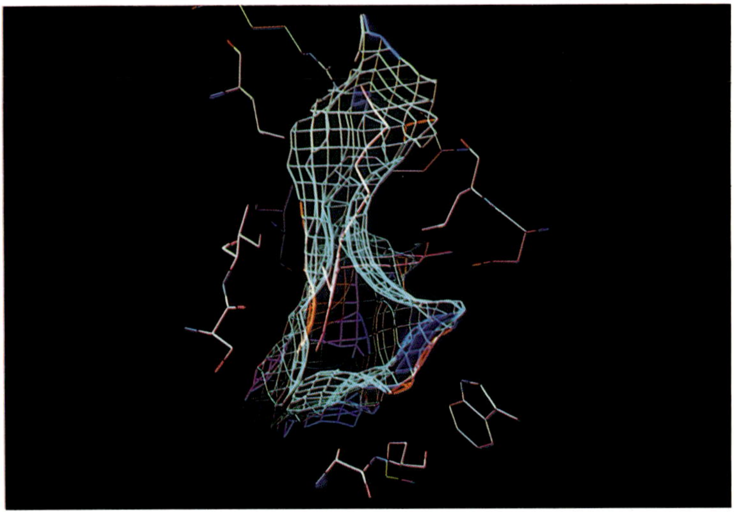

Figure 2.2 (c) Model of the binary complex of active-site residues of dihydrofolate reductase with TMP derivatives as in (b) above superimposed on double radius surface computed for enzyme alone, indicating the space available for the skeletally represented ligand molecule and the H-bonding and van der Waals properties of the enzyme

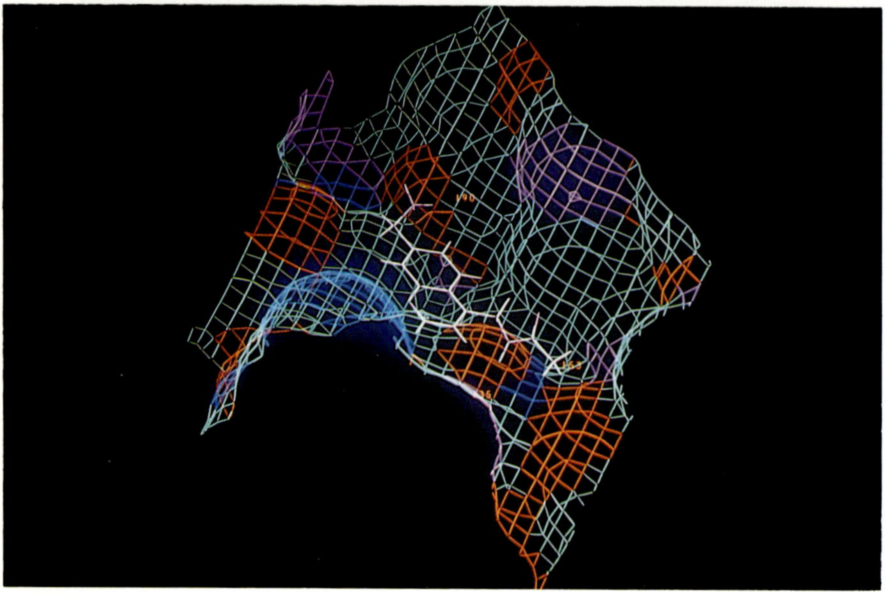

Figure 2.3 Double radius surface of the sialyllactose binding site of influenza haemagglutinin, computed from Brookhaven Data Bank coordinates 4HMG, with proposed substituted naphthalene ligand. The latter is postulated to lie in the same shallow binding cleft in which binding of NANA was observed by Weis *et al.*, making intermolecular contacts including Glu 190 and main chain carbonyl of residue 135

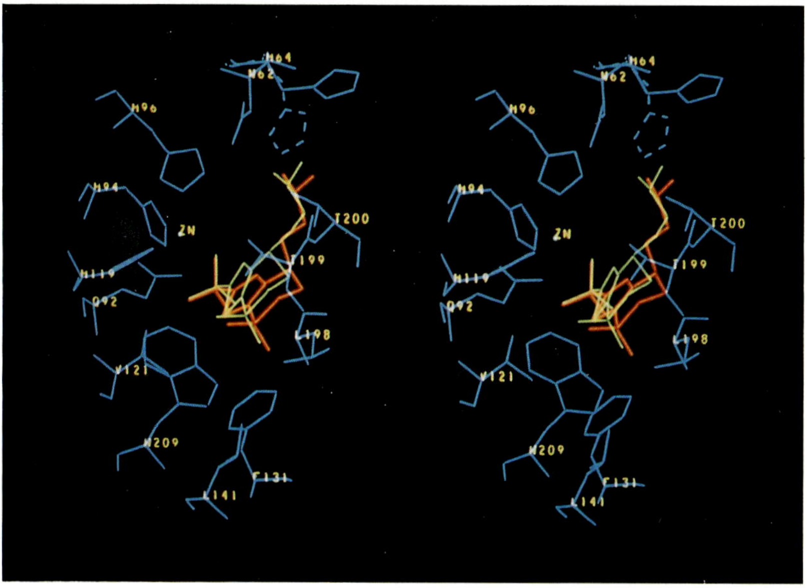

Figure 4.14 Stereo view showing a superposition of the bound conformations of the R and S isomers of **18** in the active site of human CAII. (Reprinted with permission from Baldwin, J.J., Ponticello, G.S., Anderson, P.S., Christy, M.E., Murcko, M.A., Randall, W.C., Schwam, H., Sugrue, M.F., Springer, J.P., Gautheron, P., Grove, J., Mallorga, P., Viader, M.-P., McKeever, B.M. and Navia, M.A., (1989). *J. Med. Chem.* **32**, 2510–13. Copyright (1989) American Chemical Society

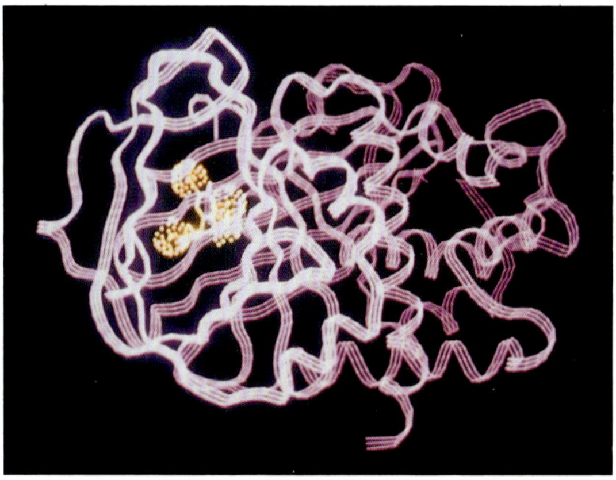

Figure 5.7 A model of TBG showing, in yellow, T_4 bound in the hypothetical binding site

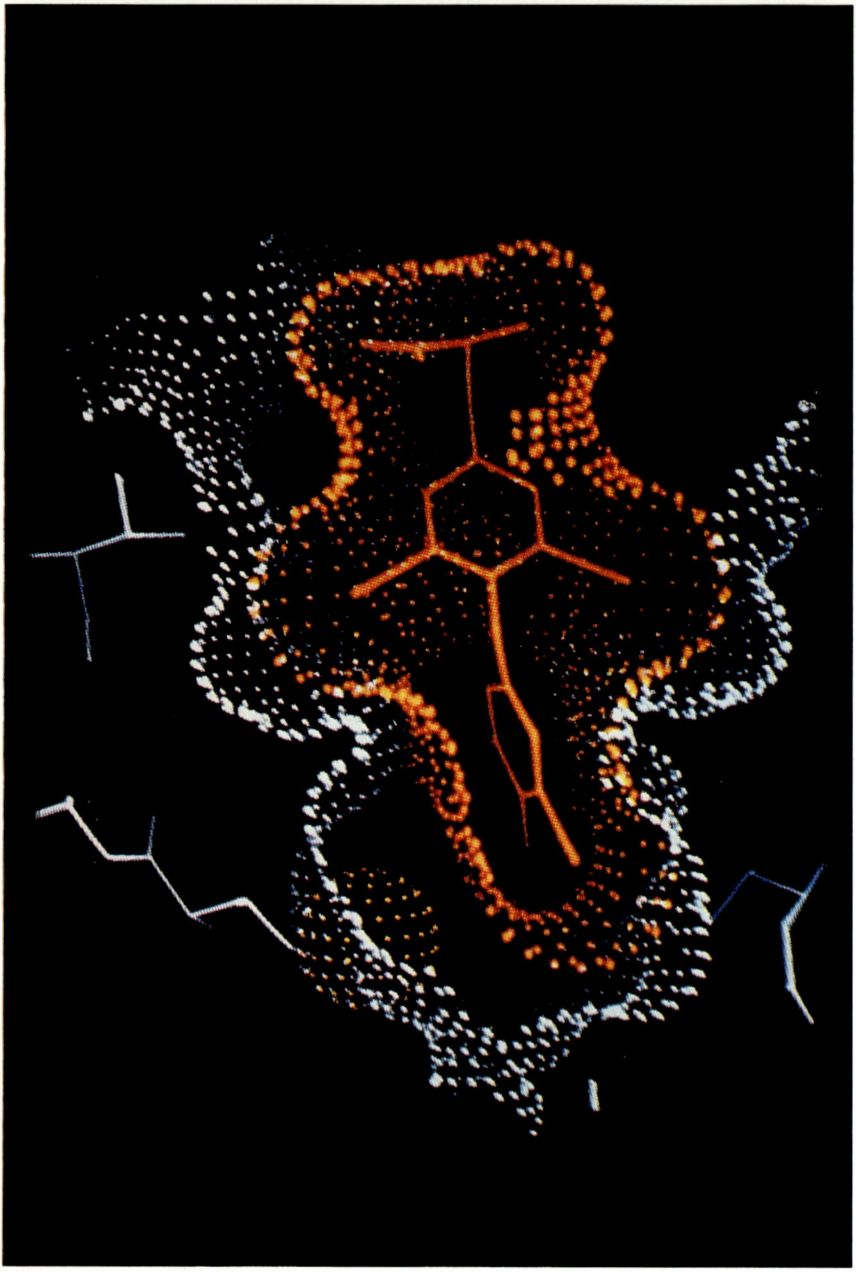

Figure 5.14 A computer graphics representation of the solvent accessible surfaces of the TTR:T_4 complex. The protein surface is in white, the hormone surface in orange and the yellow ball is the bound water molecule (W1)

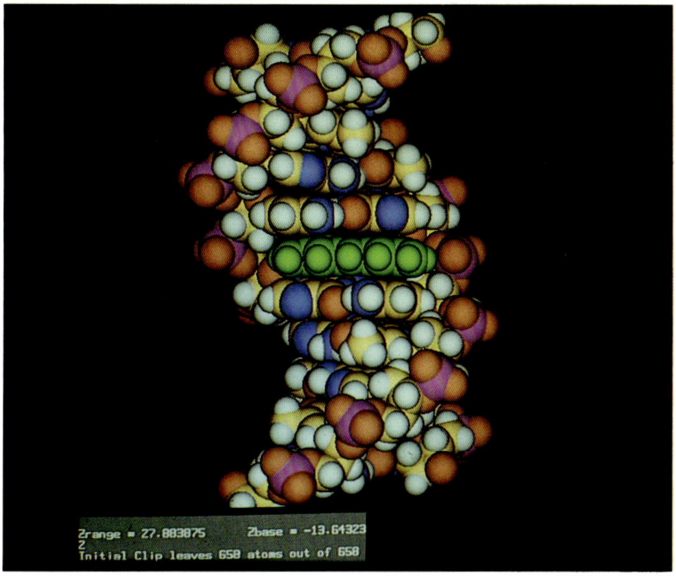

Figure 6.6 A computer-generated molecular model for proflavine intercalated into a decamer sequence shown in space-filling representation

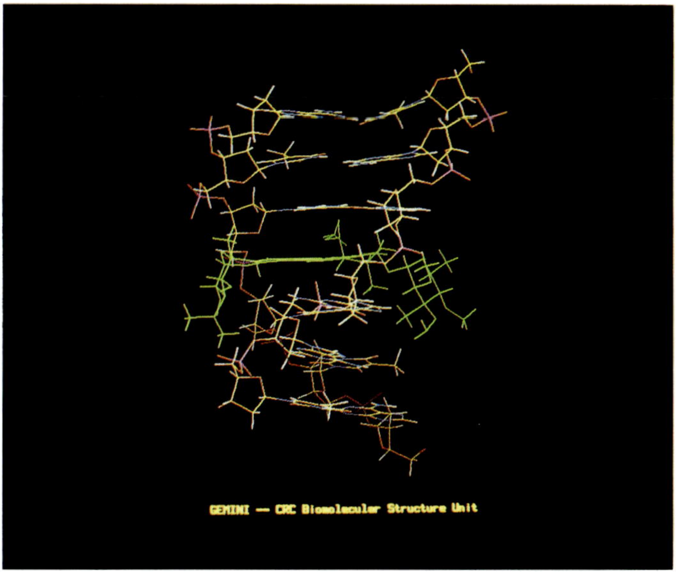

Figure 6.16 A computer-generated model for nogalamycin intercalated into the CG site of the sequence TACGTA

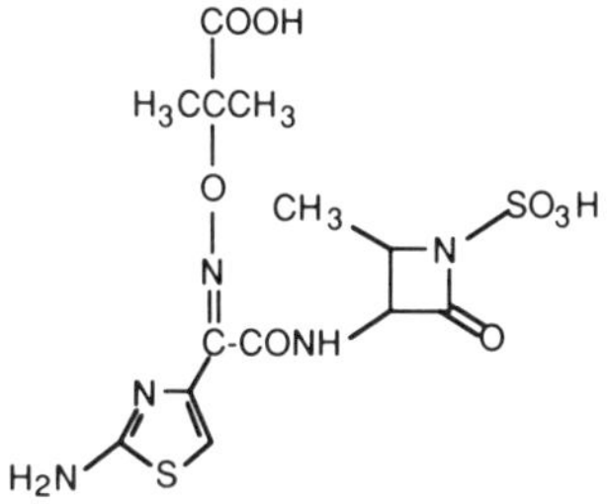

(**20**)

the β-lactam (Lee, 1971), they would be expected to bind more tightly than the substrate. It may indicate that parts of the drug are making unfavourable contacts with the enzyme. It might therefore be possible to use the model to optimize interactions without decrease in the rate of acylation, and also to further reduce the access of water to the active site and hence decrease the deacylation rate and prolong the lifetime of the acyl-enzyme complex.

REFERENCES

Antonjuk, D. J., Birdsall, B., Cheung, H. T. A., Clore, G. M., Feeney, J., Gronenborn, A., Roberts, G. C. K. and Tran, T. Q. (1984). *Br. J. Pharmac.*, **81**, 309–15.

Austin, T. W., Brooks, G. F., Bethel, M., Roberts, F. L., Turck, M. and Holmes, K. K. (1973). *J. Infect. Dis.*, **128**, S666–72.

Baccanari, D. P., Daluge, S. and King, R. W. (1982). *Biochemistry*, **21**, 5068–75.

Baccanari, D. P., Tansik, R. L., Paterson, S. J. and Stone, D. (1984). *J. Biol. Chem.*, **259**, 12291–8.

Baker, D. J., Beddell, C. R., Champness, J. N., Goodford, P. J., Norrington, F. E. A., Smith, D. R. and Stammers, D. K. (1981). *FEBS Lett.*, **126**, 49–52.

Baker, D. J., Beddell, C. R., Champness, J. N., Goodford, P. J., Norrington, F. E., Roth, B. and Stammers, D. K. (1983). In *Chemistry and Biology of Pteridines.* (Ed. J. A. Blair), deGruyter, Berlin, New York, pp. 545–9.

Baldwin, J. J., Ponticello, G. S., Anderson, P. S., Christy, M. E., Murcko, M. A., Randall, W. C., Schwam, H., Sugrue, M. F., Springer, J. P., Gautheron, P., Grove, J., Mallorga, P., Viader, M-P., McKeever, B. M. and Navia, M. A. (1989). *J. Med. Chem.*, **32**, 2510–13.

Beddell, C. R. (1984). In *X-ray Crystallography and Drug Action.* (Eds A. S. Horn and C. J. De Ranter), Clarendon Press, Oxford, pp. 169–93.

Bertino, J. R., Sawicki, W. L., Moroson, B. A., Cashmore, A. R. and Elslager, E. F. (1979). *Biochem. Pharmacol.*, **28**, 1983–7.

Birdsall, B., Roberts, G. C. K., Feeney, J., Dann, J. G. and Burgen, A. S. V. (1983). *Biochemistry.*, **22**, 5597–604.

Birdsall, B., Bevan, A. W., Pascual, C., Roberts, G. C. K., Feeney, J., Gronenborn, A. and Clore, G. M. (1984a). *Biochemistry.*, **23**, 4733–42.

Birdsall, B., Feeney, J., Pascual, C., Roberts, G. C. K., Kompis, I., Then, R. L., Muller, K. and Kroehn, A. A. (1984b). *J. Med. Chem.*, **27**, 1672–6.

Bitar, K. G., Blankenship, D. T., Walsh, K. A., Dunlap, R. B., Reddy, A. V. and Freisheim, J. H. (1977). *FEBS Lett.*, **80**, 119–22.

Bolin, J. T., Filman, D. J., Matthews, D. A., Hamlin, R. C. and Kraut, J. (1982). *J. Biol. Chem.*, **257**, 13650–62.

Burchall, J. J. and Hitchings, G. H. (1965). *Mol. Pharmac.*, **1**, 126–36.

Carrington, H. C., Crowther, A. F., Davey, D. G., Levi, A. A. and Rose, F. L. (1951). *Nature*, **168**, 1080.

Cayley, P. J., Albrand, J. P., Feeney, J., Roberts, G. C. K., Piper, E. A., and Burgen, A. S. V. (1979). *Biochemistry.*, **18**, 3886–95.

Chakravarty, S., Yadava, V. S., Kumar, V. and Kannan, K. K. (1985). *J. Biosci.*, **8**, 491–8.

Champness, J. N., Kuyper, L. F. and Beddell, C. R. (1986a). In *Molecular Graphics and Drug Design* (Eds A. S. V. Burgen, G. C. K. Roberts and M. S. Tute), Elsevier Science, B.V. (Biomedical Division), Amsterdam, pp. 336–362.

Champness, J. N., Stammers, D. K. and Beddell, C. R. (1986b). *FEBS Lett.*, **199**, 61–7.

Chan, J. H., Kuyper, L. F. and Roth, B. (1985). *Abstr. 189th Am. Chem. Soc. Meet.*, Miami, p. MED187.

Charlton, P. A., Young, D. W., Birdsall, B., Feeney, J. and Roberts, G. C. K. (1979). *J. Chem. Soc. Chem. Commun.*, 922–4.

Clore, G. M., Gronenborn, A. M., Birdsall, B., Feeney, J. and Roberts, G. C. K. (1984). *Biochem. J.*, **217**, 659–66.

Cocco, L., Roth, B., Temple, C., Jr, Montgomery, J. A., London, R. E. and Blakley, R. L. (1983). *Arch. Biochem. Biophys.*, **266**, 567–77.

Cody, V. and Zakrzewski, S. F. (1982). *J. Med. Chem.*, **25**, 427–30.

Cohen, S. S. (1971). *Ann. N. Y. Acad. Sci.*, **186**, 292–301.

Eriksson, A. E., Jones, T. A. and Liljas, A. (1988a). *Proteins: Struct., Funct., Genet.*, **4**, 274–82.

Eriksson, A. E., Kylsten, P. M., Jones, T. A. and Liljas A. (1988b). *Proteins: Struct., Funct., Genet.*, **4**, 283–93.

Fontecilla-Camps, J. C., Bugg, C. E., Temple, C., Rose, J. D., Montgomery, J. A. and Kisliuk, R. L. (1979). *J. Am. Chem. Soc.*, **101**, 6114–15.

Forsman, C., Behravan, G., Jonsson, B-H., Liang, Z-W., Lindskog, S., Ren, X., Sandström, J. and Wallgren, K. (1988). *FEBS Lett.*, **229**, 360–2.

Friedland, B. R. and Maren, T. H. (1984). In *Pharmacology of the Eye* (Ed. M. L. Sears), Springer-Verlag, Berlin, pp. 279–85.

Ghuysen, J. M., Frere, J. M., Leyh-Bouille, M., Coyette, J., Dusart, J. and Nguyen-Disteche, M. (1979). *A. Rev. Biochem.*, **48**, 73–101.

Gill, E. W., Goodall, B. M. and Hancock, P. J. B. (1983). In *Quantitative Approaches to Drug Design* (Ed. J. C. Dearden), Elsevier Science Publishers B.V., Amsterdam, pp. 121–2.

Grivsky, E., Lee, S., Sigel, C. W., Duch, D. S. and Nichol, C. A. (1980). *J. Med. Chem.*, **23**, 327–9.

Gronenborn, A., Birdsall, B., Hyde, E., Roberts, G. C. K., Feeney, J. and Burgen, A. (1981). *Mol. Pharmac.*, **20**, 145–53.

Hammond, S. J., Birdsall, B., Searle, M. S., Roberts, G. C. K. and Feeney, J. (1986). *J. Mol. Biol.*, **188**, 81–97.

Hansch, C., Hathaway, B. A., Zong-ru, G., Selassie, C. D., Dietruch, S. W., Blaney, J. M., Langridge, R., Volz, K. W. and Kaufman, B. T. (1984). *J. Med. Chem.*, **27**, 129–43.

Hewett-Emmett, D., Hopkins, P. J., Tashian, R. E. and Czelusniak, J. (1984). *Ann. N.Y. Acad. Sci.*, **429**, 338–57.

Hitchings, G. H. and Roth, B. (1959). U.S. Patent 2, 909, 522.

Holmes, R. S. (1977). *Eur. J. Biochem.*, **78**, 511–20.

Howell, E. E., Villafranca, J. E., Warren, M. S., Oatley, S. J. and Kraut, J. (1986). *Science.*, **231**, 1123–8.

Howell, E. E., Warren, M. S., Booth, C. L. J., Villafranca, J. E. and Kraut, J. (1987). *Biochemistry*, **26**, 8591–8.
Hyde, R. M., Paterson, R. A., Beddell, C. R., Champness, J. N., Stammers, D. K., Baker, D. J., Goodford, P. J., Kuyper, L. F., Ferone, R., Roth, B. and Elwell, E. (1983). In *Chemistry and Biology of Pteridines* (Ed. J. A. Blair), deGruyter, Berlin, New York, pp. 506–9.
Johnson, L. N. (1967). *Proc. R. Soc. Ser B*, **167**, 439–40.
Kannan, K. K., Vaara, I., Notstrand, B., Lovgren, S., Borrell, A., Fridborg, K. and Petef, M. (1977). In *Drug Action at the Molecular Level* (Ed. G. C. K. Roberts), Macmillan, UK, pp. 73–93.
Kannan, K. K., Ramanadham, M. and Jones, T. A. (1984). *Ann. N. Y. Acad. Sci.*, **429**, 49–60.
Kelly, J. A., Knox, J. R., Moews, P. C., Hite, G. J., Bartolone, J. B., Zhao, H., Joris, B., Frere, J. M. and Ghuysen, J. M. (1985). *J. Biol. Chem.*, **260**, 6449–58.
Kelly, J. A., Dideberg, O., Charlier, P., Wery, J. P., Libert, M., Moews, P. C., Knox, J. R., Duez, C., Fraipont, C. L., Joris, B., Dusart, J., Frere, J. M. and Ghuysen, J. M. (1986). *Science*, **231**, 1429–31.
Kelly, J. A., Knox, J. R., Zhao, H., Frere, J. M. and Ghuysen, J. M. (1989). *J. Mol. Biol.*, **209**(2), 281–95.
Khalifah, R. G. (1971). *J. Biol. Chem.*, **246**, 2561–73.
King, R. W. and Burgen, A. S. V. (1976). *Proc. R. Soc. Lond. B*, **193**, 107–25.
Kumar, A. A., Blankenship, D. T., Kaufman, B T. and Freisheim, J. H. (1980). *Biochemistry.*, **19**, 667–78.
Kuyper, L. F., Roth, B., Baccanari, D. P., Ferone, R., Beddell, C. R., Champness, J. N., Stammers, D. K., Dann, J. G., Norrington, F. E. A., Baker, D. J. and Goodford, P. J. (1982). *J. Med. Chem.*, **25**, 1120–2.
Kuyper, L. F., Roth, B., Baccanari, D. P., Ferone, R., Beddell, C. R., Champness, J. N., Stammers, D. K., Dann, J. G., Norrington, F. E. A., Baker, D. J. and Goodford, P. J. (1985). *J. Med. Chem.*, **28**, 303–11.
Lee, B. (1971). *J. Mol. Biol.*, **61**, 463–9.
Lindskog, S., Henderson, L. E., Kannan, K. K., Liljas, A., Nyman, P. O. and Strandberg, B. (1971). In *The Enzymes* (Ed. P. D. Boyer), Academic Press, New York, **5**, pp. 587–665.
Masters, J. N. and Attardi, G. (1983). *Gene*, **21**, 59–63.
Matthews, D. A., Alden, R. A., Bolin, J. T., Freer, S. T., Hamlin, R., Xuong, N., Kraut, J., Poe, M., Williams, M. and Hoogsteen, K. (1977). *Science*, **197**, 452–5.
Matthews, D. A., Alden, R. A. Bolin, J. T. Filman, D. J., Freer, S. T., Hamlin, R., Hol, W. G. J., Kisliuk, R. L., Pastore, E. J., Plante, L. J., Xuong, N. and Kraut, J. (1978). *J. Biol. Chem.*, **253**, 6946–54.
Matthews, D. A., Bolin, J. T., Burridge, J. M., Filman, D. J., Volz, K. W., Kaufman, B. T., Beddell, C. R., Champness, J. N., Stammers, D. K. and Kraut, J. (1985a). *J. Biol. Chem.*, **260**, 381–91.
Matthews, D. A. Bolin, J. T., Burridge, J. M., Filman, D. J., Volz, K. W. and Kraut, J. (1985b). *J. Biol. Chem.*, **260**, 392–9.
Matthews, D. A., Smith, S. L., Baccanari, D. P., Burchall, J. J., Oatley, S. J. and Kraut, J. (1986). *Biochemistry.*, **25**, 4194–204.
Oefner, C., D'Arcy, A. and Winkler, F. K. (1988). *Eur. J. Biochem.*, **174**, 377–85.
Oefner, C., D'Arcy, A., Daly, J. J., Gubernator, K., Charnas, R. L., Heinze, I., Hubschwerlen, C. and Winkler, F. K. (1990). *Nature*, **343**, 284–8.
Orton, C. T. and Hambly, L. R. (1971). *Poultry Sci.*, **50**, 1341–6.
Poulos, T. and Finzel, B. Personal communication (1985).

Queener, S. F., Bartlett, M. S., Jay, M. A., Durkin, M. M. and Smith, J. W. (1987). *Antimicrob. Agents Chemother.*, **31**, 1323–7.

Rasmussen, J. R. and Strominger, J. L. (1978). *Proc. Natl Acad. Sci. USA*, **75**, 84–8.

Rauckman, B. S. and Roth, B. (1980). *J. Med. Chem.*, **23**, 384–91.

Roberts, G. C. K. (1987). *Biochem. Soc. Trans.*, **15**, 762–6.

Roberts, G. C. K., Feeney, J., Burgen A. S. V. and Daluge, S. (1981). *FEBS Lett.*, **131**, 85–8.

Rosowsky, A., Wright, J. E., Ginty, C. and Uren, J. (1982). *J. Med. Chem.*, **25**, 960–4.

Roth, B. (1986). *Fed. Proc.*, **45**, 2765–72.

Roth, B. (1989). *Drugs of the Future*, **14**, 249–68.

Roth, B. and Aig, E. (1987). *J. Med. Chem.*, **30**, 1998–2004.

Roth, B., Falco, E. A., Hitchings, G. H., Bushby, S. R. M. (1962). *J. Med. Pharm. Chem.*, **5**, 1103–23.

Roth, B., Aig, E., Rauckman, B., Strelitz, J. Z., Phillips, A. P., Ferone, R., Bushby, S. R. M. and Sigel, C. W. (1981). *J. Med. Chem.*, **24**, 933–41.

Roth, B., Rauckman, B. S., Ferone, R., Baccanari, D. P., Champness, J. N. and Hyde, R. M. (1987). *J. Med. Chem.*, **30**, 348–56.

Roth, B., Baccanari, D. P., Sigel, C. W., Huggell, J. P., Eaddy, J., Kao, J. C., Grace, M. E. and Rauckman, B. S. (1988). *J. Med. Chem.*, **31**, 122–9.

Russell, P. B. and Hitchings, G. H. (1951). *J. Am. Chem. Soc.*, **73**, 3763–70.

Sanyal, G. (1984). *Ann. N.Y. Acad. Sci.*, **429**, 165–78.

Sanyal, G., Swenson, E. R., Pessah, N. I. and Maren, T. H. (1982). *Mol. Pharmac.*, **22**, 211–20.

Searle, M. S., Hammond, S. J., Birdsall, B., Roberts, G. C. K., Feeney, J., King, R. W. and Griffiths, D. V. (1986). *FEBS Lett.*, **194**, 165–70.

Seeger, D. R., Cosulich, D. B., Smith, J. M. Jr and Hultquist, M. E. (1949). *J. Am. Chem. Soc.*, **71**, 1753–8.

Smith, S. L., Stone, D., Novak, P., Baccanari, D. P. and Burchall, J. J. (1979). *J. Biol. Chem.*, **254**, 6222–5.

Stammers, D. K., Champness, J. N., Dann, J. G. and Beddell, C. R. (1983). In *Chemistry and Biology of Pteridines* (Ed. J. A. Blair), de Gruyter, Berlin, New York, pp. 567–71.

Stammers, D. K., Champness, J. N., Beddell, C. R., Dann, J. G., Elipoulos, E., Geddes, A. J., Ogg, D. and North, A. C. T. (1987). *FEBS Lett.*, **218**, 178–84.

Stone, D. and Smith, S. L. (1979). *J. Biol. Chem.*, **254**, 10857–61.

Stone, D., Phillips, A. W. and Burchall, J. J. (1977). *Eur. J. Biochem.*, **72**, 613–24.

Stone, D., Paterson, S. J., Raper, J. H. and Phillips, A. W. (1979). *J. Biol. Chem.*, **254**, 480–8.

Tansik, R. L., Averett, D. R., Roth, B., Paterson, S. J., Stone, D. and Baccanari, D. P. (1984). *J. Biol. Chem.*, **259**, 12299–305.

Tipper, D. J. and Strominger, J. L. (1965). *Proc. Natl Acad. Sci. USA*, **54**, 1133–41.

Vedani, A. and Meyer, E. F. (1984). *J. Pharm. Sci.*, **73**, 352–8.

Versenyi, L. (1974). *Magy. Allatorv. Lapja*, **29**, 278–80.

Volz, K. W., Matthews, D. A., Alden, R. A., Freer, S. T., Hansch, C., Kaufman, B. T. and Kraut, J. (1982). *J. Biol. Chem.*, **257**, 2528–36.

Waxman, D. J. and Strominger, J. L. (1983). *A. Rev. Biochem.*, **52**, 825–69.

5 Multiple Modes of Binding of Thyroid Hormones and Other Iodothyronines to Human Plasma Transthyretin

P. DE LA PAZ, J. M. BURRIDGE, S. J. OATLEY AND C. C. F. BLAKE

Laboratory of Molecular Biophysics, University of Oxford, Oxford, UK

The Design of Drugs to Macromolecular Targets. Edited by C. R. Beddell
© 1992 John Wiley & Sons Ltd

5.1 INTRODUCTION

The thyroid hormones, which are synthesized and secreted by the thyroid gland of vertebrates, exhibit a broad range of activities which include the control of tissue development and differentiation, the regulation of oxygen consumption and the promotion of various metabolic processes (for reviews see Baxter *et al.*, 1979; Eberhardt *et al.*, 1980). In recent years, the concept has emerged that thyroid hormone function is mediated by the interaction of the hormones with specific receptors, primarily those located in the nucleus of target cells (Oppenheimer *et al.*, 1976; Samuels, 1978; Latham *et al.*, 1978; Oppenheimer, 1981). However, difficulties in the isolation and purification of nuclear receptor proteins have as yet frustrated the ultimate aim of these studies: to develop a molecular model for the binding interaction between the thyroid hormones and their receptors.

The approach we have adopted to this problem has exploited knowledge of another natural binding site for the hormones, that of the serum transport protein, thyroxine binding prealbumin, otherwise known as transthyretin. X-ray crystallography has been used not only to solve the full three-dimensional structure of the protein (Blake *et al.*, 1974, 1978b) but also to provide a detailed model for its interaction with the thyroid hormones (Blake *et al.*, 1978a; Oatley *et al.*, 1982). At the same time, chemical studies have been performed on the binding of a variety of hormone analogues to transthyretin in solution (Pages *et al.*, 1973; Andrea *et al.*, 1980; Somack *et al.*, 1982). The availability of a wide range of thyroactive compounds, together with the highly resolved crystal structure of transthyretin, has provided an incentive for the present structural studies on the interaction of thyroid hormone analogues with prealbumin. This work is particularly relevant in view of the recent discovery of a range of naturally occurring iodothyronine compounds in human serum, which are normal products of thyroid hormone metabolism (Chopra *et al.*, 1978).

5.2 THE THYROID HORMONES

The thyroid hormones: 3,5,3′,5′-tetraiodo-L-thyronine or thyroxine (T_4) and 3,5,3′-triiodo-L-thyronine (T_3) are iodinated derivatives of the amino acid, tyrosine (Figure 5.1). They are synthesized in, and secreted from, the thyroid gland, under the control of the thyroid stimulating hormone (TSH), secreted by the anterior pituitary, which is in turn controlled by the thyrotropin releasing hormone (TRH) from the hypothalamus (Lissitzky, 1976; Robbins and Rall, 1979). During vertebrate evolution, they have assumed a dual role in initiating development and regulating metabolism (Frieden, 1981). Table 5.1 summarizes the main biological effects of the hormones, which form the basis of a variety of assays for hormonal activity.

5.2.1 METABOLIC DERIVATIVES OF T_4

Although T_4 is the main secretory product of the thyroid gland, it is subsequently deiodinated in the peripheral tissues such as the liver, kidney and pituitary to yield T_3, which is the biologically more potent hormone (Pitt-Rivers *et al.*, 1955; Braverman *et al.*, 1970). Besides T_3 and T_4, several other iodothyronine compounds have now been recognized in the human circulation (Chopra *et al.*, 1978; Chopra, 1981). These include 3,3′,5′-triiodo-L-thyronine (reverse T_3 or RT_3), so-called because its pattern of iodine substitution is opposite to that of T_3); three diiodothyronines (3,3′-T_2, 3′,5′-T_2 and 3,5-T_2); two monoiodothyronines (3-T_1 and 3′-T_1); and the completely unsubstituted thyronine, T_0. These compounds are chiefly derived from the metabolic deiodination of T_4 (Visser, 1980; Figure 5.2). The derivatives tetraiodothyroacetic acid (TETRAC) and triiodothyroacetic acid (TRIAC) are obtained by

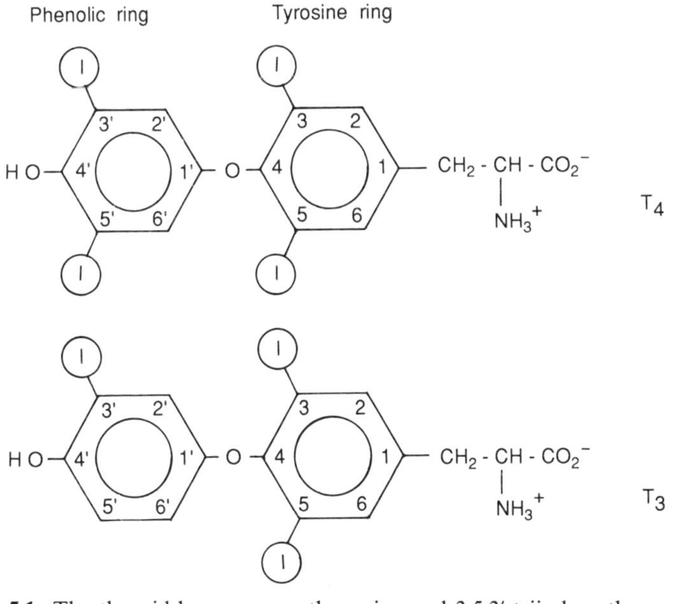

Figure 5.1. The thyroid hormones, L-thyroxine and 3,5,3′-triiodo-L-thyronine

Table 5.1 Physiological actions of the thyroid hormones

Growth promoting and developmental actions	Metabolic effects
Rate of growth of many mammalian and avian tissues, e.g. heart, liver, lung	Regulation of basal metabolic rate
Maturation of central nervous system and bone	Temperature adaptation in warm-blooded vertebrates (through effect on oxygen uptake and heat production)
Control of amphibian metamorphosis	Regulation of water and ion transport
Regulation of synthesis of some mitochrondrial enzymes, e.g. Na-K activated ATPase, adenyl cyclase, pyridine nucleotide coenzymes	Control of calcium and phosphorus metabolism Regulation of cholesterol and fat metabolism Regulation of nitrogen (urea and creatine) metabolism

After Tata (1980).

deamination of the hormone side chain. Cleavage of the ether link of the thyroid hormones results in the formation of mono- and diiodotyrosine (MIT and DIT; Balsam and Ingbar, 1979; Balsam *et al.*, 1980).

Besides T_3 and T_4, only TRIAC and TETRAC have been shown to possess calorigenic activity (Stasilli *et al.*, 1959; Money *et al.*, 1960) but their serum concentrations are too low to contribute significantly to the overall metabolic effects of the thyroid hormones. The remaining iodothyronines are inactive in many of the traditional bioassay systems (Eastman *et al.*, 1980) but at concentrations higher than physiological, and in various *in vitro* systems, these do show intrinsic biological activity. For instance, they stimulate amino acid uptake by rat thymocytes (Chopra, 1981), and growth hormone secretion by pituitary tumour cells (Papavasiliou *et al.*, 1977).

5.2.2 CHEMICAL PROPERTIES

The thyroid hormone molecules consist of three chemically distinct moieties:

(a) a hydrophobic thyronine nucleus, which accounts for the poor solubility of the compounds in water and the need for specific binding proteins to carry them to the target cells;
(b) a hydrophilic hydroxyl group attached to the phenolic ring;
(c) an amino acid side chain on the tyrosine ring.

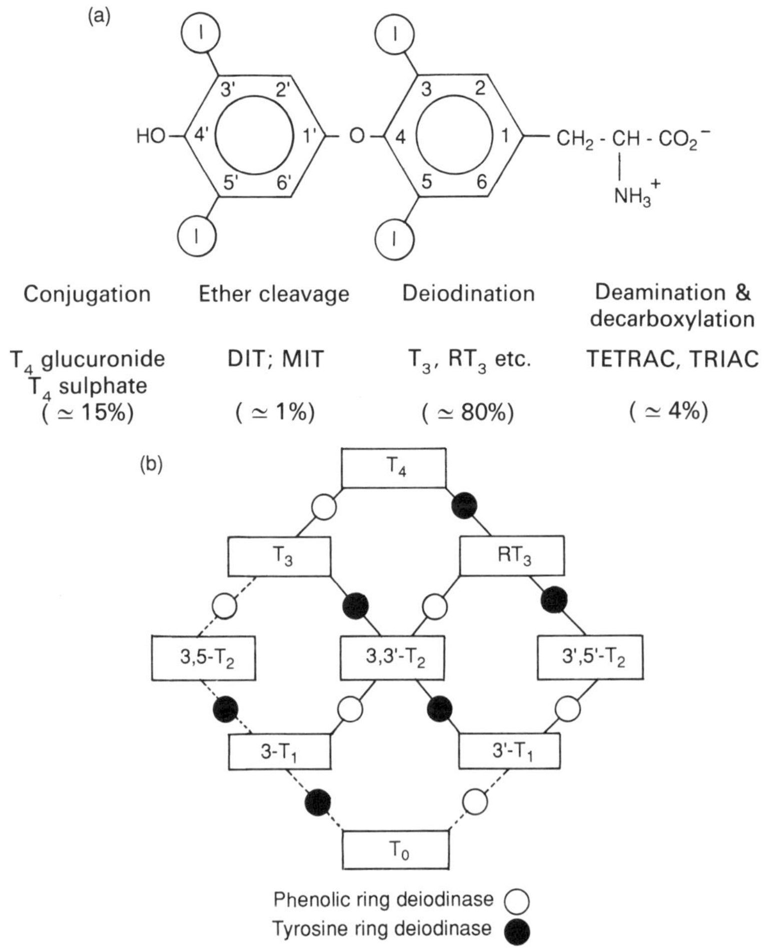

Figure 5.2. (a) Metabolic breakdown of thyroxine. (b) Deiodination cascade of thyroxine, T_4. T_4 can be deiodinated in an organized sequence of reactions to yield a series of tri-, di- and monoiodothyronines, as well as the totally unsubstituted thyronine, T_0 (after Chopra, 1981). It has been suggested that the entire cascade of reactions may be mediated by a single phenolic ring deiodinase and a single tyrosine ring deiodinase

The charge characteristics of the three ionizable groups on the hormone molecule (i.e. the α-carboxyl, the α-amino and the 4′-hydroxyl groups) are summarized in Table 5.2. This shows that at physiological pH, the amino acid side chain exists as a zwitterion. The acidity of the 4′-hydroxyl group is increased by the iodine atoms at the 3′ and 5′ positions so that for T_3 and T_4, the group is also predominantly ionized. The expected percentage of phenoxide ion for the remaining iodothyronine compounds is shown. The ionizable groups of the thyroid hormones are important in binding to proteins, not only by virtue of

Table 5.2 Charge characteristics of ionizable groups on iodothyronine compounds

Group	Compound	pK_a	% Age ionized at pH 7.4	Possible model for
α-carboxyl		2.0	100	All iodothyronines with an amino acid moiety
α-ammonium	(a) Tyrosine	9.0	90	
	(b) Diiodotyrosine	7.7	72	
4′-hydroxyl	(c) T_0	9.6	0.7	Thyronines with unsubstituted phenolic rings, e.g. 3-T_1
	(d) 3,5-T_2	9.3	1.3	
	(e) T_3	8.5	8.2	3′-iodinated thyronines, e.g. 3,3′-T_2, 3′-T_1
	(f) T_4	6.7	82.4	3′,5′-diiodinated thyronines, e.g. RT_3, 3′,5′-T_2

Data from Jorgensen (1978), based on Gemmill (1955).

their charge properties, but also as potential sites for hydrogen bonding: the ionized α-ammonium group acts as a hydrogen bond donor, the carboxylate group acts as an acceptor, the 4′-hydroxyl group can act as both a donor and an acceptor, but in the phenoxide state only as an acceptor.

5.2.3 CONFORMATIONAL PROPERTIES

The main conformations of the thyroid hormones can be described in terms of rotation about five single bonds in the thyronine nucleus (Figure 5.3). ϕ and ϕ' describe the disposition of the two ring systems about the diphenyl ether link, whose bridging angle is 120° (Jorgensen, 1976). χ_1, χ_2 and ψ describe the conformation of the amino acid side chain.

Crystallographic studies on a range of iodothyronine compounds (Cody, 1978, 1980) have shown that the bulky 3,5 iodines on the tyrosine ring restrict the flexibility of the ether link, constraining the rings to be nearly mutually perpendicular. This gives rise to two possibilities:

(a) a skewed conformation ($\phi, \phi' = 90°, 0°$),
(b) an antiskewed conformation ($\phi, \phi' = 0°, 90°$),

depending upon which ring is perpendicular to, and which is coplanar with, the ether link (Figure 5.4). The former has been widely observed in thyroid hormone

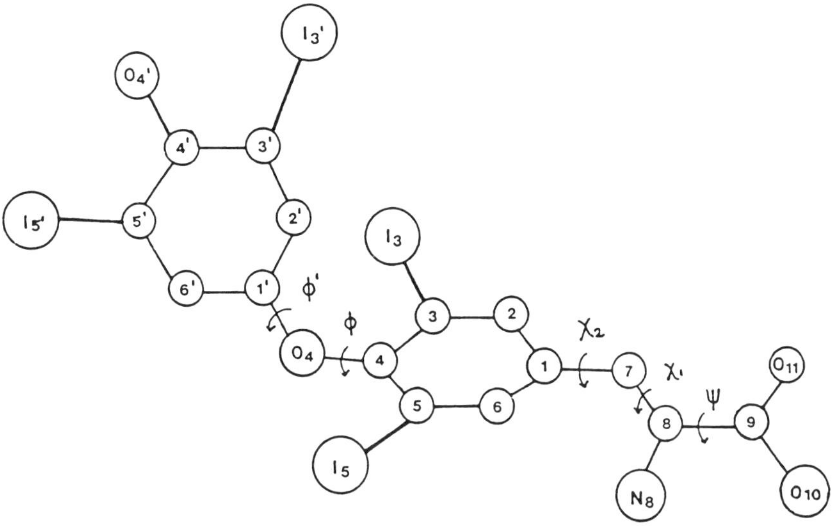

Figure 5.3. Definitions of the five single bond rotations which characterize thyroid hormone conformation (after Cody, 1980). Diphenyl ether conformation: $\phi =$ C3-C4-O4-Cl′; $\phi' =$ C4-O4-Cl′-C2′. Amino acid side chain conformation: $\chi_2 =$ C2-Cl-C7-C8; $\chi_1 =$ Cl-C7-C8-N8; $\psi =$ N8-C8-C9-O10. The torsion angle A—B—C—D is positive if, when viewed down the B—C bond, the A—B bond will eclipse the C—D bond when rotated less than 180° in the clockwise direction

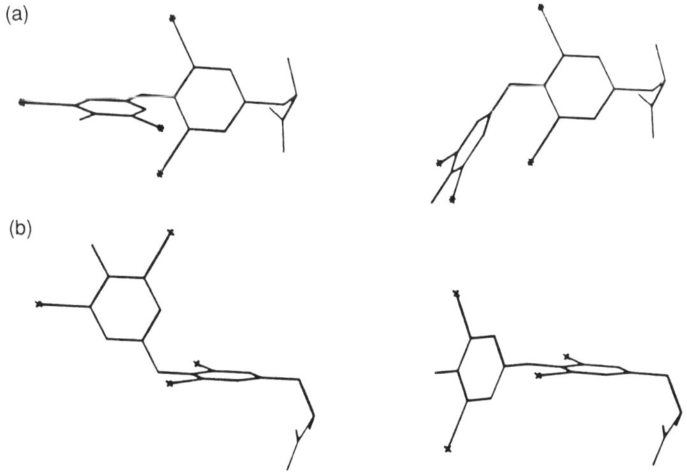

Figure 5.4. Diphenyl ether conformations for the thyroid hormones. (a) View in plane of tyrosine ring. Left: skewed, ϕ, $\phi' = 90°, 0°$; phenolic ring in plane of ether link. Right: antiskewed, ϕ, $\phi' = 0°, 90°$; tyrosine ring in plane of ether link. (b) View normal to plane of tyrosine ring. Left: skewed, ϕ, $\phi' = 90°, 0°$; phenolic ring in plane of ether link. Right: antiskewed, ϕ, $\phi' = 0°, 90°$; tyrosine ring in plane of ether link

structures (Cody, 1980). The latter, by contrast, is confined to compounds which have either only one or no bulky 3,5 substituents, e.g. RT_3 (Okabe *et al.*, 1982). The reason for this is clear from Figure 5.4, which shows that, for a 3,5 disubstituted compound, an antiskewed conformation would cause steric hindrance between the phenolic ring and one of the 3,5 substituents.

Restricted rotation about the C—O—C bridge also implies that for compounds such as T_3 and its analogues, which have only a single 3′ (or 5′) substituent on the phenolic ring, two distinct conformers are possible (Figure 5.5a).

(a) proximal, in which the 3′ substituent lies *near* the tyrosine ring;
(b) distal, in which the 3′ substituent lies *away* from the tyrosine ring.

There are also two overall conformations possible for the hormones, depending upon whether the phenolic ring and the alanine side chain are on the *same* side of the tyrosine ring plane (cisoid), or on *opposite* sides (transoid; Figure 5.5b). Conversion from cisoid to transoid forms is governed by a ϕ rotation, and from proximal to distal forms by a ϕ' rotation. The precise diphenyl ether conformation adopted by an analogue depends, however, on the nature of substituents on the thyronine nucleus and on the crystal environment (Andrea, 1977; Andrea *et al.*, 1979).

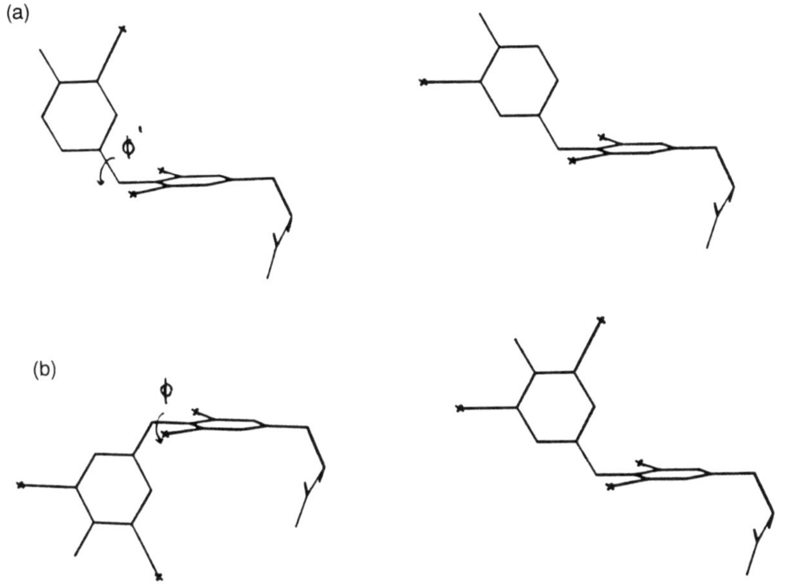

Figure 5.5. Diphenyl ether conformations for the thyroid hormones: (a) proximal (left) versus distal (right), governed by ϕ' rotation; (b) cisoid (left) versus transoid (right), governed by ϕ rotation

The amino acid moiety of the hormone tends to be oriented so that its Cβ atom is perpendicular to the plane of the tyrosine ring. Molecular energy calculations (Andrea *et al.*, 1979) and X-ray studies (Cody, 1980) both emphasize, however, that the side chain itself is highly flexible and can adopt a range of conformations which would allow it to maximize binding interactions.

5.3 THYROID HORMONE BINDING PROTEINS

The binding of the thyroid hormones to specific proteins is central, not only to hormone transport and tissue distribution, but also to hormone metabolism and biological activity. The thyroid hormones bind to two main classes of proteins (Figure 5.6):

(a) plasma proteins, which transport the hormones through the general circulation (Robbins, 1975, 1976);

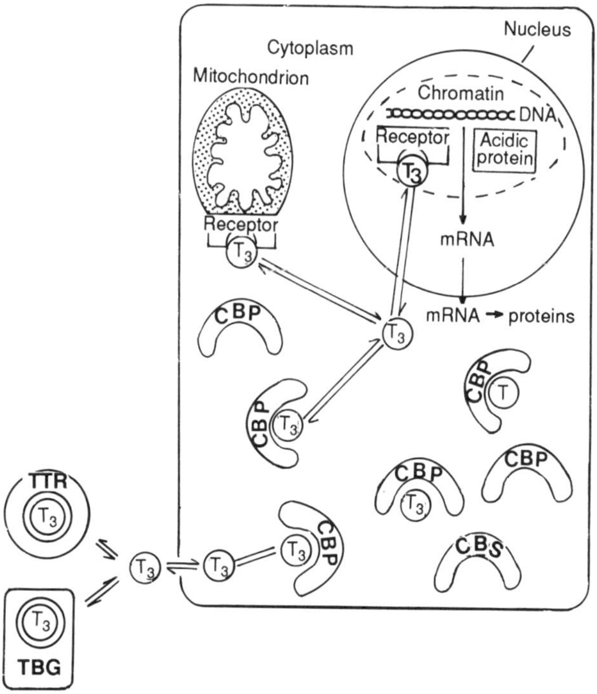

Figure 5.6. A cartoon showing the major binding proteins for the thyroid hormones. TTR is transthyretin, TBG is thyroxine binding globulin, and CBPs are cytosol binding proteins

(b) cellular proteins, located in the nuclei (Oppenheimer *et al.*, 1972), cytoplasm (Dillman *et al.*, 1974), plasma membrane (Pliam and Goldfine, 1977; Cheng *et al.*, 1980) and mitochondria (Sterling *et al.*, 1978) of target tissues. The nuclear proteins act as specific receptors which may mediate thyroid hormone function through an effect on DNA transcription. The other hormone binding proteins may be involved in additional biological effects of the thyroid hormones (Tata, 1975; Sterling, 1979a, b).

5.3.1 TRANSPORT PROTEINS

The thyroid hormones are transported in the blood by three main plasma proteins: thyroxine-binding globulin (TBG), transthyretin (TTR) and serum albumin (SA). Table 5.3 compares the main physical and chemical properties of these proteins, while Table 5.4 summarizes their relative affinities for a selection of iodothyronine compounds. Table 5.3 shows that despite its low serum concentration TBG carries the bulk of the thyroid hormones because of its high affinity for these compounds. Transthyretin, with a 20-fold higher concentration, has a 50-fold lower affinity and acts as a secondary carrier for the hormone. Serum albumin, despite its very high serum concentration, carries a relatively small proportion of T_4 and T_3 because its major thryoxine binding site is a further 50-fold weaker than that of transthyretin.

The transport proteins are not directly involved in thyroid hormone action since the absence of either TBG or serum albumin on a genetic basis produces no apparent abnormality (Robbins, 1976). Nevertheless, they influence the distribution of hormone in body fluids and tissues by contributing to the level of biologically active free hormone in the blood (Ekins, 1982; Pardridge, 1981). Although over 99% of the thyroid hormones in the plasma are bound to transport proteins, the dissociation rates of the hormones from these proteins are sufficient to maintain the concentration of free hormone at a relatively constant level, with transthyretin contributing as much as TBG to the hormone flux (Robbins, 1975). Moreover, the proteins act as buffers to maintain a large store of thyroid hormone in the blood in a non-diffusible form. This prevents large fluctuations in hormone availability at the target cell, obviating the need for very rapid hormone synthesis and protecting non-target sites from excessive hormone effects (Robbins and Rall, 1979). Finally, the transport proteins are of clinical importance because any variations in thyroid hormone binding will affect the detected levels of thyroid hormone in the blood. Such variations can arise from changes in protein synthesis and metabolism, or from the presence of other ligands which compete with the hormone for the binding sites.

5.3.2 THYROXINE BINDING GLOBULIN (TBG)

TBG is an acid glycoprotein, rich in sialic aid, with a molecular weight of ~54 000. Partial desialylation yields a protein with decreased electrophoretic

Table 5.3 Physical and chemical properties of thyroid hormone transport proteins

Property	TBG	TTR	SA
Serum concentration (mg/l plasma)	15	300	45 000
Binding capacity (μg/l)	200	2300	100
Affinity for T_4 of strongest site (M^{-1})	$\cong 10^{10}$	$\cong 10^{8}$	$\cong 10^{6}$
% Age T_4 carried	75	15	10
% Age T_3 carried	50	25	25
Physical properties			
Molecular weight	54 000	54 000	66 000
No. of peptide chains	1	4	1
β structure (%)	40	45	15
α Helix (%)	30	1	48
Chemical properties			
Half-cystines	2	4	35
S–S bridges	0	0	17
Total carbohydrate (%)	23	0	0
Sialic acid (%)	10	0	0
State of structure determination	Entire sequence known structurally homologous to α_1-antitrypsin	Entire sequence known Crystal structure solved	Entire sequence known Crystal structure solved

Data from Robbins and Rall (1979).

Table 5.4 Relative affinities of transport proteins for thyroid hormone analogues

Compound	TBG[a]	TTR[b]	SA[c]
T4	100.0	100.0	100.0
T3	9.0	9.2	49.0
RT3	38.0	32.9	—
TETRAC	1.7	680.0	20.0
TRIAC	0.3	20.0	—

[a]Snyder *et al.* (1976); K_A $T_4 = 2.5 \times 10^9\ M^{-1}$
[b]Andrea *et al.* (1980); K_A $T_4 = 3.5 \times 10^7\ M^{-1}$
[c]Sterling (1964); K_A $T_4 = 5.0 \times 10^5\ M^{-1}$

mobility and a lower T_4 affinity but with the same binding capacity as native TBG (Marshall *et al.*, 1972). The recent determination of the complete amino acid sequence (Flink *et al.*, 1986) has shown that human TBG is a single chain of 395 amino acids, which shows a 40% sequence identity with α_1-antitrypsin. Surprisingly, therefore, TBG is a member of the serpin superfamily of serine proteinase inhibitors (Carrell and Boswell, 1986), but shows no homology with transthyretin. This observation gives indirect access to the three-dimensional structure of TBG, since the structure of the homologous α_1-antitrypsin has already been determined by X-ray analysis (Löbermann *et al.*, 1984). The structure of the serpins is divided into two parts, a large flat β-sheet region flanked by α-helices, and a β-barrel structure. This latter region has a tantalizingly similar overall organization to the thyroxine binding site in transthyretin (*vide infra*), and the location of the thyroxine binding site in this region is supported by specific labelling studies (Tabachnick and Perret, 1987). On the basis of this information, a computer model building study of T_4 binding to TBG has been carried out (Terry and Blake, unpublished) which shows that thyroxine can be bound along the axis of the β barrel (see Figure 5.7 in the colour section) in a totally hydrophobic environment rich in aromatic amino acids. The location of this binding site near the reactive centre of the serpins suggests that a single amino acid mutation could have been responsible for converting a serine proteinase inhibitor into a thyroxine binding protein.

5.3.3 TRANSTHYRETIN (TTR)

Transthyretin is the most fully characterized of the thyroid hormone transport proteins because its stability and relative abundance have made isolation and purification easy to accomplish. The protein, which derived its original name of prealbumin from its ability to migrate faster than serum albumin on electrophoresis of whole plasma, is important not only in the transport of the thyroid hormones but also of vitamin A (Raz and Goodman, 1969; Goodman, 1981). The complete amino acid sequence of transthyretin is known (Kanda *et al.*, 1974) and its full three-dimensional structure has been solved by X-ray crystallography. Transthyretin has a molecular weight of 55 000 and is composed of four identical polypeptide chains, each of 127 amino acid residues. The subunits, which have an extensive β-sheet structure, are linked together to form a compact molecule with a channel running through its centre. This contains two symmetry-related binding sites of the thyroid hormones. Chemical data have shown, however, that only one molecule of T_4 binds to transthyretin with high affinity (Table 5.4), indicating that the binding of T_4 to one site inhibits binding to the other, a negative cooperative effect. The mechanism of this effect is still not understood.

5.3.4 SERUM ALBUMIN

Although serum albumin binds a wide variety of ligands (Krag-Hansen, 1981), its interaction with the thyroid hormones has been relatively little studied. The protein possesses one high affinity site for T_4, together with a number of weaker secondary sites (Steiner *et al.*, 1966; Tabachnick, 1967). Spin-label studies in solution indicate that the high affinity site is sufficiently deep to accommodate a molecule that is $\cong 23$ Å long (Cheng *et al.*, 1981). Binding is reduced by the presence of anions such as chloride, as well as by chemical modification of the terminal aspartate of the protein (Sterling *et al.*, 1962; Tritsch and Tritsch, 1963).

5.4 TRANSTHYRETIN

Transthyretin acts as a secondary carrier for the thyroid hormones, after the main transport protein, TBG. Thus, Davis *et al.* (1972) demonstrated that in the serum of patients deficient in TBG, the percentage of T_3 bound to transthyretin increased threefold. Although early chemical studies suggested a single site for T_4 with an affinity of about 10^8 M^{-1} (Raz and Goodman, 1969; Nilsson and Peterson, 1971; Pages *et al.*, 1973), X-ray evidence (Blake *et al.*, 1971, 1974) pointed to the existence of two symmetry-related hormone binding sites on the transthyretin molecule. Moreover, several compounds, structurally similar to T_4, were shown to bind at two equivalent sites on transthyretin. These included the fluorescent probes 8-anilinonaphthalene-1-sulphonate (ANS; Ferguson *et al.*, 1975) and 1-dimethylaminonaphthalene-5-sulphonate (DNS; Cheng *et al.*, 1975), as well as the thyroxine analogue, 4-hydroxy-3,5-diiodobenzaldehyde (Pages *et al.*, 1973). Finally, T_4 was found to compete with ANS for the same binding site(s) on transthyretin as shown by fluorescence quenching (Branch *et al.*, 1972; Nilsson *et al.*, 1975). Further binding studies, using higher concentrations of T_4, and equilibrium dialysis to measure directly the T_4-ANS competition, revealed the existence of a second T_4 binding site with an affinity of $\cong 10^6$ M^{-1} (Cheng *et al.*, 1977a). T_3 and a number of hormone analogues were shown to behave in the same way, having two sites of differing affinity.

In view of the X-ray data that the binding sites are identical, these results are most simply explained in terms of a negative cooperative effect, such that the binding of ligand at one site reduces the binding affinity at the second site. This negative cooperativity could, in turn, arise from two possible sources:

(a) ligand–ligand interaction (e.g. by steric or electrostatic effects);
(b) site–site interaction (mediated, for instance by conformational change within the protein).

5.4.1 INTERACTION WITH RETINOL BINDING PROTEIN; VITAMIN A TRANSPORT

Transthyretin plays an important role in the transport of vitamin A (retinol), from its main storage site in the liver to its target cells. This function is independent of T_4 binding and is carried out in a different manner because retinol does not itself bind directly to transthyretin. Rather, it combines with its specific carrier, retinol binding protein (RBP), which in turn interacts with transthyretin to form a 1 : 1 molar protein–protein complex (Goodman, 1981). The amino acid sequence of human plasma RPB (molecular weight 21 000) is known (Rask *et al.*, 1981) and its three-dimensional structure has been determined (Newcomer *et al.*, 1984). It has an eight-stranded β barrel which binds vitamin A at its centre.

5.4.2 INTERACTION WITH OTHER MOLECULES

Transthyretin binds a variety of other small molecules. The fluorescent probe, ANS, competes with T_4 for common sites on the protein (see above), and crystal structure determinations of ANS (Cody and Hazel, 1977) indicate that it can adopt a conformation similar to that of the thyroid hormones. Transthyretin also interacts with a number of drugs containing a ring structure. These include sodium barbiturate (Ingbar, 1963), penicillin (Surks and Oppenheimer, 1963) and salicylate (Larsen, 1972). The interaction with salicylate is of clinical importance because it affects the binding of the thyroid hormones at pharmacologically attained levels of the drug (Robbins, 1975).

5.4.3 CRYSTAL DATA AND MOLECULAR STRUCTURE

Human transthyretin crystallizes from a variety of sulphate and phosphate precipitants in the pH range 3.9–9.0 (Purdy *et al.*, 1965; Haupt and Heide, 1966), to form diamond-shaped flat plates with dimensions of about $0.5 \times 1.0 \times 2.0\ \text{mm}^3$. Preliminary X-ray studies (Blake *et al.*, 1971) showed that the crystals belong to the orthorhombic space group $P2_12_12$ with cell dimensions $a = 43.5$ Å, $b = 85.7$ Å, $c = 66.0$ Å. Two molecules, each composed of four identical subunits related by tetrahedral symmetry, are arranged in the unit cell such that one molecular twofold axis (the z-axis) coincides with the crystallographic twofold axis along c.

The molecular structure of transthyretin has been determined by isomorphous replacement, followed by crystallographic refinement, extending the resolution in three stages:

(a) to 6 Å on the basis of five isomorphous derivatives (Swan, 1970);
(b) to 2.5 Å on the basis of three isomorphous derivatives (Geisow, 1973; Blake *et al.*, 1974);

(c) to 1.8 Å by refinement of the 2.5 Å model against data collected to 1.8 Å, the practical limit of the diffraction pattern (Oatley, 1976; Blake *et al.*, 1978b).

The high resolution model has been further refined, resulting in an *R* factor of 0.19 for all terms with spacings less than 10 Å (S. J. Oatley, unpublished results). The refinement has confirmed the amino acid sequence determined by Kanda *et al.* (1974), while providing a very detailed and accurate description of the native structure (Blake and Oatley, 1977; Blake *et al.*, 1974, 1978b).

The tertiary structure of transthyretin is shown as a schematic diagram of the dimer in Figure 5.8. This shows that in each TTR monomer, about half of the 127 amino acid residues are organized into eight β strands, labelled A–H according to their position in the sequence. These form two four-stranded sheets (DAGH and CBEF in Figure 5.8) to give a wedge-shaped β-barrel structure stabilized by antiparallel hydrogen bonding. The β-strands are linked by loops of varying length, one of which (loop EF) contains a nine-residue stretch of α helix.

The dimer is formed by joining two monomers edge-to-edge along their F and H strands, in an interaction stabilized by extensive antiparallel hydrogen

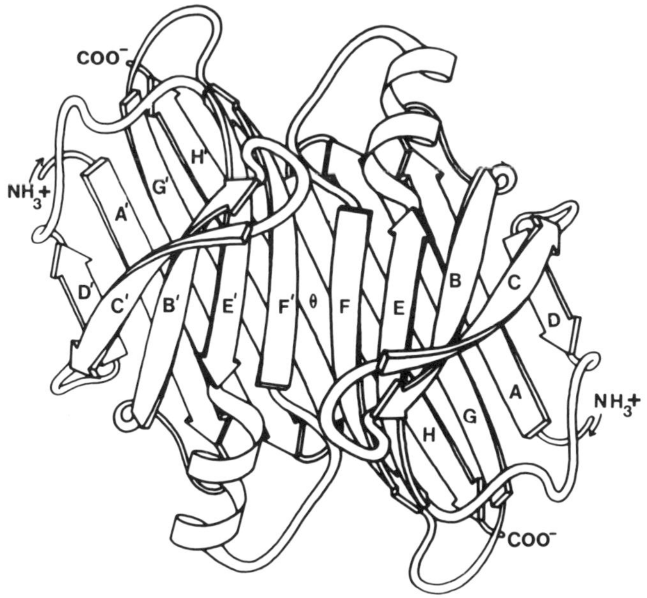

Figure 5.8. A schematic drawing of the transthyretin dimer (after Jane Richardson). The eight β strands in one monomer are labelled A–H, those in the second monomer A′–H′

bonding. This yields a pair of eight-stranded sheets: DAGHH′G′A′D′ (termed the 'inner' sheet) and CBEFF′E′B′C′ (termed the 'outer' sheet). The twist of the β sheets in each monomer, coupled with molecular symmetry, causes the eight-stranded sheets of the dimer to be markedly concave.

Two dimers assemble into a tetramer by opposing their inner sheets face-to-face about the molecular z-axis. The concave nature of the sheets causes the dimer–dimer interactions to occur mainly along the sheet edges, leaving a central channel about 8 Å in diameter and 50 Å long running horizontally through the molecule. This has been identified as the site for hormone binding.

The resulting transthyretin tetramer (Figure 5.9) therefore consists of four eight-stranded sheets, a pair of inner sheets sandwiched between two outer sheets, and is characterized by the following properties:

(a) It has 222 symmetry, i.e. three twofold axes coinciding at the centre of the molecule. The z-axis, which coincides with the crystallographic twofold axis, runs down the centre of the channel.
(b) The compact arrangement of the subunits is stabilized by both hydrogen bonding and hydrophobic interactions, which accounts for the unusual stability of the transthyretin molecule (Raz and Goodman, 1969).
(c) The inner sheets together form a 16-stranded cylindrical framework for the hormone binding channel (see Figure 5.10).
(d) The twist of the outer sheets, by contrast, forms two deep semicylindrical depressions on the outer surface of the molecule. Computer modelling suggests that these regions, together with their associated BC loops, are structurally complementary to the double helical B form of DNA (Blake and Oatley, 1977).

5.5 HORMONE BINDING INTERACTIONS

Studies of the interactions of transthyretin with thyroid hormones, iodothyronines and other ligands were carried out by the conventional X-ray procedure of preparing crystalline complexes and using their diffraction patterns to calculate the differences in the electron density distributions between the native and complexed form of the protein. The major difficulty with the use of this approach on complexes of transthyretin and thyroid hormones or iodothyronines was that the very low solubility of the ligands in the ammonium sulphate mother liquor of the crystals resulted in very long soaking times for the formation of crystalline complexes. This was compensated to some degree by the very high scattering power of the iodine atoms of the ligands which enabled these atoms of the bound ligands to be placed with a high degree of precision in relation to the protein. All hormones and other ligands that were examined showed electron density features associated with binding within the long

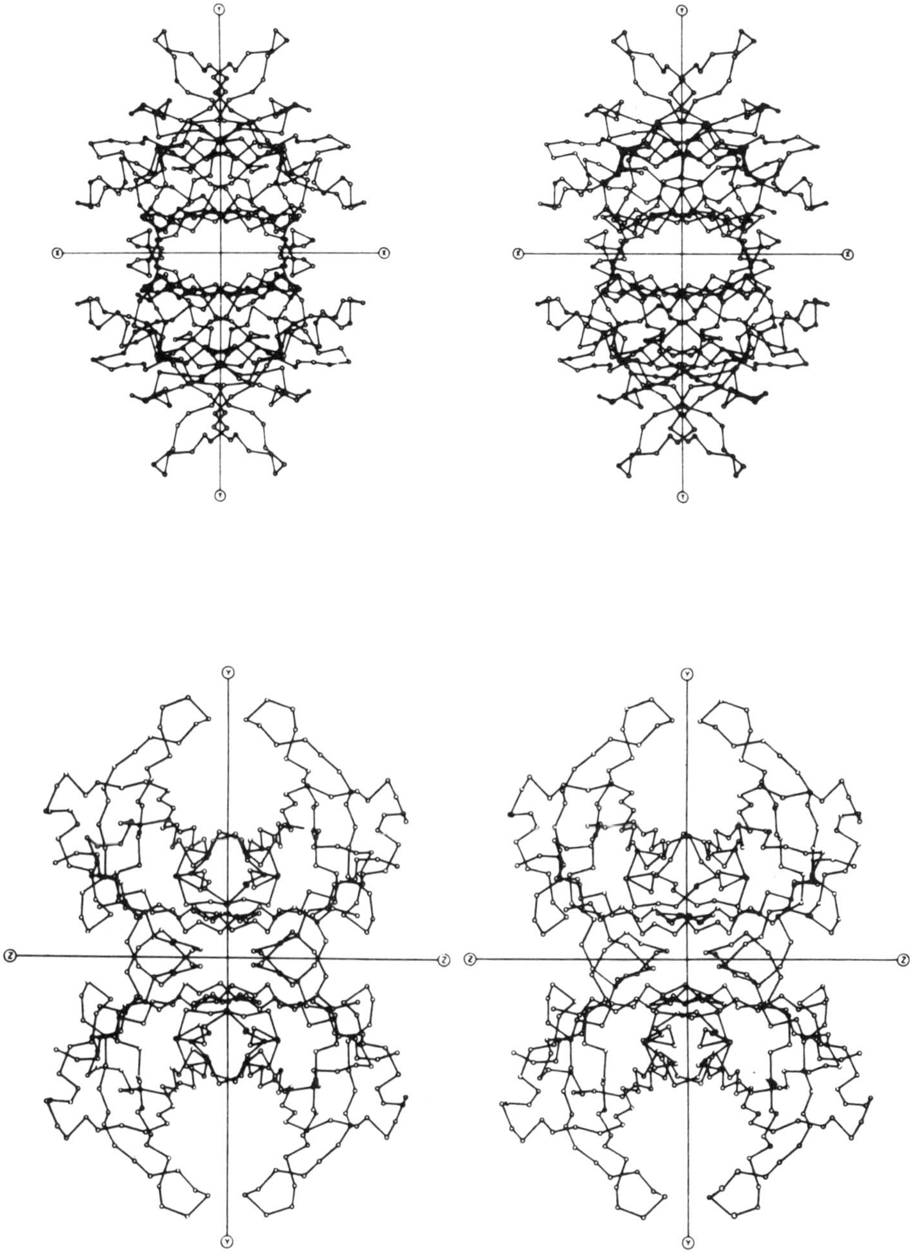

Figure 5.9. Stereoscopic drawings of the TTR molecule, looking down the molecular *Z*-axis (top) and *Y*-axis (bottom). The hormone binding channel is located along the molecular *Z*-axis

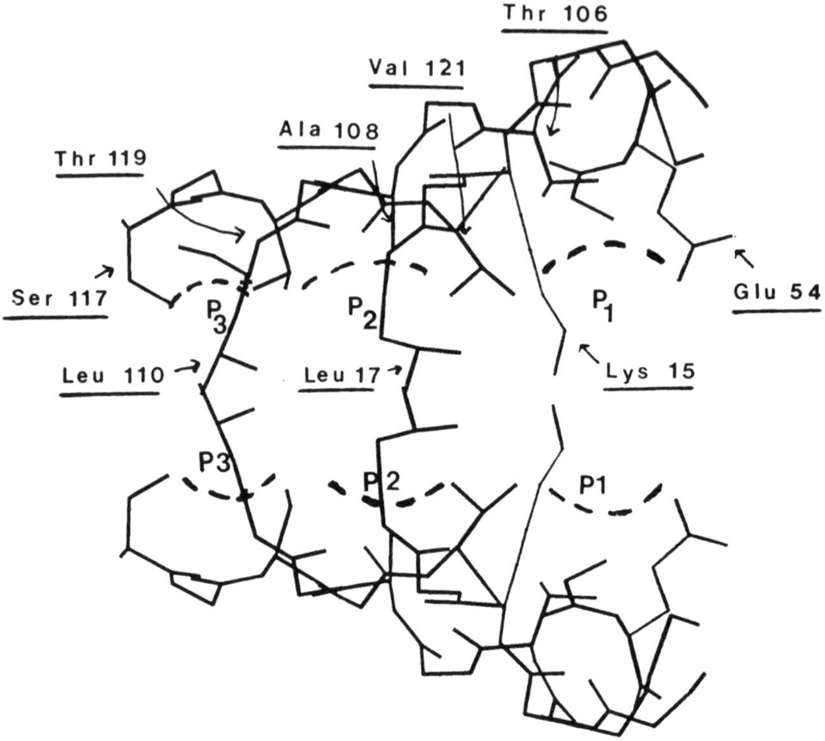

Figure 5.10. Diagram of the hormone binding site in TTR showing the side chains which project into the hormone channel. Entry into the channel is on the right. Broken lines represent the three symmetry related pairs of iodine binding pockets; P_1, P_2 and P_3 (see also Figure 5.17)

channel that runs through the centre of the transthyretin molecule, with no indication of binding elsewhere. This channel can therefore be unequivocally identified as the hormone binding site of transthyretin.

5.5.1 STRUCTURE OF THE HORMONE BINDING SITE

The main features of the central hormone channel in transthyretin are as follows:

(a) It is a 16-stranded β barrel about 50 Å long and 8 Å in diameter. The grooves between neighbouring β strands form a series of 'pockets' into which the iodine substituents of the hormones may be accommodated.
(b) The molecular 222 symmetry of the prealbumin tetramer divides the channel into two symmetry-related halves, each containing one hormone

binding site. The first is located between the monomers A and C, and the second between monomers B and D.

(c) The crystallographic twofold axis parallell to *z* runs down the centre of the channel, giving each binding site twofold symmetry.

(d) The side chains which project into the binding site are arranged in symmetry-related pairs and give rise to three chemically distinct regions along the channel (Figure 5.10):
   (i) a *hydrophilic* patch at the centre of the molecule, derived from the hydroxyl groups of the Ser 112, Ser 115, Ser 117 and Thr 119 pairs of residues;
   (ii) a *hydrophobic* patch, formed from the paired residues of Leu 17, Thr 106, Ala 108, Leu 110 and Val 121;
   (iii) a *charged* region near the channel entrance, formed by the paired side chains of Lys 15, and Glu 54.

Several well defined water molecules have been located at the centre of the binding channel, which form a hydrogen-bonded network linking the residues Ser 115, Ser 117 and Thr 119 across the two binding sites.

Recent solution studies (Cheng *et al.*, 1981) using spin-labelled probes have estimated the length of each binding site to be $\cong 21$ Å, which is consistent with the crystal structure data and suggests that the solution and crystal conformations of the protein are very similar.

5.5.2 X-RAY STUDIES ON HORMONE BINDING TO TRANSTHYRETIN

The first successful attempt to bind the thyroid hormones to transthyretin crystals was achieved by Geisow (1973), using protein crystals soaked in saturated solutions of the hormones, made up in the mother liquor of crystallization. The experiments were carried out at pH 7.4 for T_4 and 8.6 for T_3, in order to capture the 4′-hydroxyl group of the ligand in its ionized form, which was known to promote binding to prealbumin. Soak times of 1–2 months were found to be necessary, however, presumably because of the low solubility of the compounds in the transthyretin mother liquor.

Difference electron density maps, calculated at 6 Å resolution, were very similar for both compunds. They showed that the hormones were bound at two symmetry-related sites within the central channel, each site being located between two of the four subunits. In view of the chemical data showing that only one mole of hormone binds per mole of transthyretin, the simplest explanation is that the hormone can enter either of the two equivalent sites at random, so that the X-ray maps show an averaged situation with $\cong 50\%$ occupancy of each site.

Difference maps were calculated at resolutions of 1.8 Å for T_4 and 1.9 Å for T_3 (Burridge, 1982; Blake *et al.*, 1978a) representing the electron-dense iodine substituents of the hormones. Figure 5.11 shows the maps of T_3 and T_4 binding, projected down the *x*-axis of the transthyretin molecule. The pattern of peaks in both maps was consistent with the hormones binding with their long axes parallel to the long axis of the hormone channel, and the phenolic hydroxyl group buried closest to the channel centre. This interpretation allowed the transthyretin binding site to match the hydrophilic–hydrophobic–ionic character of the thyroid hormones in a striking manner (Figure 5.12).

The chief difference features in the maps correspond solely to the iodine atoms, with very little indication of the light-atom carbon skeleton. This results

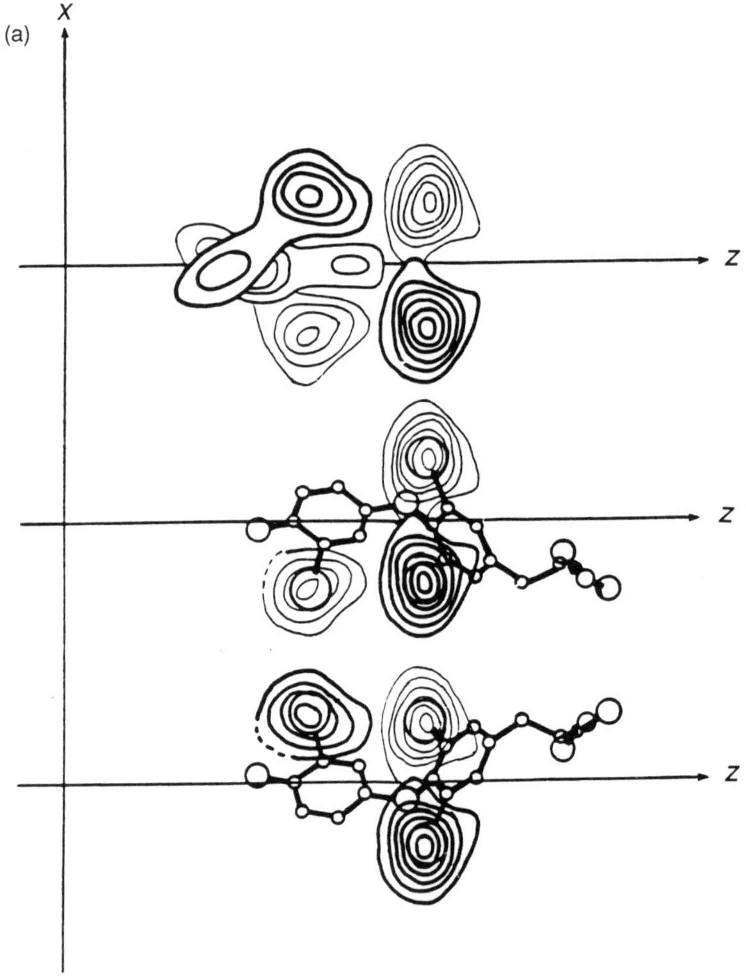

Figure 5.11. Drawings of the difference electron density representing the bound hormone, and its fitting by two alternative symmetry related hormone molecules, for (a) T_3; (b) T_4.

from a combination of three factors. First, the negative cooperativity of hormone binding to transthyretin implies that only one molecule binds per transthyretin tetramer, giving an average occupancy of 0.5 ligand molecules in each of the two identical binding sites. Secondly, the twofold axis which runs through the centre of each site divides the ligand into two symmetry-related orientations, reducing the difference features by a further factor of two. Even at full occupancy, the ligand features are therefore reduced to one-quarter of their normal weight. Finally, the poor solubility of the thyroid hormones in the transthyretin mother liquor limits their actual occupancy to about 60%, so that the ligand features in each site appear at only about 15% of their normal height. This is sufficient to drop the features of the light atoms below the error level of

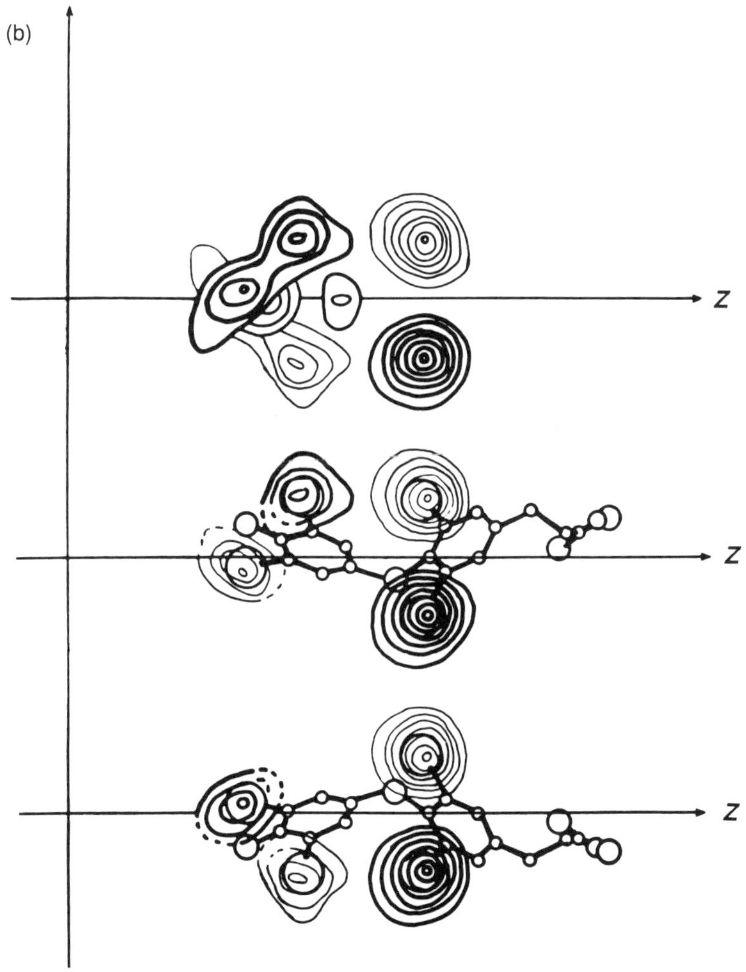

Figure 5.11 (*continued*)

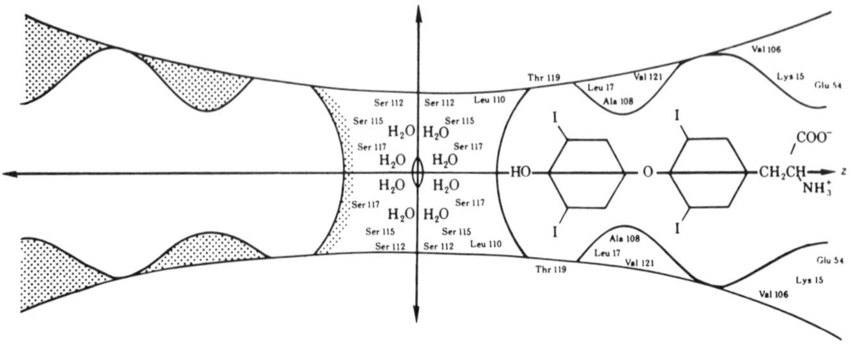

Figure 5.12. A schematic drawing of the relative positions of the protein side chains and the bound hormone in the transthyretin binding site

the maps (Burridge, 1982). However, given the limited conformational variability within the thyronine nucleus for T_3 and T_4, it has been possible to interpret the maps by fitting the iodine atoms to the major difference features. This has led to a detailed interpretation for hormone binding which is presented below.

5.5.3 T_4 BINDING TO TRANSTHYRETIN

In the initial interpretation for hormone binding to transthyretin (Burridge, 1982), the thyroxine molecule lies along the two-fold axis of the channel, with the phenolic hydroxyl group buried at the centre and the amino acid side chain near the channel entrance (Figure 5.13). This orientation was chosen on the

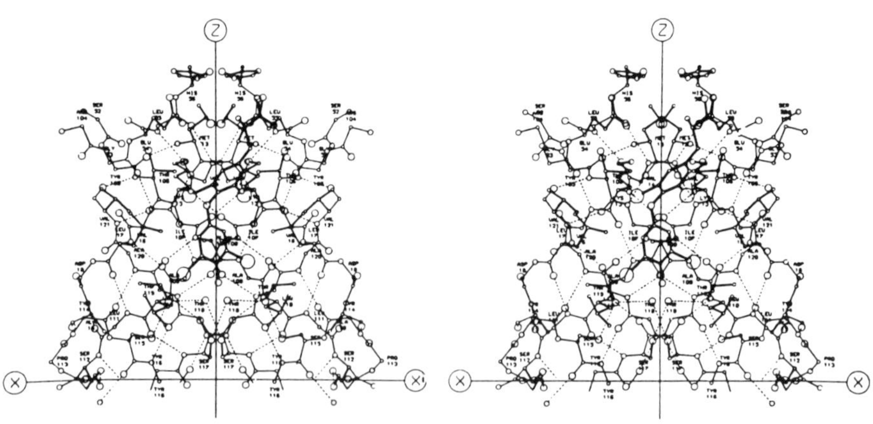

Figure 5.13. Stereoscopic view of the location of T_4 in the binding site

basis of steric considerations since the side chains near the molecular centre, particularly Leu 110, Ser 117 and Thr 119, would, in their proposed native conformations, prevent entry of the hormone with its amino acid moiety lying innermost.

In each hormone binding site, there are six pockets capable of binding an iodine substituent (Figures 5.10 and 5.13). These are formed from the grooves between neighbouring β strands and are arranged in three sets of symmetry-equivalent pairs: an outer pair of pockets (P_1 and P_1'), lying about 13 Å from the molecular centre; and two sets of inner pockets, a 'proximal' pair (P_2 and P_2') and a 'distal' pair (P_3 and P_3'), lying about 9 Å and 6 Å respectively from the channel centre. The distal pockets are each occupied in the native structure by a well defined water molecule (W1 or W1′), which is hydrogen bonded to the hydroxyl groups of two residues, Thr 119 and Ser 117.

When T_4 binds, four of these six pockets become occupied by the iodines of the hormone molecule, as seen in Figure 5.14 (colour section), in which the solvent-accessible surfaces of the T_4 complex have been calculated (Blaney *et al.*, 1982a). The close steric fit between the ligand and the binding site is strikingly demonstrated. The main protein–hormone interactions that arise from the final fitting are as follows:

(a) the *3,5 iodines* of the tyrosine ring bind to the hydrophobic outer pockets P_1 and P_1', making similar, but not quite symmetrical interactions with the methyl and methylene groups of Lys 15, Leu 17, Thr 106, Ala 108 and Val 121. Residues from both monomers of the binding site contribute to each outer pocket (Figure 5.15a).

(b) the *3′,5′ iodines* of the phenolic ring fit into pockets which are not related by symmetry and each iodine interacts exclusively with one monomer of the binding site (Figure 5.15b). The 3′ iodine atom, proximal to the tyrosine ring, binds to the so-called 'proximal' pocket P_2, defined by the β strands of Ala 108, Ala 109 and Leu 110 on the one hand and Lys 15 and Leu 17 on the other. The 5′ iodine atom, distal to the tyrosine ring, binds to the 'distal' pocket P'_3, formed by the neighbouring β strands of Ala 108, Ala 109 and Leu 110 on the one side and Ser 117 and Thr 119 on the other. The binding of the 5′ iodine to the distal pocket causes the well defined water molecule (W1′) in this pocket to be displaced.

The proximal and distal inner pockets are more polar than the outer pockets, not only because they involve hydrophilic serine and threonine residues, but also because they contain contributions from the nitrogen and oxygen atoms of the peptide backbone. The contacts made with the iodine atoms of the hormone are also tighter and more numerous as a direct result of the architecture of the binding site which flares out in a trumpet shape as the distance from the channel centre increases. Thus, contacts for the outer pockets occur mainly through the side chains of

(a)

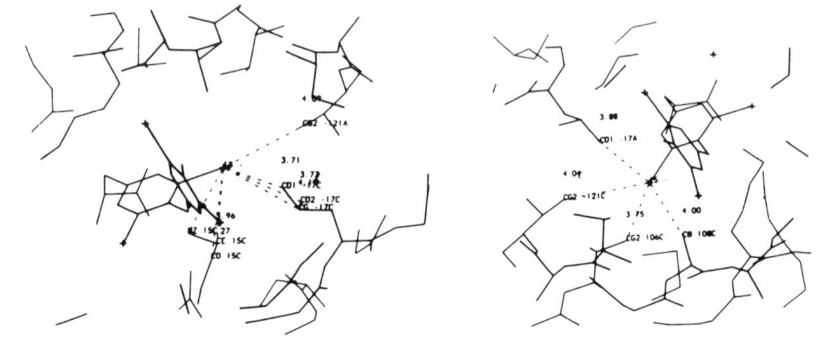

(b)

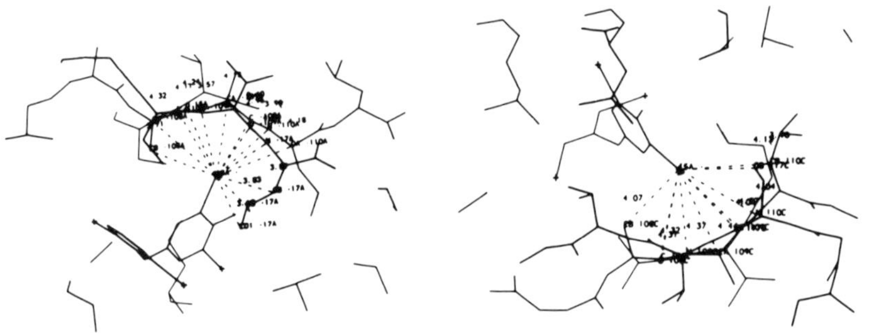

Figure 5.15. Interactions of the iodine substituents of bound T_4: (a) the 3,5 iodines; (b) the 3′,5′ iodines. All interactions less than 4.5 Å are indicated by dotted lines

amino acid residues, while those for the inner pockets often involve the peptide backbone itself.

A protein–iodine interaction of interest is the close contact observed between the proximal 3′ iodine of the hormone and the carbonyl oxygen of Ala 109, which at about 3.0 Å is closer than that normally expected for non-bonded iodine–oxygen interactions (3.6 Å). However, there is evidence from small molecule structures that iodine–oxygen distances exhibit a pronounced shortening to as little as 2.8 Å as the carbon–iodine–oxygen angle approaches 180° (Murray-Rust and Motherwell, 1979). The carbon–iodine–oxygen angle in this instance (about 160°) conforms to such a pattern. Indeed, the iodine–oxygen distances noted from crystal structures of the thyroid hormones range from 2.99 Å to 3.45 Å, with an average value of 3.19 Å (Cody, 1978). This could reflect the polarizability and charge-transfer ability of the iodine atom and may contribute to the importance of iodine in thyroid hormone–protein interactions.

(c) the *phenolic hydroxyl group* binds close to the distal pocket P_3 but in the current (TTR : T_4) model appears to make largely hydrophic contacts, not interacting directly with Ser 117 or Thr 119 (Figure 5.16).

(d) the *amino acid side chain* can interact with the charged Glu 54 and Lys 15 pairs of residues at the mouth of the channel (Figure 5.16). There appears to be a balance of attractive and repulsive forces, involving the carboxylate and ammonium groups of the ligand and the Lys 15 and Glu 54 residues of the protein. This is consistent with chemical studies which show that the carboxylate group of the amino acid moiety in thyroxine analogues contributes positively, while the ammonium group contributes negatively, to transthyretin binding (Andrea *et al*., 1980).

5.5.4 T_3 BINDING TO TRANSTHYRETIN

The general orientation for T_3 modelled in the transthyretin binding site is similar to that for T_4, with the phenolic hydroxyl binding closest to the channel

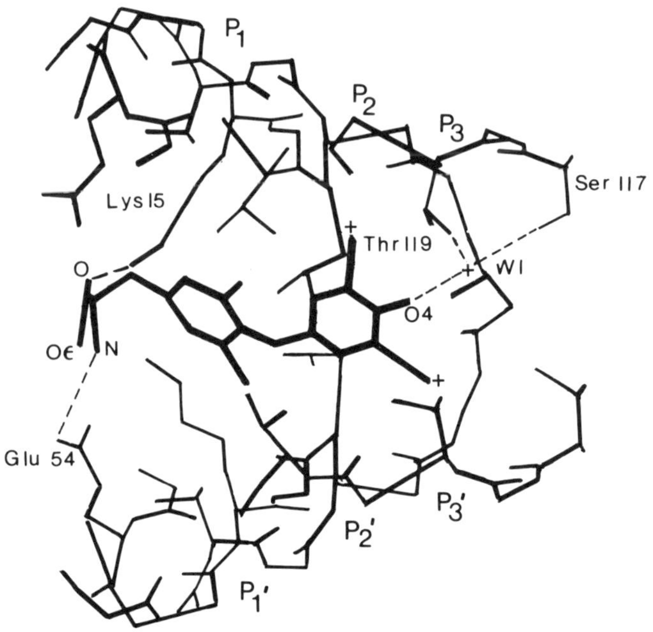

Figure 5.16. Interactions of the non-iodine substituents of bound T_4. Hydrogen bonds involving the amino acid group and the 4′-hydroxyl group are shown by dotted lines.

centre. In order to explain the presence in the difference maps of two sets of inner iodine peaks for T_3, which contains only one iodine on its phenolic ring, two modes of hormone binding have been proposed, a '*proximal*' and a '*distal*' T_3 entry, depending upon whether the hormone is in its proximal or distal conformation and, hence, upon whether the single phenolic ring iodine occupies the proximal (P_2) or distal (P_3) inner pocket of prealbumin (Figure 5.17).

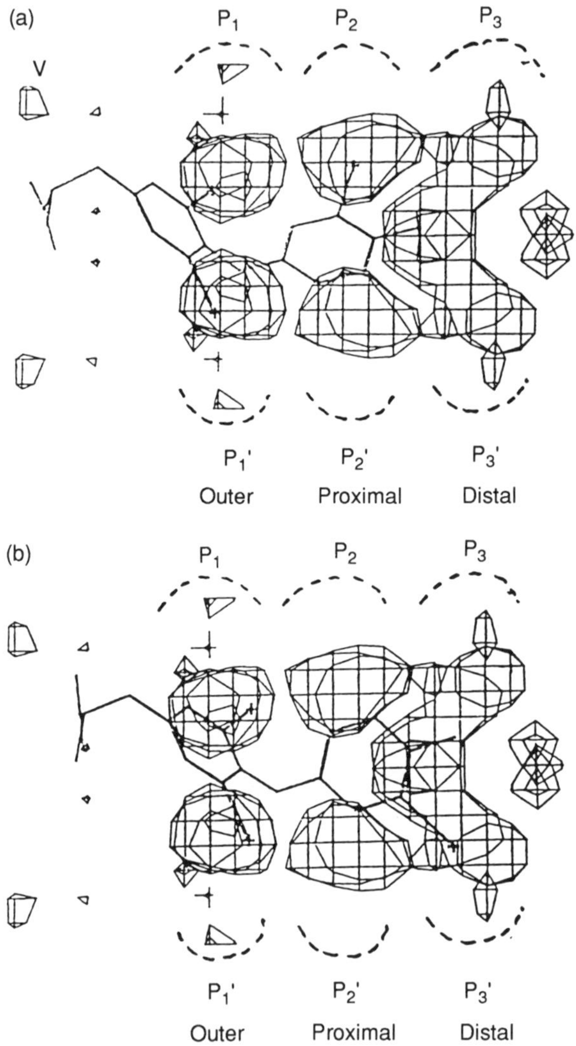

Figure 5.17. Fitting of the difference density of T_3 by (a) the proximal conformation and (b) the distal conformation; P_1, P_2 and P_3 represent the three pairs of iodine binding pockets

The overall pattern for protein–hormone interaction in both modes of T_3 binding is similar to that obtained for T_4, the chief differences arising in the exact position of the hormone in the binding site and the resulting interactions for the 4′-phenolic hydroxyl group at the centre of the channel.

The proximal T_3 model places the hormone molecule about 1 Å further away from the channel centre than T_4, with the following consequences:

(a) the contacts with the protein for the 3′ iodine in the proximal pocket are less close than those for the corresponding iodine in T_4, with no unusually short iodine–oxygen distances being observed;
(b) the 4′-hydroxyl group of the hormone continues to form hydrophobic interactions with the protein, as in T_4;
(c) very little protein conformational change is needed to accommodate the ligand into the binding site.

By contrast, the distal T_3 model places the hormone about 0.4 Å closer towards the channel centre than for T_4, with the 3′ iodine displacing the water molecule from one distal pocket (P'_3), while the phenolic hydroxyl group of the hormone is brought within hydrogen-bonding distance of the symmetry-related water molecule in the other distal pocket (P_3). The protein–hormone interactions for the distal iodine are thus closer than those observed for T_4 itself, while the hydrogen-bonding potential of the phenolic hydroxyl group is satisfied, in direct contrast to the situation for T_4 and proximal T_3.

The apparent preference for a proximal, as opposed to a distal, mode of T_3 entry (as deduced from the greater strength of the proximal iodine peaks in the T_3 difference map) has been attributed to the conformational change required in the protein for the two different modes of entry (Burridge, 1982). A proximal mode of binding apparently requires little conformational change, while for a distal mode of entry, movement of the paired Ala 108-bearing β strands is necessary to accommodate the hormone satisfactorily. It has been argued that such strand movement in the relatively rigid framework of the prealbumin molecule may be less favoured.

5.5.5 ADVANTAGES AND LIMITATIONS OF THE HORMONE BINDING MODEL

The overall scheme for hormone binding which has emerged from this initial interpretation reflects the geometrical and chemical constraints imposed by both the ligand and the binding site. Thus, the hormone binds with its long axis parallel to the long axis of the prealbumin channel. Moreover, chemically distinct regions on the hormone bind to complementary regions on the binding site: the iodines and hormone rings are located in the hydrophobic region of the channel; the amino acid side chain is located near charged residues at the channel entrance; while the phenolic hydroxyl binds close to (but does not apparently interact with) groups at the hydrophilic channel centre. The resulting model provides a structural explanation for some of the biochemical data on the binding of thyroid hormone analogues to transthyretin:

(a) The close steric fit between ligand and binding site contributes to the high association constant observed for T_4 binding to the protein. The fitting of the bulky iodine substituents into 'pockets' on the binding site is evidently a major feature of the interaction. The contribution of a given substituent to binding may be related to the closeness of fit it achieves within the appropriate transthyretin 'pocket'. The closer and more extensive contacts for the 3′,5′ iodines relative to the 3,5 iodines may therefore account for their greater contribution to binding (Andrea *et al.*, 1980; Blaney *et al.*, 1982a).

(b) The presence of an unusually close protein–iodine interaction in the T_4 complex may reflect the role of electronic factors in contributing to the binding of halogen substituents to the protein (Andrea *et al.*, 1980).

(c) The interactions modelled for the amino acid side chain have corroborated affinity labelling studies which indicate that Lys 15 is part of the T_4-binding site (Cheng *et al.*, 1977b). Moreover, molecular mechanics calculations based on the X-ray model have qualitatively reproduced the observed differences in prealbumin affinity for the D- and L-analogues of the hormones as well as for the des-amino and des-carboxy compounds (Blaney *et al.*, 1982b).

One unsatisfactory feature of the current interpretation for T_4 binding, however, is that it provides no explanation in structural terms for the role of the phenolic hydroxyl group in binding to transthyretin. Chemical studies have shown that the thyroid hormones bind more strongly to transthyretin at pH values corresponding to the ionization of the phenolic hydroxyl (Nilsson and Peterson, 1971). Such evidence would suggest some form of electrostatic interaction between the phenoxide ion of the ligand and a suitable group on the protein molecule, presumably one of the hydroxy amino acids or the water molecules located at the heart of the binding site. The hormone binding studies were carried out at pH values which promoted the ionization of the phenolic hydroxyl group. In the current model for T_4 binding, however, not only is the electrostatic charge of the phenoxide ion not neutralized, but in the case of proximal T_3, it is placed in a generally hydrophobic environment.

5.6 MULTIPLE BINDING MODES

5.6.1 PROJECTION MAPS OF LIGAND BINDING

Because of the inability of the model of hormone binding just described to account for the role of the 4′-hydroxyl group of the hormone, and the lack of complete certainty on the precise mode of hormone binding due to our inability to detect its carbon atom skeleton, it seemed desirable to examine the interaction of a wider range of ligands. These included a number of iodothyronines, such as reverse T_3, the three possible T_2s, the two T_1s, and some T_4 analogues with altered substituents. There were 13 ligands in all whose interaction with transthyretin we wished to examine. Since, a three-dimensional data set to 1.8–1.9 Å resolution, such as had been collected for T_4 and T_3, required the collection of a minimum of 20 000–23 000 X-ray intensities, a strategy was evolved to make a preliminary survey of the full range of ligands, using X-ray photographs of the Okl zone of the TTR crystals, requiring some 600 intensities at 2.9 Å resolution, and using maps projected down the crystallographic X-axis to determine the binding mode. Any interactions that looked particularly interesting in this survey would be examined in detail at high resolution in three dimensions.

Figure 5.18 shows the difference electron density features, averaged over the two binding sites, projected down the crystallographic x-axis at 2.9 Å resolution. Only features associated with the binding of the ligands are shown. The panels can be thought of as representing the binding channel whose centre is on the broken line, and the electron density peaks represent the iodine atoms of the ligands. As Table 5.5 indicates the iodine peaks are located broadly at three distances from the centre of the channel: 6 Å, 9 Å and 13 Å, corresponding respectively, to the expected distances for the distal 3′,5′, proximal 3′,5′ and 3,5 iodine atoms, as described previously. Figures 5.18 and Table 5.5 are divided into three sections depending on whether the binding of the ligands can be related to the mode of binding (forward mode) already described (Group A), or not (Group B). Group C corresponds to those ligands whose mode of binding is not clear.

The compound that most strikingly reveals an anomalous mode of binding is 3′,5′-T_2, which, while having no iodine substituents at the 3,5 positions, shows major electron density peaks at the position of the 3,5 pockets of the transthyretin binding site. In a similar way, RT_3 with two iodines in the 3′,5′ position, and only one at the 3,5 positions, produces a difference map showing larger peaks in the 3,5 pocket than in the 3′,5′ pockets. These results open up the question of the mode of binding of ligands in the hormone binding site of transthyretin, and suggest that there is a real possibility that some ligands can bind with their amino acid groups innermost in the site (reverse binding) while others appear to bind with their 4′-hydroxyl group innermost (forward binding). Two other ligands, 3′,5′-diisopropyl-3,5-T_2 and 3,5-T_2, although apparently using a forward mode of binding, have their 3,5 iodine atoms positioned closer to the centre of the binding by 1–3 Å than is normal, and may also represent anomalous modes of binding.

5.6.2 THREE-DIMENSIONAL DOUBLE DIFFERENCE MAPS

In order to examine the exact interaction of these apparently anomalous ligands, full high-resolution three-dimensional X-ray data sets were collected and

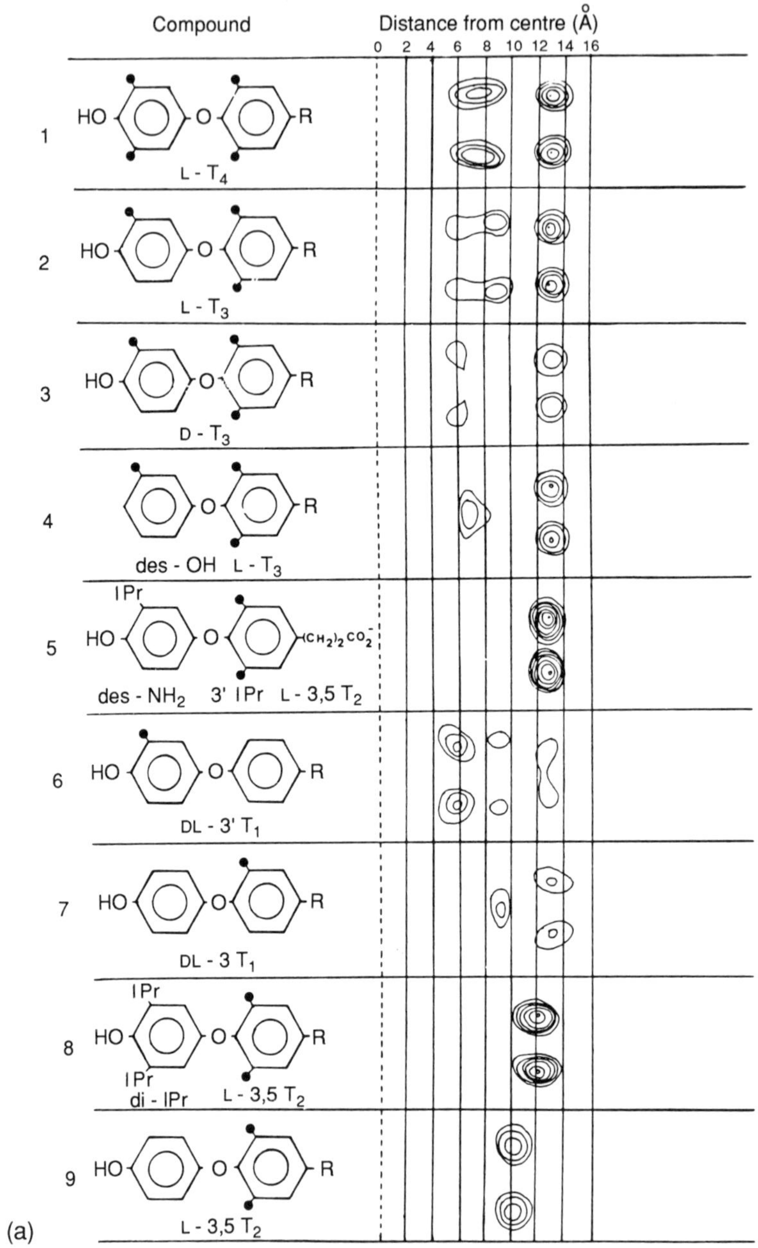

Figure 5.18. Difference electron density maps showing features in the binding site, calculated at 2.9 Å resolution down the crystallographic *a*-axis. (a) The Group A compounds showing forward modes of binding; (b) the Group B compounds showing anomalous modes of binding; (c) the Group C compounds showing indeterminate modes of binding

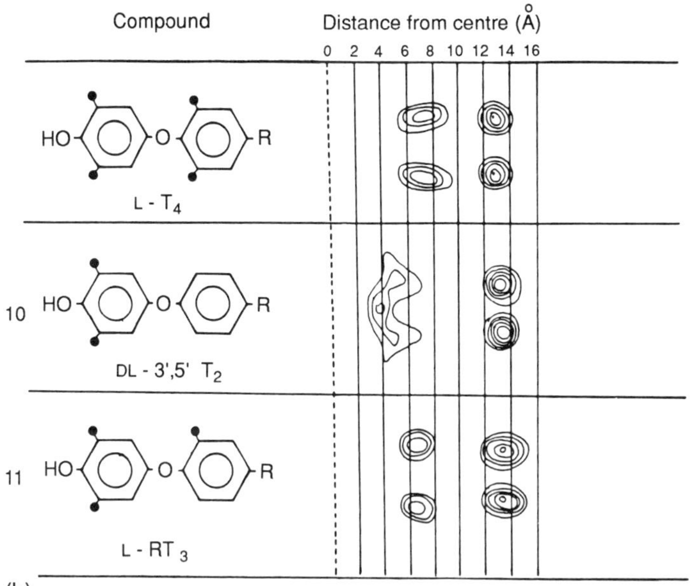

(b)

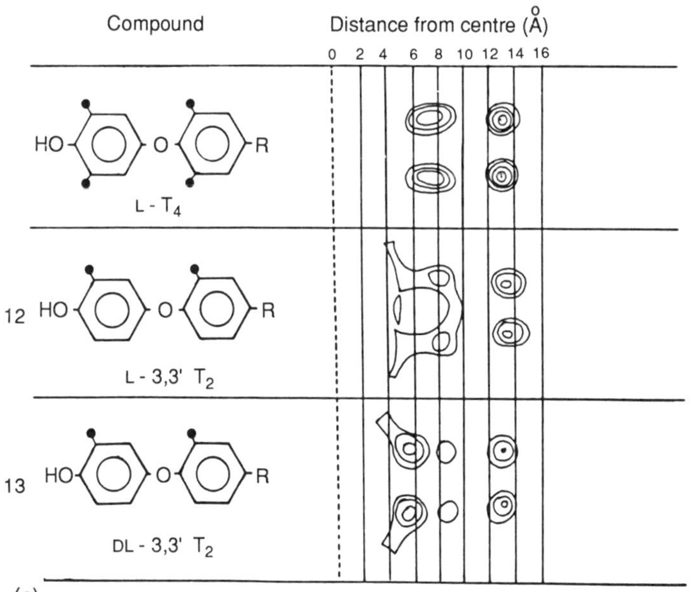

(c)

Figure 5.18 (*continued*)

Table 5.5 Analogue projection results, grouped according to different modes of binding

	Compound		Soak time (mths)	MFID* (%)	Iodine position from centre (Å)			Prox/ distal†	Iodine peak strength	Confor change	% 4′-OH ionised pH 9.0
					Distal	Prox	Outer				
Group A: forward mode of binding											
1	T_4	L	5	28.6	7.5		13.3	Both	+ +	Small	100
2	T_3	L	5	25.5	5.7	9.5	13.2	P	+ +	Small	80
3	T_3	D	1.5	23.5	6.0	—	13.4	D	+	Small	80
4	des-OH T_3	L	3	26.0	7.0		13.2	D	+	Small	80
5	des-NH_2 T_2	L	3	29.9	—	—	13.2		+ + +	Small	33
6	3′-T_1	DL	3	30.9	5.8	(9.0)	(13.0)	D	+ +	Large	80
7	3-T_1	DL	3	25.7	—	9.0	13.0	P	+	Small	33
8	DIT_2	L	1.5	29.1	—	—	12.0		+ + +	Small	33
9	3,5-T_2	L	1.5	24.3	—	10.3	—	P	+ +	Small	33
Group B: anomalous mode of binding											
10	3′,5′-T_2	DL	3	23.4	6.0	—	13.2	D	+ +	Large	100
11	RT_3	L	2	29.7	6.8	—	13.5	D	+ +	Large	100
Group C: indeterminate mode of binding											
12	3,3′-T_2	L	3	36.8	(6.0)	8.5	13.3	P	+ +	Large	80
13	3,3′-T_2	DL	3	35.0	5.5	(8.5)	13.3	D	+ +	Large	80

*MFID, mean fractional isomorphous difference, $= 100 \sum |F_{PL} - F_P| / \sum F_P$

†Column indicates which of the inner peaks (P, proximal; D, distal) is the more significant feature in the projection difference map.

interpreted for the following ligands; 3′,5′-diisopropyl-T_2 (DIT_2), 3′,5′-T_2; RT_3 and 3,3′T_2. The difference maps were calculated at resolutions between 1.8 Å and 2.0 Å (see Table 5.6). As with the earlier T_4 and T_3 maps, the iodine substituents of the ligands showed up as very clear discrete peaks of electron density, with little indication of the carbon skeleton of the ligands nor any substantial conformation changes in the transthyretin molecule. Marked differences in the pattern of iodine peaks were evident in the three-dimensional maps, as indicated by Figure 5.19 which shows the same section of the difference maps of DIT_2, 3,3′-T_2 and 3′,5′-T_2, with that for T_4 shown for reference.

A very valuable technique to obtain information on the precise differences between two complex difference maps, such as those shown in Figure 5.19, is the double difference method. If PX represents the complex between the protein and ligand X, and PY the complex between protein and ligand Y, then, assuming that the difference Fourier maps are on the same scale, in a double difference map (PX − PY), positive features represent elements in PX absent in PY,

Table 5.6 Final merging statistics for 3-D data sets

	T_4	T_3	RT_3	DIT_2	3,3′-T_2	3′,5′-T_2
Maximum resoln (Å)	1.8	1.9	1.8	2.0	1.8	1.8
Number reflns	28 530	24 423	29 178	20 880	88 268	89 286
Number equivalents	1	1	1	1	4	4
Unique reflns	23 875	20 376	23 319	17 478	20 624	20 947
Negative reflns (%)	14	15	14	7	10	13
Merging R (intensity) (%)	6.0	6.7	9.1	4.9	9.4	10.0
MFID* (%)	25.7	26.5	29.7	22.5	25.9	27.3
Soak time (mths)	5	5	3	1.5	3	3
Data collection method	5-Circle diffractometer				Rotation camera	

*MFID, mean fractional isomorphous difference.
$R_{merg} = 100 \sum_{hkl} \sum^{N}_{n=1} |I_i - \bar{I}| \sum_{hkl} N\bar{I}$
where I_i is the value of the ith of N measurements of an intensity; $\bar{I}$ is the mean value of all N measurements of that intensity.
Data for T_3 and T_4 from Burridge (1982).

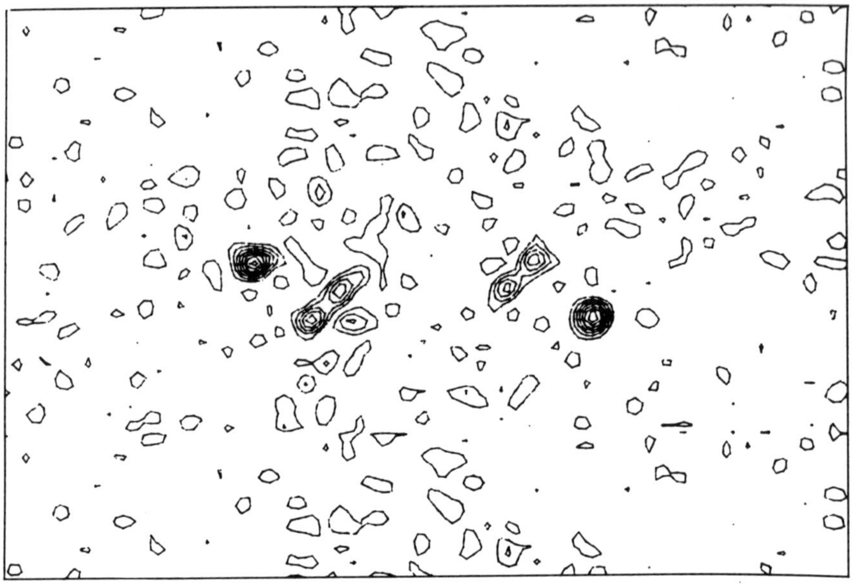

(a)

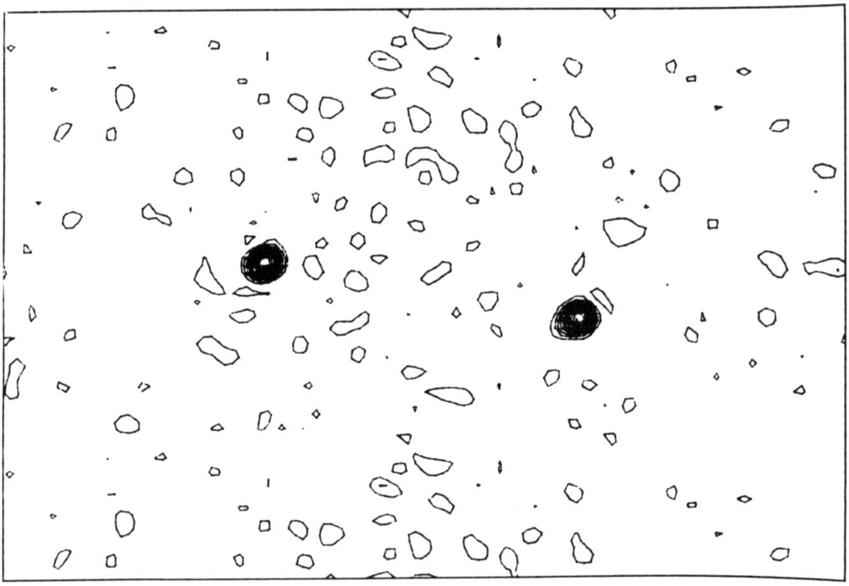

(b)

Figure 5.19. The same section of the three-dimensional difference maps of (a) T_4, (b) 3′,5′-diisopropyl-T_2, (c) 3,3′-T_2, (d) 3′,5′-T_2, calculated at 1.8–2 Å. Contours at equal but arbitrary levels

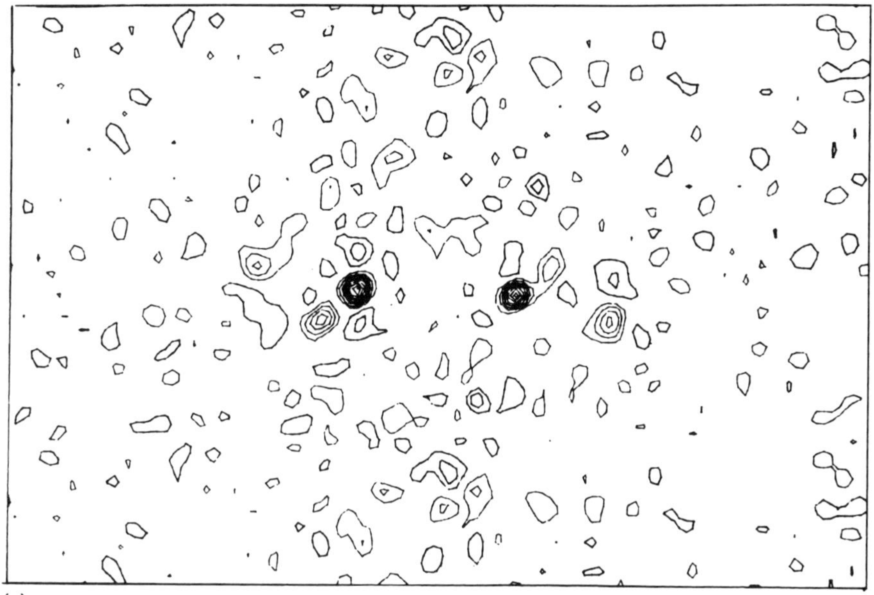

(c)

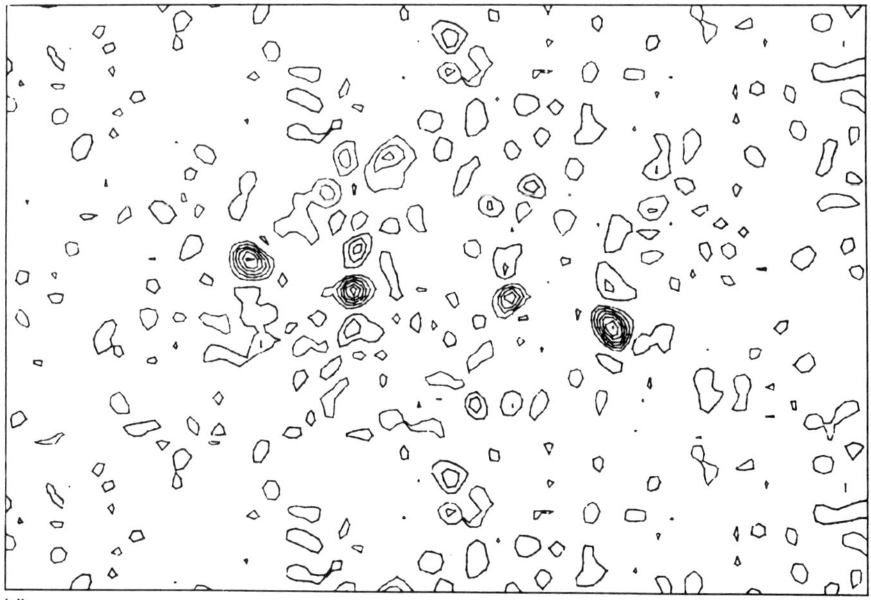

(d)

Figure 5.19 (*continued*)

negative feature elements absent in PX but present in PY, while absence of features represent similarity of elements in PX and PY. In calculating double difference maps it was essential to take into account the differing occupancy of the bound ligands. Assuming, for instance, that the occupancy of T_4 is Q_{T_4}, and that of a given analogue, X, is Q_X, then the coefficients in a (T_4 – X) double difference map were of the form:

$$[(T_4 - \text{native}) - Q_{T_4}/Q_X(X - \text{native})]\exp i\alpha_P$$

for centric terms and

$$2[(T_4 - \text{native}) - Q_{T_4}/Q_X(X - \text{native})]\exp i\alpha_P$$

for acentric terms. Double difference maps were extremely useful in translating a structural difference between any two compounds into a difference between their corresponding complexes when bound to transthyretin. Thus, for example, T_3 differs from T_4 in the loss of one iodine; a ($T_3 - T_4$) double difference map indicated how the loss of this iodine manifested itself on transfer from a (TTR:T_4) to a (TTR:T_3) complex. This gave a clearer idea, not only of subtle differences in the mode of binding of various analogues but also of the relative degree of conformational change induced in the protein.

5.6.3 MODES OF LIGAND BINDING

Figure 5.20 shows the iodine features from both difference and double difference maps for all the compounds studied at high resolution. The positions and relative heights of the iodine peaks in the proximal, distal and outer sites of the transthyretin channel are summarized in Table 5.5. The modes of ligand binding deduced for these analogues are shown in Figure 5.21.

5.6.3.1 Diisopropyl-3,5-T_2

DIT_2 clearly enters in a T_4-like mode, with its phenolic ring innermost, since its 3,5 iodine peaks are observed in the outer sites of the transthyretin channel. The centres of the iodine peaks are displaced about 1 Å closer towards the channel centre than in T_4, however, confirming the result obtained earlier with Okl projection maps.

This finding is borne out by a (T_4 – DIT_2) double difference map, which indicates that on transfer from a (TTR:T_4) to a (TTR:DIT_2) complex:

(a) there is a loss of iodine atoms from both the inner sites of prealbumin, one from the proximal and one from the distal pocket. This is consistent with DIT_2 entering with its non-iodinated ring innermost.
(b) there is an inward shift of the iodines located in the outer pockets.

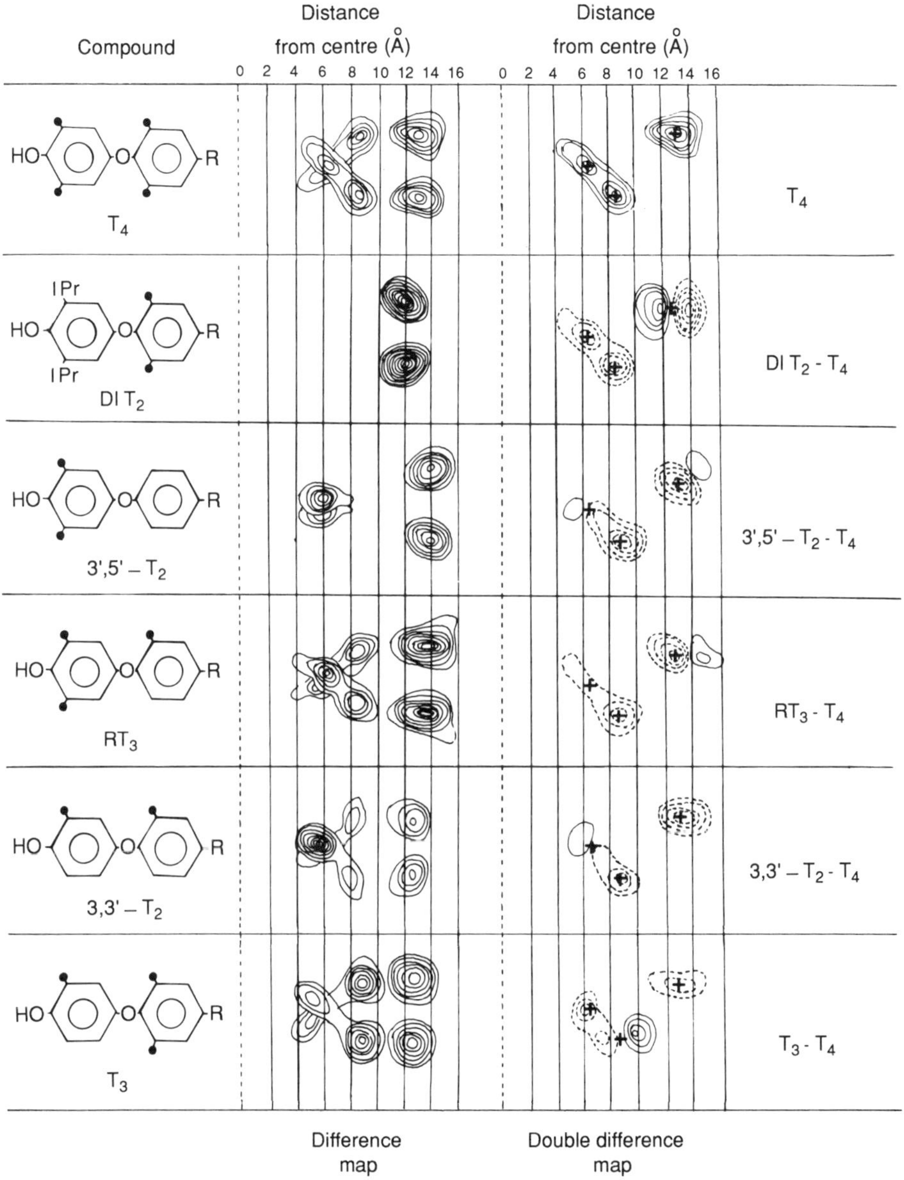

Figure 5.20. Difference and double difference electron density features for TTR : ligand complexes at high resolution

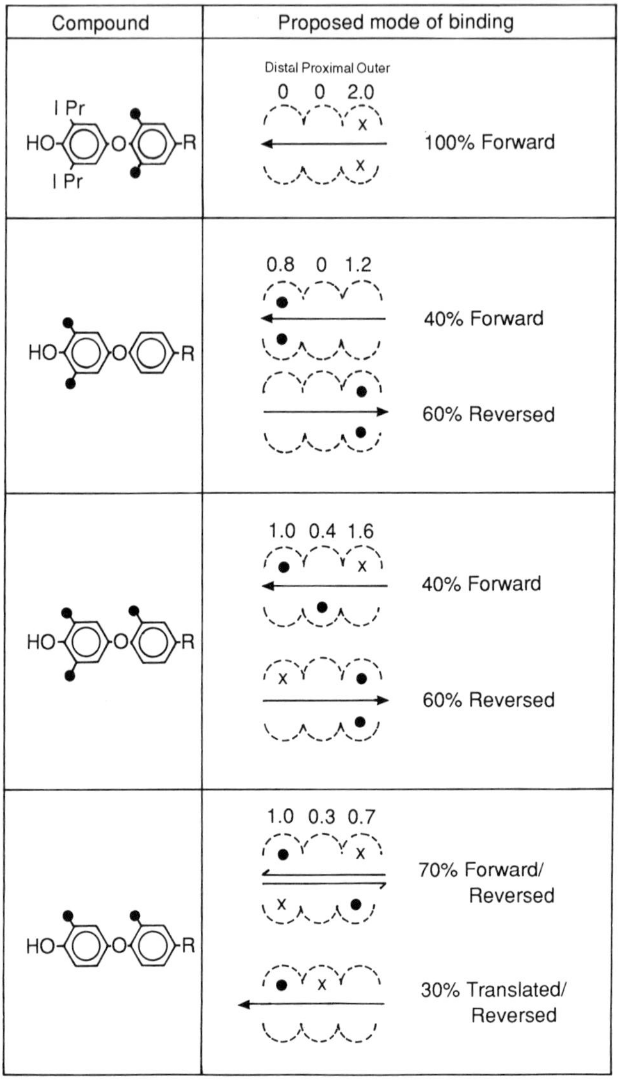

Figure 5.21. Interpretation of binding modes for hormone analogues, based on relative heights of iodine peaks in distal, proximal and outer pockets of transthyretin. ←, forward binding; →, reversed binding; x, tyrosine ring iodine (I_3; I_5); ●, phenolic ring iodine (I_3', I_5'). Numbers in boxes show the observed ratio of iodine peak heights

5.6.3.2 3′,5′-T_2

The binding of this compound cannot be explained by a simple, T_4-like mode of entry on two counts:

(a) the inner iodine peaks are confined only to the distal sites of the prealbumin channel, with none in the proximal sites, as for T_4. This immediately implies that in a forward mode of binding, the 3′ and 5′ iodines of the analogue fit symmetrically into the distal pockets, in contrast to T_4, where one iodine enters the proximal pocket, while the other enters the distal.

(b) There is an extra set of outer iodine peaks in both isomorphous and anomalous difference maps, which, assuming that impurities do not compete effectively for binding to the protein, clearly points to a second mode of binding. This is borne out by the double difference maps which show that on transferring from a (TTR:T_4) to a (TTR:3′,5′-T_2) complex, loss of iodines occurs not only from the outer sites of transthyretin (as would be expected from a T_4-like forward mode of entry) but also from the inner sites.

The extra set of iodine peaks (which are displaced $\cong$1 Å further from the channel centre than the outer peaks in T_4) could be explained either by partial entry into the binding site or by a reversed mode of binding. The former appears less likely on three counts:

(a) there are no indications, in difference maps calculated at 1.8 Å, 3 Å and 6 Å resolution, of any ligand density protruding outside the binding channel, as would be expected for a partial forward mode of entry.

(b) for ligands with iodines on both aromatic rings (e.g. T_4, T_3, RT_3 and 3,3′-T_2) the effect of a partial forward entry would be to increase the strength of the outer iodine peaks relative to the inner peaks. In three out of four of these compounds, the inner iodine peaks are either as strong as the outer peaks (e.g. T_4) or stronger (e.g. T_3, 3,3′-T_2).

(c) a partial forward mode of entry would leave one of the hydrophobic aromatic rings of the ligand unfavourably exposed to the solvent.

Assuming reversed entry then, the relative peak heights of 0.8 inner iodines: 1.2 outer iodines (Table 5.5) suggest that about 40% of the 3′,5′-T_2 molecules bind in a forward mode, while about 60% bind in the reversed mode (Figure 5.21).

5.6.3.3 Reverse T_3

RT_3 binding cannot solely be explained by a T_4-like mode of entry for two reasons:

(a) the outer peaks in RT_3 (as in 3′,5′-T_2) are displaced ≅1 Å further out of the channel centre than in T_4;
(b) a T_4-like mode of entry predicts that the inner iodine peaks should be twice as strong as the outer peaks. Almost the reverse is in fact observed, with the outer peaks being somewhat stronger: this immediately suggests the presence of a reversed component of binding, as noted for 3′,5′-T_2. The observed ratio of peak heights is in fact 1.4 inner iodines: 1.6 outer iodines, which would be consistent with approximately 40% forward entry and 60% reversed entry (Figure 5.21). This would explain the ($T_4 - RT_3$) double difference map which shows that, on passing from a (TTR:T_4) to a (TTR:RT_3) complex, loss of iodine does not occur solely from the outer sites (as expected from a simple forward mode of entry) but also from the inner sites (as would be expected from reversed entry).

5.6.3.4 3,3′-T_2

Since this compound possesses a symmetrical distribution of iodines across its two aromatic rings, both a reversed and a T_4-like forward mode of entry would predict equal numbers of iodines for the inner and outer sites on a 3,3′-T_2 difference map. However, there are roughly twice as many iodines in the inner than the outer sites, with the distal inner pockets being the most highly occupied (Table 5.5). The distal iodine peaks also shift about 0.4 Å closer to the channel centre than in T_4.

The above results are clearly seen in a (3,3′-T_2 − T_4) double difference map which indicates that on transfer from a (TTR:T_4) to a (TTR:3,3′-T_2) complex:

(a) loss of iodines does not occur equally from both inner and outer sites (as expected from a T_4-like mode of entry) but predominantly from the outer sites.
(b) loss of iodines from the inner sites is confined solely to the proximal pocket, while the distal pocket remains fully occupied by one iodine, as in a (TTR:T_4) complex.
(c) there is a slight inner shift of the distal iodine peak in a (TTR:3,3′-T_2) complex relative to a (TTR:T_4) complex.

The strength of the distal inner peaks relative to the proximal and outer ones cannot be explained in terms of impurities in the experimental sample used. This is because the chief contaminant (8% 3-T_1) appears from projection data to bind with its single iodine chiefly at the outer and proximal sites and would not contribute to the observed distal features. Moreover, the relative affinity of the two compounds for prealbumin (K_A 3-T_1 = 0.01; K_A 3,3′-T_2 = 0.6, where K_A T_4 = 100) suggests that the contaminating monoiodothyronine would not compete effectively with 3,3′-T_2 for binding to the protein.

The simplest interpretation for the observed peak electron counts, therefore, is that a single iodine occupies the distal pocket of transthyretin, while the other distributes itself between the proximal and outer pockets. This could be achieved by a combination of two modes of entry, as shown in Figure 5.21:

(a) mode (1) places one iodine in the distal and one in the outer pocket (this could be achieved *either* by a conventional forward entry *or* a reversed entry);
(b) mode (2) is unusual in that the iodines from both aromatic rings simultaneously occupy the inner sites of transthyretin, one binding to the distal and one to the proximal pocket. This can only be achieved if the entire ligand molecule slipped further into the binding site. Such a mode of entry may be possible for this analogue because of the decrease in the number of bulky iodine substituents at the phenolic ring. This new binding mode may also account for the inner shift of the distal iodine peak relative to T_4.

5.6.3.5 Reassessment of T_3 Binding

In view of the success of the double difference technique in comparing the mode of binding of different analogues relative to T_4, the same method was used to re-examine T_3 binding. T_3 differs from T_4 solely in the loss of an iodine at the 5′ position. If it binds to transthyretin in the manner modelled for T_4, then in a $(T_3 - T_4)$ double difference map, loss of an iodine should occur solely from the inner sites of the hormone channel. However, results show that:

(a) Loss of iodine occurs from both *outer* and *inner* sites (Figure 5.20), giving a ratio of 1.45 outer iodines: 1.55 inner iodines, instead of the expected ratio of 2 outer iodines: 1 inner iodine (Table 5.5).
(b) Of the two inner iodine sites, the proximal pocket is preferentially occupied when T_3 binds.
(c) There is an *outward* shift of the proximal iodine peak and an *inner* shift of the distal iodine peak in a (TTR:T_3) relative to a (TTR:T_4) complex.

These results have been confirmed by a (RT_3–T_3) difference map (Figure 5.22). Bearing in mind that since both molecules are triiodothyronines, there is no net gain or loss of iodines on transfer from one complex to the other, the map shows that:

(a) There are slightly more iodines in the outer sites of a (TTR:RT_3) than a (TTR:T_3) complex. This is precisely the opposite of what would be expected if both ligands bound solely in the T_4-like mode. (The outer iodines of RT_3 are also displaced further away from the channel centre than those of T_3).

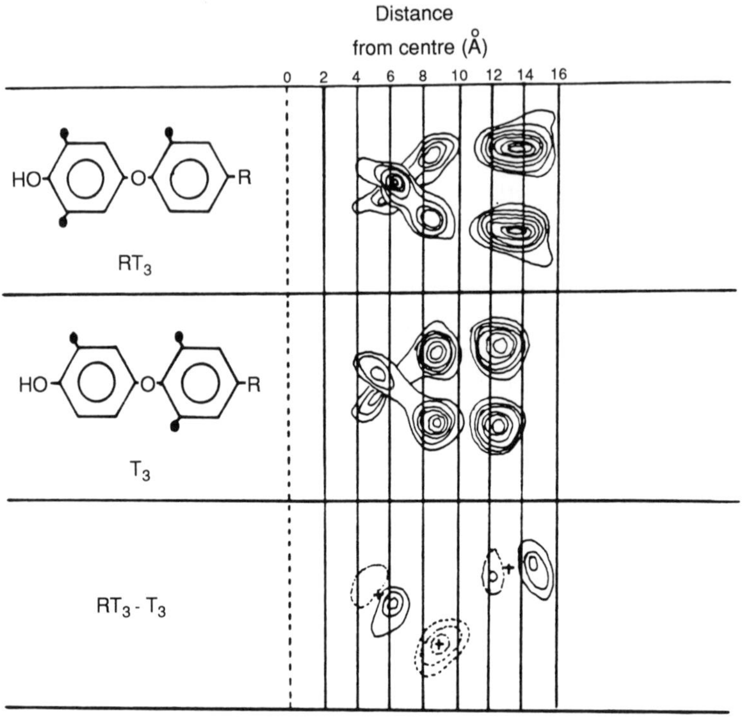

Figure 5.22. Comparison of difference features for the binding of T_3 and RT_3 to prealbumin. In ($RT_3 - T_3$) difference map, bold lines indicate a *gain*, and dotted lines a *loss*, of iodines, on transfer from a (TTR:T_3) to a (TTR:RT_3) complex

(b) Of the inner sites in transthyretin, it is the proximal pocket which is preferential occupied by T_3, and the distal one by RT_3.
(c) There is a slight inward shift of the distal iodine peak in a (TTR:T_3) relative to a (TTR:RT_3) complex.

There are two possible explanations for the increased strength of the inner iodine peaks relative to the outer ones in a T_3-containing complex:

(a) *Either* T_3 binds in a combination of forward (45%) and reversed (55%) modes (Figure 5.23a);
(b) *Or* T_3 binds in a combination of *forward* modes, including $\cong$30% of a translated forward binding (Figure 5.23b), as proposed for 3,3′-T_2.

In both instances, the 3,5 iodines of T_3 must be accommodated *symmetrically* in the proximal pockets, in order to account for the observed peak heights. The projection map for the binding of 3,5-T_2 (Figure 5.18), where the iodine features are observed solely in the region of the proximal pockets, indicates that such a

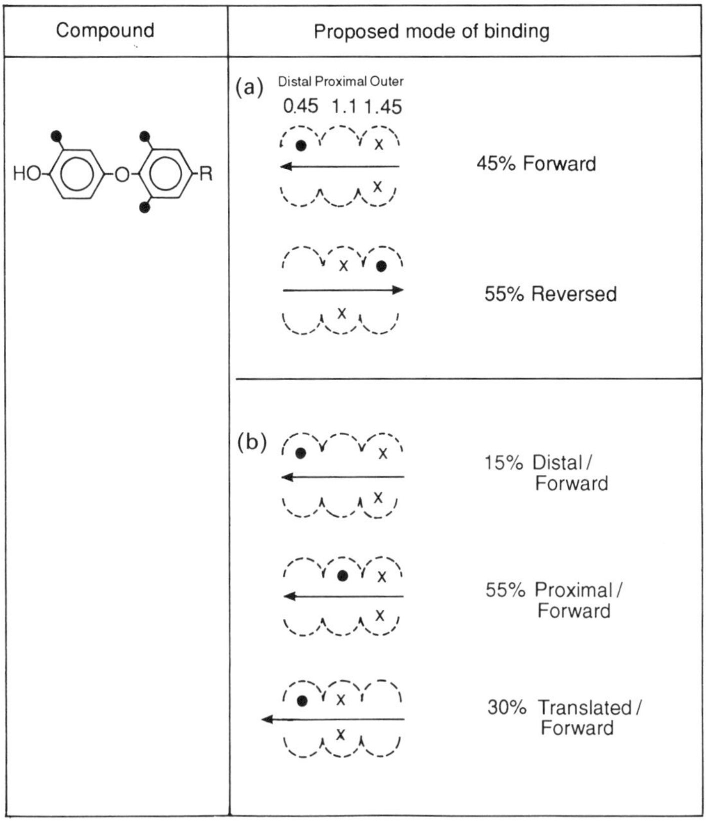

Figure 5.23. Interpretation of binding modes for T_3, based on relative heights of iodine peaks in distal, proximal and outer pockets of prealbumin. ←, forward binding; →, reversed binding; x, tyrosine ring iodine (I_3; I_5); ●, phenolic ring iodine (I_3'; I_5'). Numbers in boxes show the observed ratio of iodine peak heights

mode of binding is in fact possible. Computer graphics modelling of the (TTR:T_3) complex has also supported this view.

5.6.3.6 Reassessment of T_4 Binding

(a) Forward binding

The analogue studies have shown that in a forward mode of binding, the phenolic hydroxyl group of the ligand interacts with the Ser 117 and Thr 119 residues of prealbumin, either directly, or via a bound water molecule W1. In the earlier interpretation for T_4 binding (Section 5.5.3), however, the hydrogen bonding potential of this group is not satisfied, despite spectroscopic evidence suggesting that this group is stabilized by the protein environment (Ferguson *et al.*, 1975). The only pattern of forward binding accessible to T_4 is as modelled by DIT_2 (Mode 1). This compound binds unambiguously in the forward mode, but about 0.5 Å closer to the channel centre than T_4, allowing the 4′-hydroxyl group to hydrogen bond to the water molecule W1. The DIT_2 model shows clearly that a compound which is constrained to adopt the same conformation as T_4 can bind sufficiently close to the channel centre to allow the hydrogen-bonding potential of the phenolic hydroxyl group to be satisfied. Moreover, this can be achieved with 3′ and 5′ substituents of approximately the same size and hydrophobicity as iodine atoms. Finally, the observed position of the outer iodine peaks in DIT_2 coincides precisely with that already modelled for T_3 in its 'distal' mode of entry, where the hydrogen-bonding potential of the hydroxyl group is again satisfied. These findings strongly suggest that, for Mode 1 forward entry, the outer iodine positions delineated by DIT_2 binding are necessary to permit hydrogen bonding between the phenolic hydroxyl group and the bound water molecule W1. A comparison of the outer peaks for T_4 and DIT_2 shows that, while the centres of the T_4 peaks are displaced 1 Å further out than those for DIT_2, the elongated shape of the T_4 peaks results in a significant overlap with those for DIT_2. This gives rise to the possibility that T_4 itself may bind in the same position as DIT_2 within the hormone channel.

A refitting of the T_4 molecule has shown that the outer iodines of T_4 can be made to coincide more closely with those for DIT_2, while still being fully contained within their observed density peaks. This allows the hydrogen-bonding requirements of the 4′-hydroxyl group to be satisfied, without sacrificing any favourable interactions for the remaining substituents on the ligand. In particular, the amino acid side chain can still interact with the charged Glu 54 and Lys 15 residues at the channel entrance, and the short iodine–oxygen contact for the 3′ iodine in the proximal pocket is retained.

Most striking of all, however, is the fact that this new mode of binding can be accommodated without inducing any significant conformational changes in the Ser 117 and Thr 119 residues at the heart of the binding channel. Since this new model fails to account fully for the observed T_4-induced conformational changes, or indeed, for the displaced positions of the outer iodine peaks in T_4 relative to DIT_2, the possibility arises of an additional mode of binding for T_4, which is not present for DIT_2. This mode is clearly favoured by the increased acidity of the phenolic hydroxyl group which occurs when the 3′,5′-isopropyl groups of DIT_2 are replaced by iodines in T_4. Studies with analogues containing 3′,5′ iodine substituents (3′,5′-T_2; RT_3) have shown that these are capable of binding in a reversed mode. In both cases, this is associated not only with displaced outer iodine peak positions, but also with distinct conformational changes at the heart of the binding site. The simplest explanation, therefore, is that T_4 may itself bind in the reversed mode.

(b) Reversed binding

Two constraints must be taken into account when fitting T_4 in the reversed mode:

(i) the ligand can only adopt a skewed conformation. This precludes the type of reversed binding observed for antiskewed structures.
(ii) the ligand binds with two of its iodines in the outer transthyretin pockets, while the remaining two are equally distributed between the proximal and distal inner sites.

Figure 5.24 shows a model of reversed entry for T_4, which satisfies the above requirements. It has been achieved with a diphenyl ether conformation of ϕ, $\phi' = 77°$, $3°$, which falls within the range of energetically allowed values determined by Andrea *et al.* (1979). Although there is no density observed for the fitted amino acid side chain in the T_4 map, the group lies within the boundary delineated by the corresponding density features in the RT_3 map, where indications for the side chain are clearest. The main features of the final fit are as follows:

(i) the *3′ and 5′ iodines* bind asymmetrically in the outer pockets, P_1 and P_1'.
(ii) the *3 and 5 iodines* bind asymmetrically, one in the proximal pocket P_2,

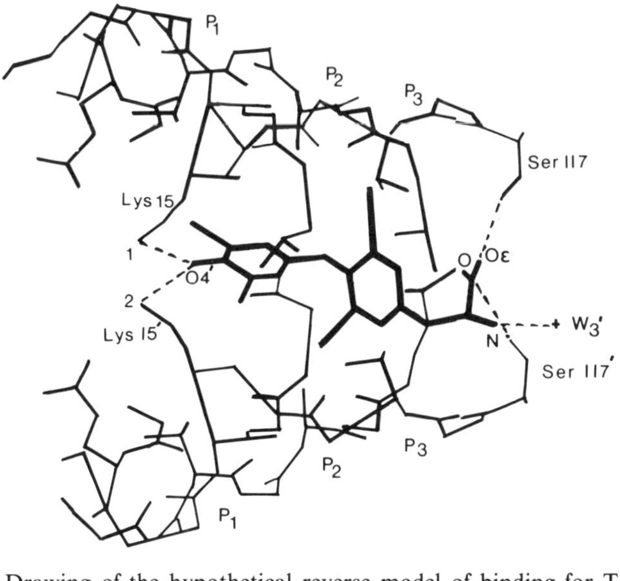

Figure 5.24. Drawing of the hypothetical reverse model of binding for T_4. Hydrogen bond interactions for the amino acid group and the 4′-hydroxyl group are shown with dotted lines (cf. Figure 5.16)

and one in the distal pocket P_3. This mode of reversed binding reflects the only compromise available for fitting four iodine substituents into the available transthyretin pockets.

(iii) the *amino acid moiety* binds in the symmetry-related distal pocket P_3', where it can interact with the paired Ser 117 residues.

(iv) the *phenoxide ion* lies within hydrogen bonding distance of the paired lysine 15 residues at the mouth of the channel.

Although the amino acid side chain makes some short contacts with the Ser 117 residues at the centre of the binding site, these are acceptable within a margin of about 0.2 Å.

The model accounts for two hitherto unexplained features in the difference maps:

(i) the elongated shape of the T_4 iodine peaks and their outward displacement relative to DIT_2;

(ii) the observed shifts and rotations in the Thr 119 and Ser 117 residues, which are due to the binding of the amino acid side chain at the centre of the binding site.

Finally, reversed binding has two characteristics which are consistent with chemical observations on hormone binding:

(i) the phenolic hydroxyl lies at the mouth of the transthyretin channel, where it is exposed to solvent even when protein-bound. This may explain the observation that the pK_a of this group varies relatively little between the bound and unbound states (Nilsson and Peterson, 1971).

(ii) the amino acid side chain binds innermost within the channel. This could account for the reported failure of T_4-sepharose columns to retain transthyretin (Robbins, 1976): the hormone is attached to the column matrix by its amino acid moiety, which is therefore no longer available to interact with the protein.

5.6.4 CONFORMATIONAL CHANGES IN BINDING SITE

The conformational changes in the hormone channel which are observed on ligand binding involve three main sets of residues:

(a) the hydrophobic residues, Val 121, Leu 17, Ala 108 and Leu 110, which delineate the 'pockets' for iodine binding;

(b) the charged Lys 15 and Glu 54 residues at the mouth of the channel;

(c) the hydrophilic residues, Ser 115, Ser 117 and Thr 119 which, with their associated water molecules (labelled W1–W3 in Figure 5.25), form a complex hydrogen-bonded network at the centre of the channel.

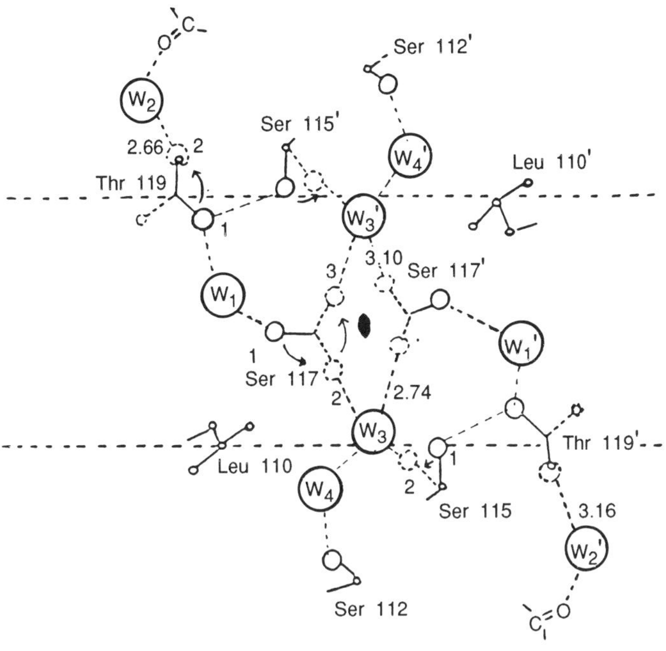

Figure 5.25. Schematic drawing of the different side chain conformations and hydrogen bond networks at the centre of the hormone binding channel. Alternative orientations of Ser 115, Ser 117 and Thr 119 are shown numbered and in dotted lines. The hydroxyl group of those side chains are shown as the larger circles. Water molecules are denoted by the larger circles and are labelled W1–W4 in one half of the channel and W1′–W4′ in the symmetry related half. Hydrogen bonds are indicated by the light dotted lines, with the lengths of the hydrogen bonds shown in Å units nearby

The effects of hormone binding in the central hydrophilic region are difficult to interpret for two reasons:

(a) its constituent residues exhibit a degree of disorder, as shown by the B factor values for their side chain atoms and can therefore adopt a variety of side chain orientations even in the native structure. Figure 5.25 shows the alternative conformations possible of Ser 117, Thr 119 and Ser 115, each associated with different patterns of hydrogen bonding.

(b) in each binding site, the water molecule W1, which may be involved in hydrogen bonding to the phenolic hydroxyl group of the hormone, is related by symmetry to a second water molecule (W1′) which is located close to the position where the distal iodine of the ligand binds. The density features due to these water molecules are thus masked by the features due to the electron dense iodines.

Nevertheless, it has been possible to correlate certain conformational changes in this region with the mode of ligand binding, by exploiting the information derived from the analogue DIT_2. This had two advantages over the other compounds:

(a) it displayed a unique mode of binding to transthyretin;
(b) no electron dense iodines were found to bind at the inner sites of the transthyretin channel, making it possible to observe directly the status of the water molecule W1 on ligand binding.

When DIT_2 binds with its phenolic ring innermost, the density for W1 strengthens, while Ser 117, Thr 119 and Ser 115 adopt conformation 1. Computer graphics modelling has shown that this disposition of residues, together with the inward translation of DIT_2 relative to T_4, permits hydrogen bonding between the phenolic hydroxyl group of the hormone and the residues Ser 117 and Thr 119, via the water molecule W1. By contrast, all remaining analogues, which exhibited mixed modes of entry, showed subtle but distinct differences in the side chain rotations of the above residues (Table 5.7). In RT_3 and $3',5'$-T_2, for instance, which binds in the reversed mode, two features have been noted:

(a) a rotation of Thr 119 to conformation 2, with the associated strengthening of the density feature for W2;
(b) a rotation of Ser 117 to conformation 3.

If these two rotations are concerted and lead to the expulsion of W1, a new pocket is created at the heart of the channel, which would allow enough room to accommodate the bulky amino acid side chain in a reversed mode of entry.

Table 5.7 Summary of side chain conformations observed in the central hydrophilic region of the prealbumin channel

	Leu 110	Ser 115	Ser 117	Thr 119	W1	W2
T_4	Shifts	1 & 2	1 & 2	2	?	Strengthens
			Main chain shift			
T_3	Shifts	1	Weak density	2	?	Strengthens
DIT_2	No shift	1	1	1	Strengthens	Weakens
$3',5'$-T_2	Shifts	1	Weak density	2	?	Strengthens
RT_3	Shifts	1 & 2	3	1 & 2	?	Strengthens
			Main chain shift			
$3,3'$-T_2	Shifts	1	3	2	?	Strengthens

Conformations are as depicted in Figure 5.25.

5.6.5 SUMMARY OF OBSERVED MODES OF INTERACTION

Difference Fourier maps have clearly shown that the mode of ligand interaction with transthyretin is much more complicated than previously envisaged:

(a) DIT_2 unambigously showed a simple forward mode of binding, with its phenolic ring located nearest to the channel centre.
(b) 3′,5′-T_2 and RT_3 showed mixed modes of entry, with about 40% of the molecules binding in the forward mode and 60% in the reversed mode.
(c) In addition to a normal forward (and possibly, a reversed) binding, 3,3′-T_2 showed a different mode of forward entry, involving an inward translation of the entire ligand molecule relative to T_4.
(d) A reinterpretation of the T_3 and T_4 results showed that they could also be explained in terms of mixed modes of entry with T_3. These involved *either* a combination of 45% forward and 55% reversed binding *or* a variety of forward binding modes, including $\cong$30% of a 'translated' forward entry, similar to that envisaged for 3,3′-T_2. T_4 binding probably combines the forward DIT_2 mode and reverse mode but in unknown proportions.

These conclusions emphasize the absence of a single mode of binding for the thyroid hormone analogues and point to four distinctly different patterns:

Mode 1: a conventional T_4-like mode of *forward* binding (e.g. for DIT_2 and probably T_4);
Mode 2: a *symmetrical forward* binding (e.g. for 3′,5′-T_2);
Mode 3: a *translated forward* binding (e.g. for 3,3′-T_2);
Mode 4: a *reversed* mode of binding (e.g. for 3′,5′-T_2 and RT_3 and probably T_4).

These modes share two common features. First, the fitting of the bulky iodine substituents of the ligand into hydrophobic pockets on the protein is a major aspect of the interaction: the final positions of these iodines in the different binding patterns vary by no more than about 1 Å from their mean value (i.e. only one-quarter of the diameter of an iodine atom). Secondly, symmetry-related features on the ligand (i.e. the 3,5 and the 3′,5′ substituents) are preferentially located in symmetry-equivalent pockets of the protein. This allows chemically identical substituents to participate in energetically equivalent interactions and demonstrates the importance of symmetry considerations in ligand binding to prealbumin.

Nevertheless, results from a range of hormone analogues have shown that:

(a) there are different positions for the binding of the 3,5 iodines, depending upon the size of the substituents on the phenolic ring.
(b) there are different binding patterns for the 3′,5′ iodines, depending upon the nature of the substituents on the tyrosine ring and on the diphenyl ether conformation of the ligand (skewed versus antiskewed).

(c) there may be different modes of binding for analogues with differing amino acid configuration.

Most strikingly, two compounds (RT_3, $3'5'$-T_2) have demonstrated that changes in the pattern of iodine substitution can lead to a change in the orientation of the ligand within the binding site, leading to a reversed mode of binding which brings the amino acid moiety closest to the channel centre. The ligand is located $\cong 1$ Å further out of the binding site than in the forward mode, and conformational changes are observed at the centre of the channel to accommodate the amino acid side chain. These include rotations in the Thr 119 and Ser 117 residues and expulsion of a crystallographically well defined water molecule to create an additional pocket for the binding of the side chain moiety. At the experimental pH (9.0), forward and reversed binding have been observed in approximately equal proportions for these compounds, indicating that they are of comparable energy.

Modelling studies suggest that two factors are involved in determining the balance between forward and reverse binding:

(a) the pK_a of the phenolic hydroxyl group (as determined by the nature of the 3′,5′ substituents).
(b) the conformational flexibility at the ether link (as determined by the nature of the 3,5 substituents).

Reversed entry is favoured by compounds with an ionized phenolic hydroxyl group since this can interact electrostatically with the paired Lys 15 residues at the mouth of the binding channel. Flexibility at the ether link is important, however, in accommodating the iodine substituents within the binding site when the amino acid moiety is situated in the more sterically constrained region around the channel centre.

A re-examination of the difference electron density maps for hormone binding, in terms of iodine peak heights, peak positions, and observed conformational changes in the protein, indicates that T_3 and T_4 most probably bind in a mixture of forward and reversed modes.

5.7 IMPLICATIONS FOR HORMONE BINDING

The remarkable ability of transthyretin to accommodate ligands in a variety of binding modes may reflect its function as a transport protein for the thyroid hormones and their deiodination products. Under physiological conditions, these compounds differ in their conformational flexibility at the ether link and in the charge properties of the phenolic hydroxyl group. The structure of the transthyretin binding site has presumably evolved to enhance the probability of binding, while tolerating these differences, as indicated by the following features:

(a) The diameter of the binding channel (8 Å) can accommodate iodothyronine compounds in both a skewed and an antiskewed conformation;
(b) There is a redundancy in the number of hydrophobic pockets capable of binding an iodine substituent. (There are six pockets on prealbumin compared to a maximum of four iodines in T_4).
(c) The twofold axis which runs down the centre of the binding channel allows the asymmetric hormones to bind in two energetically equivalent orientations, which further increases the probability of binding to the protein.
(d) The binding site pockets themselves are not rigid since they are composed of flexible side chain residues (lysine, threonine, valine, leucine) which are free to change conformation to maximize binding interactions. Moreover, each of the paired distal pockets contains a crystallographically well defined water molecule which may be expelled to create an additional site for binding a ligand substituent.
(e) The central region of the binding site contains a complex network of water molecules and hydrophilic residues (serine, threonine) which are capable of acting as both hydrogen bond donors and acceptors. This allows both forward and reversed modes of binding to be accommodated.

The ability of the transthyretin binding site to detect, and respond to, differences in iodothyronine structure strikingly demonstrates how, even in a comparatively simple system, involving a ligand with limited conformational variability and a defined and relatively rigid binding site, a subtle discrimination between closely related ligands can be achieved. This is of considerable importance in formulating an understanding of the specificity of other physiologically important binding sites for the thyroid hormones, notably the thyroid hormone nuclear receptor.

Finally, these studies have emphasized the importance of structural information as a complement to biochemical structure–activity data. A quantitative estimate of the contribution of a particular ligand substituent to the overall mode of ligand binding can only be made in the case of a uniform and defined mode of binding to the receptor site. The work on transthyretin shows that, in a situation where multiple modes of ligand binding exist, the validity of such an estimate becomes questionable.

ACKNOWLEDGEMENTS

We gratefully acknowledge the Medical Research Council for their financial support of this work. P. de la P., J.M.B. and S.J.O. also acknowledge the Medical Research Council for their individual financial support, and J.M.B. also acknowledges the financial support of the Science and Engineering Research Council.

REFERENCES

Andrea, T. A. (1977). PhD Thesis, University of California, San Francisco

Andrea, T. A., Dietrich S. W., Wallace, J. M., Kollman, P. A. and Jorgensen, E. C. (1979). *J. Med. Chem.*, **22**, 221–32.

Andrea, T. A., Cavalieri, R. R. Goldfine, I. D. and Jorgensen, E. C. (1980). *Biochemistry*, **19**, 55–63.

Balsam, A. and Ingbar, S. H. (1979). *Clin. Res.*, **27**, 247A.

Balsam, A., Surks, M. I. and Ingbar, S. H. (1980). *Clin. Res.*, **28**, 255A.

Baxter, J. D., Eberhardt, N. L., Apriletti, J. W., Johnson, L. K., Ivarie, R. D., Schachter, B. S., Morris, J. A., Seeburg, P. H., Goodman, H. M., Latham, K. R., Polansky, J. R. and Martial, J. A. (1979). *Rec. Prog. Horm. Res.*, **35**, 97–153.

Blake, C. C.F. and Oatley, S. J. (1977). *Nature*, **268**, 115–20.

Blake, C. C. F., Swan, I. D., Rerat, C., Berthou, J., Laurent, A. and Rerat, B. (1971). *J. Mol. Biol.*, **61**, 217–24.

Blake, C. C. F., Geisow, M. J., Swan, I. D. A., Rerat, C. and Rerat, B. (1974). *J. Mol. Biol.*, **88**, 1–12.

Blake, C. C. F., Burridge, J. M. and Oatley, S. J. (1978a). *Biochem. Soc. Trans.*, **6**, 1114–18.

Blake, C. C. F., Geisow, M. J., Oatley, S. J., Rerat, C. and Rerat, B. (1978b). *J. Mol. Biol.*, **121**, 339–56.

Blaney, J. M., Jorgensen, E. C., Connolly, M. L., Ferrin, T. E., Langridge, R., Oatley, S. J., Burridge, J. M. and Blake, C. C. F. (1982a). *J. Med. Chem.*, **25**, 785–90.

Blaney, J. M., Weiner, P. K., Dearing, A., Kollmann, P. A., Jorgensen, E. C., Oatley, S. J., Burridge, J. M. and Blake, C. C. F. (1982b). *J. Am. Chem. Soc.*, **104**, 6424–34.

Branch, W. T., Robbins, J. and Edelhoch, (1972). *Arch. Biochem. Biophys.*, **152**, 144–51.

Braverman, L. E., Ingbar, S. and Sterling, K. (1970). *J. Clin. Invest.*, **49**, 855–64.

Burridge, J. M. (1982). DPhil Thesis, University of Oxford.

Carrell, R. W. and Boswell, D. R. (1986). In *Proteinase Inhibitors.* (Eds A. J. Barrett and G. Salvesan), Elsevier Science Publishers, Amsterdam, pp. 403–20.

Cheng, S.-Y., Cahnmann, H. J., Wilchek, M. and Ferguson, R. N. (1975). *Biochemistry*, **14**, 4132–6.

Cheng, S-Y., Pages, R. A., Saroff, H. A., Edelhoch, H. and Robbins, J. (1977a). *Biochemistry*, **16**, 3707–13.

Cheng, S-Y., Wilchek, K., Cahnmann, H. J. and Robbins, J. (1977b). *J. Biol. Chem.*, **252**, 6076–81.

Cheng, S-Y., Maxfield, F. R., Robbins, J., Willingham, M. C. and Pastan, I. H. (1980). *Proc. Natl Acad. Sci.*, **77**, 3425–9.

Cheng, S-Y., Rakhit, G., Erard, F., Robbins, J. and Chignell, C. F. (1981). *J. Biol. Chem.*, **256**, 831–6.

Chopra, I. J. (1981). In *Physiopathology of Endocrine Diseases and Mechanisms of Hormone Action* (Eds R. J. Soto, A. de Nicola and J. Blaquier), Alan Liss, New York, pp. 67–80.

Chopra, I. J., Solomon, D. H., Chopra, U., Wu, S-Y., Fisher, D. A. and Nakamura, Y. (1978). *Rec. Prog. Horm. Res.*, **34**, 521–67.

Cody, V. (1978). *Rec. Prog. Horm. Res.*, **34**, 437–75.

Cody, V. (1980). *Endocr. Rev.*, **1**, 140–66.

Cody, V. and Hazel, J. (1977). *Acta Cryst.*, **B33**, 3180–4.

Davis, P. J., Handwerger, B. S. and Gregermann, R. J. (1972). *J. Clin. Invest.*, **51**, 515–21.

Dillmann, W. H., Surks, M. I. and Oppenheimer, J. H. (1974). *Endocrinology*, **95**, 492–8.

Eastman, C. J., Corcoran, J. M. and Smith, H. C. (1980). *6th Int. Congr. Endocrin.*, Program & Abstracts, S-43.

Eberhardt, N. L., Apriletti, J. W. and Baxter, J. D. (1980). In *Biochemical Actions of Hormones*, (Ed. G. Litwack), Academic Press, New York, Vol. VII, pp. 311–94.
Ekins, P. J. (1982). In *Recent Advances in Endocrinology and Metabolism* (Ed. J. L. H. O'Riordan), Churchill Livingstone, Edinburgh, pp. 289–327.
Ferguson, R. N., Edelhoch, H., Saroff, H. A. and Robbins, J. (1975). *Biochemistry*, **14**, 282–9.
Flink, I. L., Bailey, T. J., Gustafoon, T. A., Markham, B. E. and Norkin, E. (1986). *Proc. Natl Acad. Sci. USA*, **83**, 7708–12.
Frieden, E. (1981). *Trends Biochem. Sci.*, **6**, 50–3.
Geisow, M. J. (1973). DPhil Thesis, University of Oxford.
Gemmill, C. L. (1955). *Arch. Biochem. Biophys.*, **54**, 359–67.
Goodman, D. S. (1981). *Ann. N.Y. Acad. Sci.*, **359**, 69–78.
Haupt, H. and Heide, K. (1966). *Specialia*, **15** (vii), 449–51.
Ingbar, S. H. (1963). *J. Clin. Invest.*, **42**, 143–8.
Jorgensen, E. C. (1976). *Pharmac. Ther. B*, **2**, 661–82.
Jorgensen, E. C. (1978). *Hormonal Proteins and Peptides* (Ed. C. H. Li), Vol. 6, Academic Press, New York, pp. 57–105.
Kanda, Y., Goodman, D. S., Canfield, R. E. and Morgan, F. J. (1974). *J. Biol. Chem.*, **249**, 6796–805.
Larsen, P. R. (1972). *J. Clin. Invest.*, **51**, 1125–34.
Latham, K. R., MacLeod, K. M., Papavasiliou, S. S., Martial, J. A., Seeburg, P. H., Goodman, H. M. and Baxter, J. D. (1978). In *Receptors and Hormone Action* (Eds B. W. O'Malley and L. Birnbaumer), Vol. 3, Academic Press, New York, pp. 76–100.
Lissitzky, S. (1976). *Pharmac. Ther. B*, **2**, 219–46.
Löbermann, H., Tokuoka, R., Deisenhofer, J. and Huber, R. (1984). *J. Mol. Biol.*, **177**, 531–56.
Marshall, J. S., Pensky, J. and Green, A. M. (1972). *J. Clin. Invest.*, **51** 3173–81.
Money, W. L., Kumaoka, S., Rawson, R. W. and Kroc, R. C. (1960). *Ann. N.Y. Acad. Sci.*, **86**, 512–44.
Murray-Rust, P. and Motherwell, W. (1979). *J. Am. Chem. Soc.*, **101**, 4374–6.
Newcomer, M. E., Jones, T. A., Aquist, J., Sundelin, J., Eriksson, U., Rask, L. and Peterson, P. A. (1984). *EMBO J.*, **3**, 1451–4.
Nilsson, S. F. and Peterson, P. A. (1971). *J. Biol. Chem.*, **246**, 6098–105.
Nilsson, S. F., Rask, L. and Peterson, P. A. (1975). *J. Biol. Chem.*, **250**, 8554–63.
Oatley, S. J. (1976). DPhil. thesis, University of Oxford.
Oatley, S. J., Burridge, J. M. and Blake, C. C. F. (1982). In *Hormone Antagonists*, (Ed. M. K. Agarwal), Walter de Gruyter and Co., Berlin, pp. 705–15.
Okabe, N., Fujiwara, T., Yamagata, Y. and Tomita, K. (1982). *Biochim. Biophys. Acta*, **717**, 179–81.
Oppenheimer, J. H. (1981). In *Physiopathology of Endocrine Diseases and Mechanisms of Hormone Action* (Eds R. J. Soto, A. de Nicola and J. Blaquier), Alan R. Liss Inc., New York, pp. 45–55.
Oppenheimer, J. H., Koerner, D., Schwartz, H. L. and Surks, M. I. (1972). *J. Clin. Endocr. Metab.*, **35**, 330–3.
Oppenheimer, J. H., Schwartz, H. L., Surks, M. I., Koerner, D. and Dillmann, W. H. (1976). *Rec. Prog. Horm. Res.*, **32**, 529–65.
Pages, R. A., Robbins, J. and Edelhoch, H. L. (1973). *Biochemistry*, **12**, 2773–9.
Papavasiliou, S. S., Martial, J. A., Latham, K. R. and Baxter, J. D. (1977). *J. Clin. Invest.*, **60**, 1230–9.
Pardridge, W. M. (1981). *Endocr. Rev.*, **2**, 103–23.
Pitt-Rivers, R., Stanbury, J. B. and Rapp, B. (1955). *J. Clin. Endocr. Metab.*, **15**, 616–20.

Pliam, N. B. and Goldfine, I. D. (1977). *Biochem. Biophys, Res. Commun.*, **79**, 166–72.

Purdy, R. H., Woeber, K. A., Holloway, M. T. and Ingbar, S. H. (1965). *Biochemistry*, **4**, 1888–95.

Rask, L., Anundi, H., Bohme, J., Eriksson, U., Ronne, H., Sege, K. and Peterson, P. A. (1981). *Ann. N.Y. Acad. Sci.*, **359**, 79–90.

Raz, A. and Goodman, D. S. (1969). *J. Biol. Chem.*, **244**, 3230–7.

Robbins, J. (1975). In *Thyroid Hormone Metabolism* (Eds W. E. Harland and J. S. Orr), Academic Press, New York, pp. 1–22

Robbins, J. (1976). In *Trace Components of Plasma: Isolation and Clinical Significance* (Eds G. A. Jamieson and T. J. Greenwalt), Alan R. Liss Inc., New York, pp. 331–50.

Robbins, J. and Rall, J. E. (1979). In *Hormones in Blood* (Eds C. H. Gray, and V. H. T. James), 3rd edn, Vol. 1, Academic Press, New York, pp. 576–688.

Samuels, H. H. (1978). In *Receptors and Hormone Action* (Eds B. W. O'Malley and L. Birnbaumer), Vol. 3, Academic Press, New York, pp. 35–74.

Snyder, S. M., Cavalieri, R. R., Golafine I. D., Ingbar, S. H. and Jorgensen, E. C. (1976). *J. Biol. Chem.*, **251**, 6489–94.

Somack, R., Andrea, T. A. and Jorgensen, E. C. (1982). *Biochemistry*, **21**, 163–70.

Stasilli, N., Kroc, R. L. and Meltzer, R. I. (1959). *Endocrinology*, **64**, 62–82.

Steiner, R. F., Roth, R. J. and Robbins, J. (1966). *J. Biol. Chem.*, **241**, 560–7.

Sterling, K. (1964). *J. Clin. Invest.*, **43**, 1721–29.

Sterling, K. (1979a). *New Engl J. Med.*, **300**, 117–23.

Sterling, K. (1979b). *New Engl J. Med.*, **300**, 173–7.

Sterling, K., Rosen, P. and Tabacknick, M. (1962). *J. Clin. Invest.*, **41**, 1021–30.

Sterling, K., Lazarus, J. H., Milch, P. O., Sakurada, T. and Brenner, M. A. (1978). *Science*, **201**, 1126–9.

Surks, M. I. and Oppenheimer, J. H. (1963). *Endocrinology*, **72**, 567–73.

Swan, I. D. A. (1970). DPhil Thesis, University of Oxford.

Tabachnick, M. (1967). *J. Biol. Chem.*, **242**, 1646–50.

Tabachnick, M. and Perret, V. (1987). *Biochem. Int.*, **15**, 409–17.

Tata, J. R. (1975). *Nature*, **257**, 18–23.

Tata, J. R. (1980). In *Cellular Receptors for Hormones and Neurotransmitters* (Eds D. Schulster and A. Levitski), Wiley, New York, pp. 127–46.

Tritsch, G. L. and Tritsch, N. E. (1963). *J. Biol. Chem.*, **238**, 138–42.

Visser, T. J. (1980). *Trends Biochem. Sci.*, **5**, 222–4.

6 Computer Modelling of Drug–DNA Intercalative Interactions

STEPHEN NEIDLE
The Institute of Cancer Research, Sutton, Surrey, UK

6.1 INTRODUCTION

The human genome encodes some tens of thousands of gene products. These proteins and enzymes generally play highly specific structural and functional roles in cellular and organellular processes. Their functional specificity is due to the very large number of tertiary structural possibilities that can result from the folding of polypeptide chains composed of the 20 amino acids. Thus, an enzyme active site for a particular catalytic function has evolved to have an unique three-dimensional geometry (which is only sometimes conserved across species). It is this feature that is taken advantage of by many drugs, which can competitively inhibit or irreversibly bind to the active site of a particular enzyme or receptor that may be implicated in a disease process. The increasing body of knowledge of the tertiary structures of these pharmacological targets, largely from X-ray crystallographic studies, has been the major driving force behind the current approaches to rational drug design that are having a major impact in

The Design of Drugs to Macromolecular Targets. Edited by C. R. Beddell
© 1992 John Wiley & Sons Ltd

many areas of medicine. These are discussed in other chapters in this volume. The principal exceptions to this picture of drug-receptor specificity (with all its implications for selective toxicity) are in the area of cancer chemotherapy. The generic term 'cancer' covers well over 100 distinct diseases with a commonality of uncontrolled cell growth. This feature over all others has been utilized over the past 40 years in the development of drugs that inhibit cell growth and division at its most general level—this is by interference with DNA function (Waring, 1981; Neidle and Waring, 1983). Historically, the first anti-tumour drugs were the nitrogen mustards, whose descendants are still of major use in the clinic. These compounds interact directly with DNA, forming covalent cross-links.

They illustrate the fundamental feature of all such agents: they appear to have non-selective reactivity with *all* susceptible DNA sequences, and hence are apparently inherently non-selective to tumour cells. The fact that such compounds can indeed have an acceptable and clinically useful therapeutic index is due to at least four factors:

(a) Rapidly-proliferating cells are inherently more sensitive to DNA damage, and thus rapidly growing cancers such as some leukaemias can be highly responsive to drug chemotherapy.
(b) Pharmacokinetic and metabolic properties of specific tumour cells can differ sufficiently from those of normal cells to confer chemotherapeutic advantage.
(c) It is increasingly apparent that even structurally simple nitrogen mustards interact non-randomly with target DNA sequences. Studies using DNA-sequencing methodologies have, for example, shown that nitrogen mustards do not simply bind to random guanines (their chemically preferred site of nucleophilic attach), but interact preferentially with long runs of guanines (Kohn *et al.*, 1987a).
(d) Many DNA-interactive compounds produce DNA strand breaks *in vivo* as a result of their stabilization of a 'cleavable complex' between DNA and the enzyme DNA topoisomerase II (Nelson *et al.*, 1984; Kohn *et al.*, 1987b; Lock and Ross, 1987; Tan *et al.*, 1987). Levels of this enzyme are elevated in many tumour cells as a result of increased gene expression, and the non-repairable (and hence lethal) double-strand breaks that are produced, are at non-random sequences.

Thus, these factors are increasingly seen to be interrelated, and the classic picture of DNA interaction being a totally random event, is a misleading oversimplification at best. The further question of whether there are indeed fundamental differences between normal and neoplastic cells at the DNA level has recently been answered in the affirmative by the discovery of families of oncogenes that when expressed can produce malignant changes (Barbacid,

1986; Sigal *et al.*, 1987; Pai, 1991). In principle, these DNA sequences could constitute highly specific targets for DNA-reactive drug intervention, as well as the oncogene proteins themselves being potential drug targets.

The apparent structural simplicity of DNA has led to the common conception of drug-DNA interactions being inherently non-specific, and therefore without the subtleties and specificities of enzyme-small molecule recognition. Evidence such as that outlined above has begun to change that view. In addition, data on protein-DNA interactions, especially from X-ray crystallography (for example, Ohlendorf *et al.*, 1982; Zhang *et al.*, 1987; Wolberger *et al.*, 1991) has shown the potential for extraordinary selectivity possessed by DNA sequences.

This chapter will review studies on molecular structural approaches to the understanding and consequent design of DNA-interactive ligands and drugs that have actual and potential relevance to cancer chemotherapy. In general, such compounds can be categorized as either covalently reactive (cf. the nitrogen mustards), or capable of interacting by means of one or more types of non-covalent association. Attention is focused here on those drugs that interact non-covalently with DNA and are members of the important subclass of DNA intercalators (Baguley 1991; Denny, 1989). This category is not only important in clinical terms, with compounds such as doxorubicin (Adriamycin), daunomycin and actinomycin being among the most effective drugs currently available. Their interactions with nucleic acids have been extensively studied over the past 30 years, and thus molecular modelling and drug design has a large body of physicochemical data with which to interrelate. As in other areas, fruitful progress has most recently been made by the interplay of X-ray crystallography and computer modelling, and accordingly there will be concentration on results from, and future directions of these methodologies. A major aspect of the subject continues to be concerned with DNA structure itself which, cven some 40 years after the Watson-Crick model, is still capable of showing hitherto unexpected features of significance. The drug-DNA intercalation field has produced several thousand publications covering its fundamental aspects alone; accordingly there will be no attempt to be comprehensive, even in the molecular modelling area. Instead, this chapter will concentrate on those studies that are deemed to be of especial significance to rational drug design. There have been fewer modelling studies on covalently bound drugs until recently, not least because of difficulties in identifying specific sites of attachment and evaluating their relative biological significance—the interested reader is referred to publications by Remers *et al.* (1986), Rao *et al.* (1986) and Warpehoski *et al.* (1990) for recent examples. The rapidly expanding area of non-covalent groove binders is of interest in relation to the design of sequence-specific ligands, but as yet does not have any direct applications in the cancer chemotherapy field. Molecular modelling studies of groove binders have been surveyed by, *inter alia*, Neidle *et al.* (1987b) and Gago *et al.* (1989).

6.2 DNA STRUCTURES

It has long been known that DNA is highly polymorphic (Arnott, 1981); following the original identifications of A and B forms by Watson, Crick, Wilkins and Franklin, the extensive studies of Arnott, Fuller and their schools in the subsequent 30 years identified many other forms of both random-sequence 'natural' DNA and defined-sequence synthetic polynucleotides (Arnott *et al.*, 1982, 1983a, b; Mahendrasingam *et al.*, 1986). Futhermore, least-squares model-fitting methods have been developed that have enabled structural models to be assigned to many of the fibre diffraction patterns from these polymers with a high degree of veracity. Factors such as the inherent paracrystallinity at best of these DNAs has meant that these models are most commonly of precisely repetitious mononucleotide (or occasionally of dinucleotide) repeats. Thus, these structures do not reveal any structural aberrations within a nucleotide sequence such as nearest-neighbour effects. They do, however, provide reliable averaged structural data (Chandrasekaran and Arnott 1989), such as the dimensions of major and minor grooves in different polymorphs (Table 6.1). Groove dimensions are of fundamental importance in relation to the interactions of both small and large molecules with DNAs, since binding sites are necessarily accessed via either major or minor grooves (Figure 6.1).

Single-crystal diffraction methods have a much more recent history as applied to nucleotide sequences. The analyses by Rich and his colleagues in the early 1970s of self-complementary ribodinucleoside monophosphates unequivocally showed Watson–Crick base-pairing without any preconceived model, as is necessary in the analysis of fibre diffraction patterns.

At least equally profound was the determination of the crystal structure of the dodecanucleotide double helix with the sequence d(CGCGAATTCGCG) (Wing *et al.* 1980; Dickerson and Drew, 1981a, b). This B-type structure shows marked sequence-dependent features, such as variations in helical twist between adjacent base pairs and base-pair roll and tilt (see Saenger, 1984; Srinivasan *et al.*, 1987; Yanagi *et al.*, 1991, for a thorough discussion of these topics). Thus, local rather than global helical parameters must be used to fully describe the structure. A consequence of these sequence-dependent differences is that the

Table 6.1. Groove dimensions (in Å) and helical parameters in the two major polymorphs of nucleic acids.

	Minor groove		Major groove		Rotation per nucleotide (°)
	Depth	Width	Depth	Width	
A helix	2.8	11.0	13.5	2.7	32.7
B helix	7.5	5.7	8.5	11.7	36.0

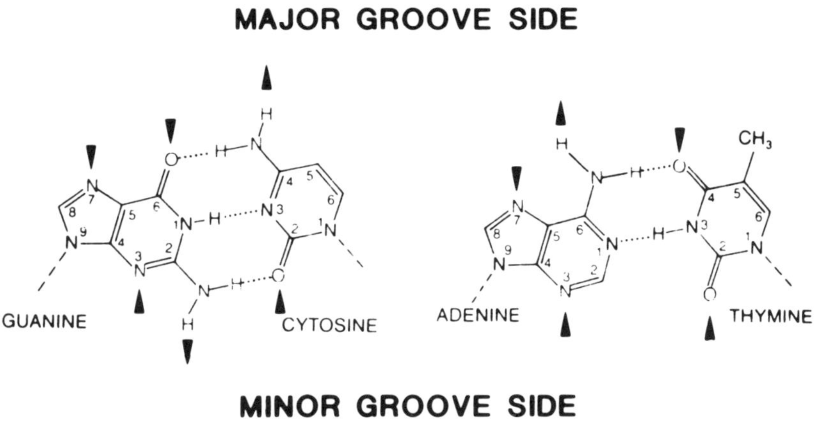

Figure 6.1. The hydrogen-bond donor and acceptor sites for GC and AT base pairs in the major and minor grooves of DNA

major and minor grooves vary in dimensions along the extent of the dodecamer, with the minor groove being narrow in the AATT section, and wider toward the ends of the helix. The major groove widths change in a compensatory manner. Related to these observations is the network of ordered water molecules found concentrated in the AATT region (Drew and Dickerson, 1981), which has been termed the 'spine of hydration'. This sequence-dependent hydration has been implicated in the B-form nature of this structure. It is also likely that the hydrogen-bonding of water molecules to bases in this region is a factor in the narrowing of the minor groove, and even in the propeller twist of some base pairs. Generalizations concerning B-type structures are now becoming possible with the analysis of further distinct sequences (Yanagi *et al.*, 1991). Certainly, the possibility of crystal-packing forces or 'end-effects' perturbing, for example, the changes in groove dimensions at each end of the dodecamer remains a real possibility. The more recent analyses of decanucleotides such as d(CCAAGATTGG) which also shows a B-like structure without the special hydration features of the central AATT sequence in the dodecamer (Prive *et al.*, 1987) promise to provide structures that have reduced crystal-packing perturbation. This decamer does not have the exceptionally narrow minor groove that the dodecamer has in its centre. A wide minor groove has also been found in the dodecamer d(CGCAAGCTGGCG) (Webster *et al.*, 1990). There is still a real need for more structural data on a wide range of B-DNA oligonucleotides in order to be able to delineate the general rules governing sequence-dependent structural behaviour. The 'Calladine' rules that have been developed to explain the variations in base-pair roll and propeller twist in terms of relief of steric clash between adjacent bases (Calladine, 1982), do have good overall predictive value in some circumstances (Drew and Travers, 1985), but are necessarily restricted to sequences related to the Dickerson dodecamer, from which the rules were derived.

Long-range sequence effects have been the subject of extensive study following the finding that AT-rich tracts can undergo bending in solution (Koo *et al.*, 1986), which appears to have important functional significance in terms of gene regulation. Poly dA.poly dT in fibres has been suggested to have a heteronomous structure with one strand A-like, and the other B-like (Arnott *et al.*, 1983b); a more recent reinvestigation of the fibre diffraction patterns from this polynucleotide at several humidities has somewhat revised this model (Park *et al.* 1987), although this sequence is still considered to have abnormal structural features even in an unbent situation. The crystal structure of the duplex from the sequence d(CGCAAAAAACGC) has a homopolymeric run of six contiguous AT base pairs (Nelson *et al.*, 1987); it has revealed a number of special and unexpected features about this sequence, which explain some of its abnormal properties. For example, the AT base pairs have large propeller twists, of up to 25°, which results in a zig-zag pattern of additional base–base bifurcated hydrogen bonds in the major groove between adenine N6 hydrogen atoms and thymine O4 atoms on their 3′ side. This, together with a high degree of purine–purine stacking, results in a narrowing of the minor groove. This structure does not directly describe the geometry of a bent region of DNA, nor how a junction between it and linear DNA would be organized. The proposals emanating from it must await further crystal structures for verification.

A number of other oligonucleotide crystal structures have been reported. Most are representative of the conformationally less flexible A form, with bases markedly tilted with respect to the overall helix axis (Shakked *et al.*, 1983; Heinemann *et al.*, 1987; Clark *et al.*, 1990). These structures vary considerably in sequence type, and it appears that the overall tendency of most sequences that have actually been crystallized to be in the A form is more a consequence of the length of sequence and its crystal packing conditions than of any inherent sequence-dependent factors. It has long been known from fibre diffraction studies that the A polymorph is favoured by an increase in hydrophobic environment. The B form family is generally considered to be the most relevant to physiological situations (Gottesfeld *et al.*, 1987), although the hydrophobicity of some interacting drugs may implicate the A form in particular circumstances.

The discovery of Z-DNA in the crystal structure of the sequence $d(CG)_3$ (Wang *et al.*, 1979) has indicated that alternating pyrimidine-purine sequences can, under appropriate environmental conditions such as high salt or torsional stress, undergo an extreme polymorphic change of a right → left-handed transition. Left-handed DNA has been confirmed to exist in solution for appropriate oligonucleotide sequences, and for such sequences inserted into plasmids (Singleton, 1987). Its existence and function in genomic DNA is, however, still a subject of controversy. Furthermore, numerous *in vitro* experiments have shown that the overwhelming majority of DNA-interactive drugs disfavour the Z form, probably on account of its abnormally high electrostatic environment.

In general, studies on the theoretical analysis, modelling and design of DNA-binding drugs have taken as their (usually implicit) basis, B-form DNA conformations. As will be discussed in subsequent sections, there is at present insufficient experimental X-ray crystallographic data for all categories of drug interaction, for this assumption to be fully verified.

6.3 METHODOLOGIES OF DRUG-DNA MODELLING

The computation of low-energy oligo- and polynucleotide conformations and inter/intramolecular energies is in general a more complex problem than that for polypeptides, on account of the much greater number of degrees of freedom per nucleotide repeating unit (Olson, 1982). In addition to the seven backbone torsional angles (Figure 6.2), the deoxyribose sugar has pseudorotational flexibility; the parameterization of base-pair twisting, rolling and bending, although clearly of considerable implication, has not as yet been explicitly included in any published force field. The majority of empirical force fields that have been developed for nucleic acids and drug-DNA complexes are partial (Olson, 1982) in the sense that they do not incorporate full geometry optimization. That of Kollman and his colleagues is probably the most extensive (Weiner *et al.*, 1984; Weiner and Kollman, 1986), with comprehensive parameterization based on the force field

$$E_{\text{total}} = E_{\text{bond stretch}} + E_{\text{angle deformation}} + E_{\text{torsional}} + E_{\text{van der Waals}} + E_{\text{electrostatic}} + E_{\text{hydrogen bonds}} \qquad (1)$$

The van der Waals term is of conventional Lennard-Jones type, and the electrostatic term utilizes partial atomic charges from, for example, CNDO/2 calculations (and a Mullikan population analysis), or better from quantum-mechanically derived electrostatic potentials fitted to a point charge model. This force field is used by a number of workers, and has been remarkably successful in

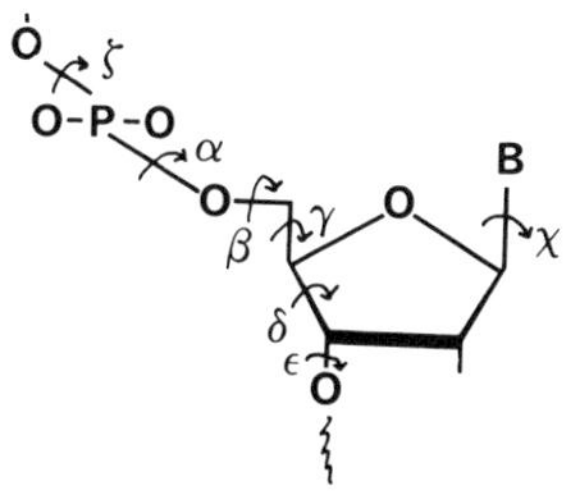

Figure 6.2. The torsion angles in a unit nucleotide (B = base)

both reproducing and predicting low-energy conformations for a number of nucleic acid duplex sequences, especially when an all-atom rather than a united-atom representation is used. All-atom force fields for nucleic acids have also been developed by Brookes *et al.* (1983) in the CHARMM program and by van Gunsteren and Berendsen (1986) in the GROMOS system. Neither has been used extensively with drug–DNA complexes. Other less extensive force fields (and computationally cheaper) can be useful for describing the particular circumstances of base–base stacking interactions (Olson, 1982; Srinivasan *et al.*, 1987).

Nucleic acids are highly charged molecules, with the phosphate groups on the exterior of a double helix carrying most of the formal negative charge. The extensive studies of the Pullmans has focused attention on the electrostatic potential and field, and has shown that contrary to expectation, the minor groove is the region of greatest negative electrostatic potential (Pullman and Pullman, 1981; Pullman, 1989), in a sequence-specific manner. The importance of potential considerations in the binding of drugs (most of which carry cationic charge), has been extensively explored by the Pullmans. These authors have developed an intermolecular energy force field which explicitly incorporates several factors that are electrostatic in origin:

$$E_{\mathrm{INTER}} = E_{\mathrm{MTP}} + E_{\mathrm{POL}} + E_{\mathrm{REP}} + E_{\mathrm{CT}} + E_{\mathrm{DISP}} \tag{2}$$

where E_{MTP} and E_{POL} represent electrostatic and polarization contributions, calculated by means of multipolar expansion of the *ab initio* wave functions of the individual structural fragments. The terms E_{REP} and E_{DISP} correspond to the standard Lennard–Jones potentials, and the E_{CT} contribution explicitly takes charge-transfer factors into account (see, for example Chen *et al.*, 1985, 1987a). These terms are implicitly incorporated in the 10–12 terms of the Kollman (and others) force field in equation (1), which have been extensively parameterized for nucleic acid constituents as such, but not for them in a variety of interacting situations, such as when polarization and charge-transfer factors can be significant.

These force fields can be used to calculate:

(a) the internal energy of a drug, E_{D};
(b) the internal energy of the oligonucleotide sequence, E_{H}, usually as B-form DNA;
(c) the internal energy of the nucleic acid helix in a drug complex, E_{HC};
(d) the drug–helix intermolecular interaction energy, E_{HD};
(e) the total of intermolecular and internal energies in the drug–DNA complex E_{COM}.

Helix destabilization energy (E_{DS}) is then defined as

$$E_{\mathrm{DS}} = E_{\mathrm{HC}} - E_{\mathrm{HD}} \tag{3}$$

and represents the energy lost as a result of producing the distortions in DNA for forming a binding site. The binding energy ΔE is then the energy of the drug–DNA complex minus the total energies of the drug and the B-DNA helix:

$$\Delta E = E_{\mathrm{COM}} - [E_{\mathrm{D}} + E_{\mathrm{H}}] \tag{4}$$

The problem of solvent treatment is common to all these molecular mechanics methods. In general, explicit structured solvent contributions have only very rarely been considered for drug–DNA complexes, not least because of the lack of relevant structural data. Thus, molecular mechanics on these systems in principle calculates potential energies and cannot normally provide data on energy values that have direct relevance to thermodynamic measurements of free energy and entropy, although relative binding energies calculated for a closely related series of drugs that do not displace different numbers of water molecules should correlate with experimental relative enthalpies and even changes in free energies. Exclusion of explicit solvent causes a further serious problem that is particular to nucleic acids; their large formal negative phosphate charges are in reality shielded by solute, and thus neglect of this factor will greatly overemphasize any electrostatic contributions in the force field. The problem has been tackled in various ways:

(a) the phosphate charges have been modified so as to neutralize the group as a whole, whilst still retaining the phosphate dipole (for example, in Tilton *et al.*, 1983);
(b) explicit counter-ions, usually of sodium, are placed between the phosphorus and its two ester oxygen atoms (Ornstein and Rein, 1979);
(c) the most straightforward, and probably most widely used, procedure is to employ a distance-dependent dielectric constant in the Coulombic electrostatic term, of the form $\varepsilon = cR_{\mathrm{ij}}$. A study (Orozco *et al.*, 1990) on an oligonucleotide–drug complex has indicated that in the situation with anionic phosphate charges and no explicit solvent, a constant $c = 4$ gives *structural* results after full energy minimization that compare rather better with relevant X-ray data than is the case with $c = 1$. The derivation of force field parameters for intercalating drugs has been discussed recently (Neidle and Jenkins, 1991).

The major problem of the conformational plurality of nucleic acids cannot be adequately explored by molecular mechanics, with its inherent limitation of being restricted to the minimization of local wells on the total multidimensional energy surface. Molecular dynamics techniques have much promise in being able to explore many more features of molecular flexibility (McCammon and Harvey, 1987), as well as being able to calculate free energies. Their application to proteins extends back to 1976; the first studies on oligonucleotides were reported some 7 years later (Levitt, 1983) not least on account of the widely perceived problems involved in the treatment of solvent and electrostatic factors, as discussed above. McCammon and Harvey (1987) have pointed out that these problems are more acute in the case of nucleic acids than for proteins, due to the former's polyelectrolyte nature and their larger surface-to-volume ratios. The successful 140 ps simulation of the oligonucleotide d(CGCGAATTCGCG), including 1927 water molecules and 22 Na^+ counter-ions (Swaminathan *et al.*, 1991), has shown that realistic dynamics simulations can provide detailed dynamic data on conformation, base-pair and base-step features. To date, few molecular dynamics simulations of drug–DNA complexes have been reported (see, for example, Prabhakaran and Harvey, 1985, 1988; Rao and Kollman, 1987; Cieplak *et al.*, 1990; Boehncke *et al.*, 1991). Undoubtedly, as techniques for solvent modelling, for example by Monte Carlo methods (Mezei *et al.*, 1983) are improved, relevant potentials for drug–solvent and drug–DNA become developed, and supercomputer-like machines increase in availability, many more such simulations will be embarked upon. The d(CpG)–proflavine crystal structure has been the subject of molecular dynamics simulations (Swaminathan *et al.*, 1990; Herzyk *et al.*, 1991) examining, in the first study, the water structure, and in the second, the dynamics of the intercalation complex itself. These two studies used the quite distinct GROMOS and AMBER force fields, and the points of difference between the results may thus be explained, at least in part. The exciting recent development of methodology that combines statistical mechanics and molecular dynamics simulations in the free energy perturbation method (Bash *et al.*, 1987; McCammon, 1987) does enable thermodynamic free energies of solvation and of the effects of conservative structural changes in a drug molecule to be accurately calculated, with results that can concur very well with experimental thermodynamic data. As yet, this technique is only in its infancy for nucleic acids and their drug complexes.

The principal alternative method in the drug–DNA area for the study of multidimensional degrees of conformational freedom involves the use of interactive computer graphics, for more than merely display and analysis purposes. It has found especial use in the estimation of plausible low-energy positions for a ligand in a macromolecular binding site, usually prior to molecular-mechanics regularization. Altough such an approach is usually unable to study all conformational and positional possibilities systematically, especially when the ligand itself has flexibility, it can be remarkably speedy in locating sterically (and therefore energetically) reasonable structures, when in the hands of a skilled molecular modeller. The graphics method is used in several laboratories for the study of drug–DNA interactions, including the author's one (see for example, Collier and Neidle, 1988; Lybrand *et al.*, 1986). Use of colour-coded surface representations (Neidle *et al.*, 1987b) have often been found to aid drug–DNA docking manoeuvres, especially when incorporating electrostatic potential display, for example to show the negative electrostatic potential on an AT-rich minor groove surface (Weiner *et al.*, 1982) or the differences between major and minor groove electrostatic potential that may be important in the location of the optimal geometry for an intercalating drug (Dean and Wakelin, 1979, 1980a, b).

6.4 FUNDAMENTAL ASPECTS OF INTERCALATIVE DRUG BINDING TO NUCLEIC ACIDS

The discovery of low-molecular-weight molecules that bind reversibly to cell nuclei predates the elucidation of the double helix itself. These were originally used as histological stains, until the discovery of the antibacterial properties of many of them, especially the acridines such as proflavine (Figure 6.3). The seminal studies of Lerman (1961, 1963) showed that these molecules, typified by proflavine and ethidium, indeed bind to DNA. It was shown that this equilibrium binding is accompanied by characteristic changes to the DNA itself, in particular a length extension of 3.4 Å per bound drug molecule and changes to

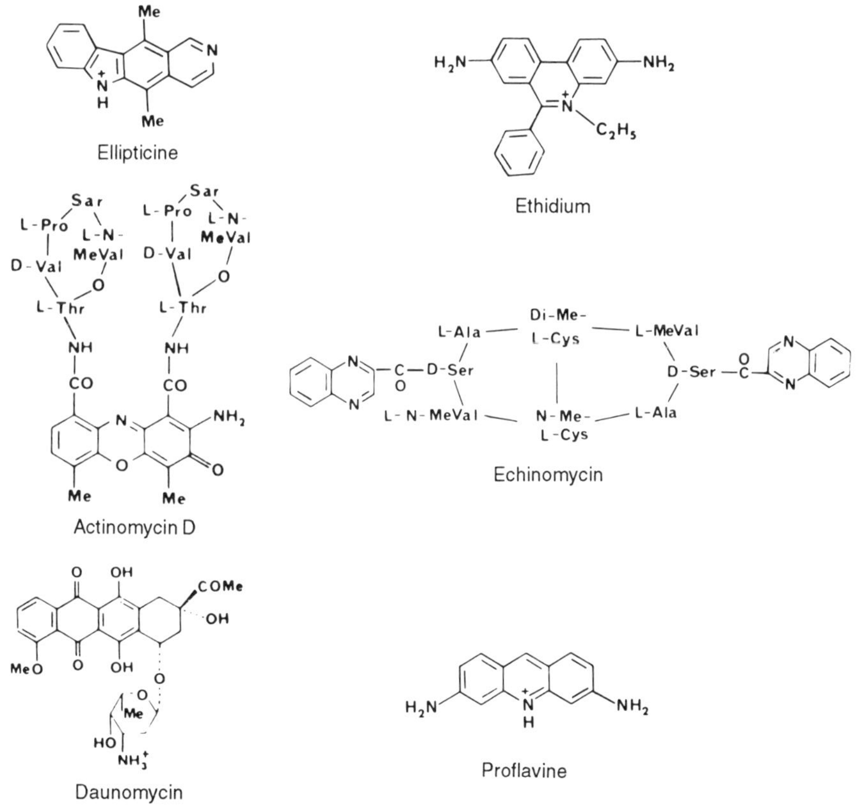

Figure 6.3. The structures of various intercalating molecules. Doxorubicin has the COMe group replaced by —CO. CH_2 OH in daunomycin

the absorption spectra of both ligand and DNA (reviewed, *inter alia*, by Neidle, 1979; Gale *et al.*, 1981; Waring, 1981). Fibre diffraction studies showed a retention of the base-pair stacking of 3.4 Å separation found in B-DNA itself, and biological studies found that these molecules acted as frameshift mutagens, i.e. they shifted the reading frame of the genetic code by an integral number of bases. These observations, together with the common structural feature of possession of a planar aromatic grouping of minimally three fused rings (Figure 6.3), led to Lerman's intercalation model (Figure 6.4), in which a drug molecule becomes inserted in between two adjacent base pairs, with their separation increasing from 3.4 to 6.8 Å. This conceptual model is still generally accepted, and has been subsequently supported by a very large body of physicochemical data. Until recently, studies of drug–DNA intercalation in solution necessarily concentrated on the bulk averaged properties of the complexes to natural and synthetic polynucleotides, using techniques such as NMR, circular and electric dichroism, unwinding of closed-circular DNA, kinetics and DNA helix-coil melting stabilization (Patel, 1979; Berman and Young, 1981; Wilson and Jones, 1981).

A major impetus for these investigations has (and continues to be) the fact that many DNA intercalators have experimental anti-tumour activity, and some, exemplified by the anthracyclines, are of very considerable clinical use. The most important member of this family of compounds, doxorubicin (Figure 6.3), continues to be the agent with the broadest spectrum of useful activity in humans (Arcamone, 1981) of all currently available drugs, and tends to be the standard against which putative new drugs are judged. A number of studies have demonstrated significant correlations between DNA binding parameters such as equilibrium binding constants, and cytotoxic activity. In some instances, there are overall correlations between experimental murine anti-tumour activity and levels of DNA intercalative binding. It should be noted that such relationships do not invariably occur, and indeed since factors of metabolism and pharmacokinetics invariably play important roles in overall drug action, it would be

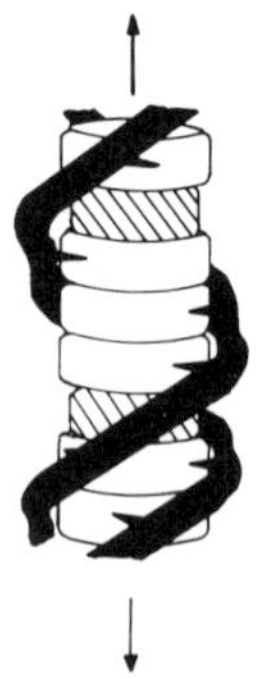

Figure 6.4. The Lerman model for intercalation into double-helical DNA, with the bound drug chromophores shown shaded

surprising if they always did. The more recent knowledge of the involvement of DNA topoisomerase II in the critical cytotoxic events, especially in neoplastic cells, strongly implies that drug–DNA intercalative interactions, being presumably early events in the overall process, are essential but in themselves insufficient for anti-tumour activity. The Lerman model for intercalation is necessarily lacking in molecular detail. It cannot provide information on:

(a) Changes induced by drugs having differences in molecular structure and substitution patterns.
(b) Whether there are preferences for particular sequences at, and adjacent to the binding sites themselves. These could have important biological implications.
(c) The molecular structure(s) of the nucleotide residues at, and around the binding sites, and how these differ from the unperturbed drug-free structures.

The restrictions on detailed knowledge of binding site preferences imposed by experimental data being confined to natural-sequence DNA and synthetic polynucleotides, have been overcome by the advent of DNA footprinting (Dervan, 1986), which is able to localize sites of drug attachment (which by implication are sites of slowest drug–DNA dissociation) on sequences of length between 100 and 1000 base pairs. The methodology used is a variant of standard gel-based DNA sequencing, with cleavage of phosphodiester DNA bonds occurring everywhere except at and adjacent to ligand binding sites. The interested reader is referred to Dervan (1986) for further details of the technique. In general, it has been found that the structurally more simple intercalators such as the acridines, ethidium and daunomycin, have failed to be footprinted, in contrast to others such as actinomycin and a number of AT-selective groove binders. However, recent careful studies (Chaires *et al.*, 1987, 1990) have shown by both footprinting and equilibrium binding experiments that daunomycin prefers a triplet sequence involving adjacent GC base pairs flanked at the 5′ end by either an AT or an additional GC one. The clear implication of this investigation is that sequence specificity, with a possible structural basis in terms of subtle differences between different regions of a DNA, is probably relevant, to at least some extent, to all DNA-binding drugs.

NMR methods have until recently provided less detailed structural information on intercalation complexes than on unperturbed oligonucleotides, in large part because of frequent fast dissociation of bound drug, as well as difficulties in dealing with the multiplicity of observed binding sites even in short sequences. The interested reader is referred to the early seminal papers in this area by Feigon *et al.* (1984) and Patel and Canuel (1978), and to the more recent analysis of a nogalamycin–octanucleotide complex (Zhang and Patel, 1990).

Sequence preferences have been largely established for many drugs as a result of footprinting experiments, together with NMR chemical shift data on short-length oligonucleotides. For the acridines, ethidium and daunomycin, it has been found that pyrimidine-3′,5′-purine sites are preferred (Patel and Canuel, 1978; Phillips and Roberts, 1980; Reinhart and Krugh, 1978).

6.5 X-RAY STRUCTURES OF DRUG–DNA INTERCALATION COMPLEXES

The shortest sequences of oligonucleotide that can conceivably act as a model for double-helical DNA, is that of a dinucleoside monophosphate. To date, some 16 crystal structures of complexes between intercalating drugs and dinucleoside 'mini helices' have been reported (detailed in Neidle, 1989). Drugs bound include proflavine (to the sequences riboCpG, deoxyriboCpG, riboCpA-UpG), ethidium, 9-aminoacridine and ellipticine. All of these complexes have a pyrimidine-3′,5′-purine sequence at the intercalation site, and have a uniform pattern of backbone conformations such that the only two backbone angles that deviate significantly from the polynucleotide-type structures of the A-form are around the O5′-C5′ and 3′ side glycosidic torsion angles (Berman *et al.*, 1978). All of these structures have the drug molecule intercalated in between base pairs, in the manner outlined by Lerman. Some drugs such as proflavine are intercalated from the major groove direction, whereas ethidium, for example, resides in the minor groove. These differences are manifest primarily in differences in base-pair roll and tilt rather than in backbone conformation. It is notable that dinucleoside complexes with different sequences, such as the actinomycin-dGpC complex (Takusagawa *et al.*, 1982), invariably do not form intercalated duplexes, but instead are found in distinctive 'opened-out' arrangements. Thus, the sequence preferences found in solution, and outlined in the above section, are paralleled in the crystalline state. The sole representatives of a crystal structure showing drug monointercalation into a sequence longer than a dinucleoside, are that of the anti-tumour drug daunomycin and several of its derivatives with hexanucleotides such as d(CGTACG) and d(CGATCG) (Quigley *et al.*, 1980; Wang *et al.*, 1987; Moore *et al.*, 1989; Frederick *et al.*, 1990). In these structures daunomycin or an analogue is intercalated from the minor groove side, with the protonated amino sugar resting along the side of the groove. The anthracycline drug nogalamycin has substituents at each end of the aglycone; the substituents lie one in each groove (Liaw *et al.*, 1990; Williams *et al.*, 1990a, b; Gao *et al.*, 1990). The anthracycline chromophores are oriented roughly perpendicular to the base-pair long axis, in overall good agreement with earlier NMR chemical shift data on daunomycin and DNA (Patel *et al.* 1981). There are two sites of intercalation in these hexamer duplexes, one between each CG base pair. Stabilization and sequence preference at the binding sites appears to be enhanced by a hydrogen bond from the drug molecule to a guanine N2 atom. It is important to note for the subsequent discussion on drug–DNA models, that these structures, like the dinucleoside complexes, do not provide information on the conformational changes produced by intercalation at *both* 5′ and 3′ sides of the binding site. The dinucleoside complexes by definition only show the nucleotide structures immediately surrounding a bound drug, and the hexamer complex defines conformation extending beyond the daunomycin binding site only in the 3′ direction, i.e. quite asymmetrically. There is thus as yet no crystal structure that shows a drug intercalated and embedded within a long stretch of sequence, so that the conformational features of intercalator-induced change at both 3′ and 5′ sides can be established as well as any longer-range structural changes. The proflavine-dCpG complex (Shieh *et al.*, 1980) is notable in that its crystal structure shows a regular arrangement of hydrogen-bonded water pentagons (Neidle *et al.*, 1980) that play a crucial role in stabilizing the complex. Subsequent Monte Carlo calculations on possible water structures in this complex have found a very similar arrangement (Mezei *et al.*, 1983).

The highly cytotoxic quinoxaline antibiotic echinomycin (Figure 6.3) and its closely related analogue triostin A (which has a modified disulphide bridge), are representatives of a more complex class of intercalators, which can have their two chromophores simultaneously intercalated into a DNA sequence. Such molecules generally have significantly higher nucleic acid binding constants than their mono-intercalative counterparts, and thus have generated interest as having potentially greater specificity at lower dosage levels than the latter (Wakelin, 1986). This bis-intercalative mode requires a separation of 10.2 Å between the chromophores (Figure 6.5), so that there are two non-opened-up base pairs between them. Solution studies of binding to a variety of polynucleotides has shown a preference for GC-rich regions (Waring and Fox, 1983), and extensive footprinting studies (Low *et al.*, 1984; Van Dyke and Dervan, 1984) have shown that echimonycin protects GC sites, with the tetranucleotide

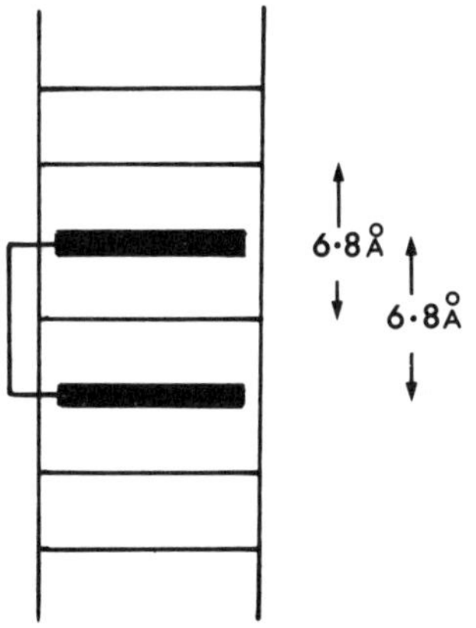

Figure 6.5. A schematic representation of bis-intercalation

sequences TCGT, ACGT, TCGA and CGCC being strong binding sites. The crystal structures of echinomycin and triostin bound to the hexanucleotide sequence dCGTACG have been determined (Wang *et al.*, 1984; Ughetto *et al.*, 1985). These show the expected bis-intercalation, in the manner described above. The drug chromophores are sandwiched between CG base pairs, and thus link together hexamer duplexes in the crystal. The octapeptide moiety of these drugs is found to lie in the minor groove of the hexamer duplex. The sequence specificity, as found by footprinting, is fully in accord with the drug binding sites in these structures, and is explained by minor groove hydrogen bonding and van der Waals interactions between an alanine residue and the exocyclic N2 amino group of a central guanine in the middle of the binding site. A remarkable and entirely unexpected feature of the structures is that the central AT base pairs that flank the outer regions of a drug binding site, have Hoogsteen rather than standard Watson–Crick base pairs, and the adenosines have *syn* glycosidic angle conformations rather than the normal *anti* type found in A- and B-form DNAs. A very similar arrangement has been found in the crystal structure of a triostin-octanucleotide complex (Quigley *et al.*, 1986), with the additional feature of some GC Hoogsteen base-pairing as well as AT ones. Quigley *et al.* suggest that steric factors may play a major role in forcing the adoption of unconventional base pairs, which if they occur in natural-sequence DNA on binding these drugs, would certainly inhibit the normal recognition processes of subsequent DNA-regulatory proteins. NMR studies (Gao and Patel, 1988), do clearly indicate that Hoogsteen base-pairing also occurs in solution with at least some oligonucleotide sequences, although chemical footprinting appears to rule it out in longer DNA sequences (McLean *et al.*, 1989).

6.6 GENERAL FEATURES OF DRUG–DNA INTERCALATION MODELS

Theoretical studies of the intercalation process have been conducted in order to answer a number of distinct questions ranging from calculations of sequence preferences for different chromophores to the establishment of stereochemically plausible backbone geometries, and in a few instances, to the design of new potential drugs. The overall greater preferences found for many 'simple' intercalators of GC-containing polynucleotides (see Feigon *et al.*, 1984 for recent experimental data) has long been rationalized on the basis of the greater polarizibility of GC base pairs compared to AT ones. The pyrimidine-3′,5′-purine sequence preferences found experimentally both in solution and in the crystal have been rationalized on the basis of molecular mechanics energy calculations with the finding that this type of sequence is energetically easier to unstack with base-pair separation increasing from 3.4 to 6.8 Å, than is the reversed purine-3′-5′-pyrimidine sequence (Broyde and Hingerty, 1979). Thus, of the ten distinct dinucleoside intercalation sites (detailed in Gale *et al.*, 1981), the CpG one is the most preferred for 'simple' intercalators. It is noteworthy that this sequence has considerable current biological interest, as it is significantly under-represented in eukaryotic genomes, and may even occur in clusters (Bird, 1986).

A number of theoretical studies of intercalation have used base-paired dinucleosides as the model for DNA itself, as have the crystallographic analyses outlined above. It is therefore surprising that the crystallographically determined backbone conformations have been used in only some of these studies, in part because it is clear that there are many ways of extending the backbone from an A or B form so as to construct a 6.8 Å pair separation and produce a stereochemically reasonable intercalation model. To date, there has been no full analysis of the relative energetics of the several models used, or indeed of longer sequences with embedded intercalation sites, for which there is no structural data. A study (Chen *et al.*, 1987a) of four computer-generated dinucleoside duplex ethidium-intercalated geometries by means of the force field (equation 2) described in Section 6.3, has suggested that the one of lowest energy is closest to the crystallographically observed one for ethidium CpG (Jain *et al.*, 1977). The fact that the numerous dinucleoside complex crystal structures all show a closely similar backbone conformation even though they crystallize in a variety of unit cells with different solvent environments, strongly suggests that this is a global minimum, and that it is invariant regardless of the nature of the bound drug. At the dinucleoside level, the nature of the sugar pucker at 5′ and 3′ ends is not a major determinant of intercalated structure (Berman *et al.*, 1978), even though this conformational variable is important for longer sequences.

The position of a bound intercalant in its binding site has been the subject of a number of theoretical studies. In general, steric differences between different drugs can produce differentiation between major and minor groove entry and disposition, especially when bulky substituents are attached. For example, ethidium is found in the crystal structures of several dinucleoside complexes to have its phenyl group lying in the minor groove, whereas proflavine, with its two exocyclic amino groups some 2–3 Å further apart than the minor groove width, is constrained to lie predominantly in the major groove. Related to this is the possibility that a chromophore can have a number of distinct equi-energetic orientations in the intercalation plane. A computational procedure has been developed to systematically explore all such possibilities (Pearl and Neidle, 1986), which has shown that the degree of flexibility that a drug has in the bound state is related to its size and that of its substituents. The nature of the substituents is of importance in determining whether they can undergo hydrogen-bonding interactions with donor/acceptors on DNA, especially on the bases. The possession of cationic charge on a chromophore has often been regarded as a necessary feature of an intercalator, but there are a number of uncharged and very effective intercalators such as actinomycin, that nevertheless have electron-deficient chromophores. Again, theoretical approaches have illuminated these topics.

6.6.1 THE MODELS

Fibre-diffraction studies of DNA–drug complexes provide relatively little observational data, even compared to nucleic acids themselves, and model-building from this basis is necessarily very speculative (see for example, Fuller and Waring, 1964). Such models incorporated the experimental finding for 'unwinding' accompanying intercalation, such that the total angle of helix twist is decreased from the 36° of B-DNA per adjacent base pair. A number of modelling analyses have erroneously ascribed these total unwinding angles (of, for example, 26° for ethidium) to the change in helix twist solely at the intercalation site. In principle, this is physically unreasonable and is unsupported by experimental data; the crystal structure of the daunomycin–hexanucleotide complex shows that the unwinding is propagated along the length of the duplex for several base pairs (Quigley *et al.*, 1980). The helix twist angles found in the crystal structures of dinucleoside intercalation complexes are thus not equivalent to polynucleotide unwinding angles. The early model-building using the dinucleoside structures as starting points (Sobell *et al.*, 1977) did make this assumption; although this work relaxed (though not by means of energetic calculations) the central intercalation geometry in an arbitrary manner it did make the important observation that all such models were necessarily no longer co-helical with standard B-DNA by virtue of the non-equivalence in helical repeating terms of the two nucleotides on each strand of the intercalation site. This concept was also developed (Berman *et al.*, 1978; Berman and Neidle, 1979) in two distinct tetranucleotide duplex models in which the central dinucleosides were restrained to their experimental crystallographic conformations. This was prior to structural work on deoxydinucleoside complexes (Shieh *et al.*, 1980), so the tetranucleotides are of RNA type. The asymmetry of the dinucleosides at the intercalation site, and in particular the high value for the glycosidic angle at the 3′ ends, force the adjacent residues to adopt conformations that are themselves non-standard—this, at least for double-helical RNA, suggested that intercalation results in considerable long-range structural disruption. Both of these tetranucleotide models have base pairs with large propeller twists, a feature that is most pronounced for the 'neighbour exclusion' one, compared to the other, 'extended-exclusion' one. (The neighbour exclusion principle of drug intercalation (Gale *et al.*, 1981) states, on the basis of much experimental data, that maximal drug binding occurs at every other potential site.) These models indicate that binding at every site is not possible on purely steric grounds, since it would result in unacceptable distortions and the 'neighbour exclusion' model was the nearest achievable one. The severe steric constraints encountered in this modelling exercise were probably due in large part to the presence of the 2′-ribohydroxyl group on the sugar residues. The problem has been recently re-examined in terms of deoxy residues (Neidle *et al.*, 1988). In this analysis, the deoxydinucleoside geometry from the crystal structure of the proflavine-dCpG complex (Shieh *et al.*, 1980) has been used as the intercalation nucleus, and a B-DNA tetranucleotide sequence was least-squares fitted to both 3′ and 5′ ends, followed by full-geometry molecular mechanics minimization on the complete decanucleotide with the drug in its binding site. This model is thus representative of the situation with a drug embedded into an oligonucleotide sequence. The simulated structure (Figure 6.6, see colour section) has marked long-range changes compared to B-DNA—the major groove has been widened along some six to eight base pairs and the propeller twist of base pairs does not follow Calladine's rules. Confirmation of such features must await a relevant crystal structure, although the changes in groove dimensions are in broad agreement with the consistent findings of footprinting experiments that drug binding induces changes in enzymatic cutting frequency that are ascribed to changes in accessibility to phosphodiester bonds via the grooves. This model has been used as a starting point in a dynamics simulation, with full solvent inclusion (Herzyk *et al.*, 1992).

Other models of intercalation do not directly relate to any of the single-crystal results; only some have been checked for stereochemical consistency (for example, Malhotra and Hopfinger, 1980). An early study of interest used the linked-atom least-squares modelling procedure to produce models for intercalation into A- and B-DNA helices (Alden and Arnott, 1975, 1977). The procedure used deliberately forced equivalence of both strands, as well as no deviation from helicity throughout the structures; both A- and B-helix complexes have two unusual (though sterically allowed) *trans* backbone conformational angles on each strand at the intercalation site, with only the latter having unwinding of base pairs. A similar method has been used to generate intercalated hexanucleotide sequences (Lybrand and Kollman, 1985) with mixed C3′endo-3′,5′-C2′endo sugar pucker geometries at the intercalation site in the centre of the helix. These hexamers closely resemble the polynucleotide intercalation models in that the torsion angles around the P—O5′ and C5′—C4′ bonds are in *trans* domains.

Three families of low-energy intercalation geometries have been generated by Miller and Pycior (1979) and Miller (1981) for tetranucleotide duplexes, with the sole constraints of base-pair coplanarity and B-DNA helicity for the 5′ and 3′ ends. These differ one from another by having distinct glycosidic torsion angles and sugar puckers as the primary variables, which result in different helix twist angles for the three models. Various intercalating drugs have been assigned to these models, on the basis of their different polynucleotide helix unwinding angles (Gale *et al.*, 1981). However, as indicated earlier in this chapter, polynucleotide unwinding is not equivalent to 'unwinding' in a short oligonucleotide sequence. These tetranucleotide models do not have any straightforward relationship with the crystallographically determined backbone conformations.

A valuable and detailed study of conformational mobility in intercalated dinucleoside sequences (Taylor and Olson, 1983) has examined all possible intercalated structures by means of calculations of non-bonded van der Waals and electrostatic potentials. All conformational angles were varied and over 400 low-energy arrangements were found. These form two classes—one with standard A- and B-like helical structures, and the other (larger) one with P—O5′ and C5′—C4′ *trans* torsion angles. The latter category resembles the original Watson–Crick structure for B-DNA (Crick and Watson, 1954) before refinements of the fibre-diffraction data, as well as the polynucleotide intercalation models of Alden and Arnott (1975). These two angles are highly correlated and themselves produce a continuum of intercalation geometries. This study has also shown that the crystallographic intercalation geometries have considerably less flexibility than most others, as well as being of significantly higher energy. However, interactions with ligands were not considered in this work, and thus it is not clear to what extent quantitative conclusions are meaningful. The notion of plurality of intercalation geometries has also been proposed on the basis of an analysis of nucleotide geometries in the tRNA crystal structure (Holbrook and Kim, 1979), where a number of distinct instances have been found of bases intercalating between base pairs so as to form the loop regions characteristic of the 'globular' structure of tRNA.

The first reported molecular dynamics study of intercalation into DNA (Prabhakaran and Harvey, 1985) used the alternating oligonucleotide sequence $d(CG)_6$ which was subjected to a 3.4 Å stretch during the simulation in order to form the intercalation cavity. It was necessary to impose a virtual van der Waals plane in the cavity during the molecular dynamics motions, in order to avoid its collapse. Unwinding was distributed over several base pairs for all models produced. Analysis of thermal motion parameters obtained from the crystallographic analysis of the proflavine–riboCpG complex (Aggarwal and Neidle, 1985) has shown that there are pronounced librational and translational effects. These are predominantly in the planes of the drug and base pairs with, for example, the principal translational eigenvector of the proflavine's thermal motion being directed towards the major groove of the complex, in accord with the proposal that this drug intercalates from the major groove direction. Thermal motion analysis is restricted to high-resolution X-ray analyses, but when this condition is met, the technique can provide valuable information and insights on molecular motions and correlated motions of component groups that complements purely computational molecular dynamics methods, and does not suffer from problems of solvent definition and relevance of energetic parameterisations.

6.7 MODELS WITH SPECIFIC DRUGS

6.7.1 PROFLAVINE AND THE ACRIDINES

An early investigation with base-paired dinucleosides (Nuss *et al.*, 1979) used the geometry found in the crystallographic analysis of a CpG–ethidium complex (Sobell and Jain, 1977) as a starting-point. This used a restricted force field with the molecular energy being minimized with respect to van der Waals and electrostatic terms, and torsion angles, for both the sequence CpG and its reverse GpC. Even so, the analysis found that the former sequence was significantly preferred in energetic terms for drug intercalation, by nearly 5 kcal/mol for proflavine, in accord with experimental evidence (Patel and Canuel, 1977). These authors also conclude that the major reason for the pyrimidine-3′,5′-purine sequence specificity is intramolecular rather than drug-base pair in origin. A subsequent study used the full force field outlined in Section 6.3 (Dearing *et al.*, 1981), albeit with a partial united-atom approach, and has examined the question of the role of the dinucleoside sugar puckers in the overall structure of the proflavine–riboCpG and deoxy-dCpG complexes, whose crystal structures (Neidle *et al.*, 1977; Shieh *et al.*, 1980) do not show the simple alternating ('mixed') pattern observed in other dinucleoside complexes (for example, in ethidium–CpG complexes (Tsai *et al.*, 1977) with C3′endo-3′,5′-C2′endo puckers). It was found that there was little energy difference between the different puckering possibilities for intercalated proflavine, whereas with acridine orange (which has *N*-dimethyl groups replacing the exocyclic amino groups of proflavine) the bulky substituents force the sugar puckers to be alternating. The intermolecular hydrogen bonding in the case of proflavine and ribo-CpG between the drug's amino groups and phosphate oxygen atoms more than compensates energetically for the slightly less stable all-C3′endo puckers that it produces. The inter-strand P...P separation in ribo duplex complexes appears to be smaller than that in the deoxy series; consequently proflavine is able to form two intermolecular hydrogen bonds with the former, and only one with the latter. Whilst sugar pucker changes at the 3′ side of an intercalation site for a dinucleoside thus have little or no effect on phosphodiester backbone conformation (see also Berman *et al.*, 1978), they would have profound consequences in a longer sequence. The conclusions of Dearing *et al.* (1981) that drug–nucleic acid interactions in addition to intercalation itself, such as hydrogen bonding to phosphates, can provide sufficient stabilization in themselves to force-change puckers, does suggest that the uniform structural pattern found experimentally for the dinucleoside complexes with a variety of drugs, may not extend to longer sequences.

Examination of a range of potential positions for proflavine to be oriented in the intercalation site (Islam and Neidle, 1984), has found overall excellent

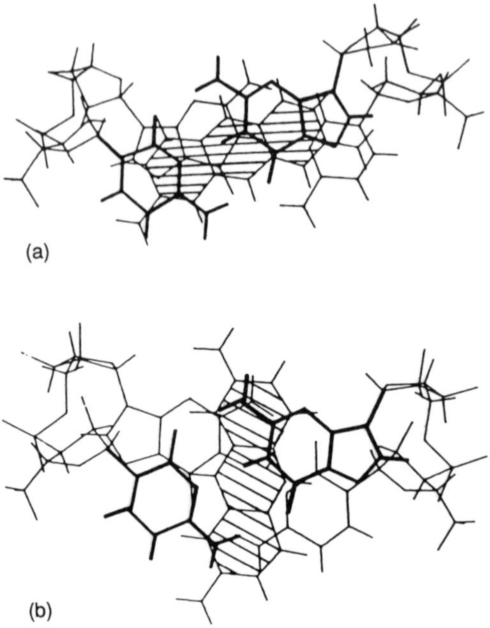

(a)

(b)

Figure 6.7. (a) The low-energy orientation of proflavine intercalated between base pairs. The drug is shown shaded. (b) A perpendicular orientation for proflavine

agreement with the crystallographic structures with the drug having its amino groups pointing out towards the major groove of the duplex (Figure 6.7a). Other orientations such as one with a 90° rotation (Figure 6.7b) have less attractive stacking interactions with base pairs.

A combined molecular mechanics and dynamics investigation of double intercalation of 9-aminoacridine into a heptanucleotide (Rao and Kollman, 1987) has examined possible theoretical origins of neighbour-exclusion binding, which is a fundamental tenet of the intercalation phenomenon. In general, experimental binding studies find that nearest-neighbour binding of an intercalator does not occur, in contrast to neighbour-exclusion site occupancy (Crothers, 1968). Since no directly relevant structural models are experimentally available, theoretical ones were constructed. It was surprising that a difference of only 5 kcal/mol was found between the drug–helix interaction energies for the molecular mechanics-refined neighbour-exclusion and neighbour-exclusion-violation models. Even this was balanced out by the former's larger helix destabilization energy. The violation model appeared to be slightly more favoured by the dynamics simulations, probably because the phosphodiester backbone is more stretched, thus reducing repulsive electrostatic interactions. On the other hand, it was found by normal-mode analysis that there is a significant entropic difference in favour of the neighbour-exclusion model. Rao and Kollman point out that counter-ion effects, which they did not fully investigate, could also be important. Of likely more significance are solvent effects—disruption of structure so as to form the sites, and relative stabilities of structured drug-DNA-solvent arrangements. The study is an important illustration of the limitations imposed by considerations solely of stereochemical energies.

The acridine derivative amsacrine (4′-[acridin-9-ylamino]-3′-methoxymethanesulphonanilide, Figure 6.8), has established anti-cancer activity in humans. Extensive structure-activity studies on over 700 analogues have been carried out (reviewed by Denny *et al.*, 1983), and have shown the relationship of DNA intercalation to cytotoxic activity (Baguley *et al.*, 1981), although unsurprisingly this factor is by itself insufficient for optimal effect. The findings that amsacrine, as well as a number of other intercalating drugs, is involved *in vivo* in a ternary complex with DNA and the enzyme DNA topoisomerase II (Minford *et al.*, 1986; Kohn *et al.*, 1987b), suggest that the structural perturbations in DNA resulting from the initial intercalating event, occur non-randomly, and that there are subtle features on the surface of the binary complex that are recognized by the enzyme; this subtlety is shown by the low propensity of the 2′-methoxy isomer of amsacrine to form a ternary complex even though it appears to bind to DNA with equal effectiveness (Wilson *et al.*, 1981). There is at present no structural data on the enzyme. Preliminary molecular modelling studies on amsacrine and several analogues (Neidle *et al.*, 1987a) has been reported. This used dinucleoside duplexes as models for intercalated sites in DNA, with a simplified force field without geometry optimization, which therefore can be taken as indicative of conformational trends rather than of reliable energetics. It had previously been shown that amsacrine itself has considerable conformational flexibility about the central —NH— linkage, and that equi-energetic conformers are separated by 180° rotations about the two bonds (Abraham *et al.*, 1985). Thus, docking of this drug into an intercalation site is rendered more complex by the addition of these two more degrees of freedom, which cannot be

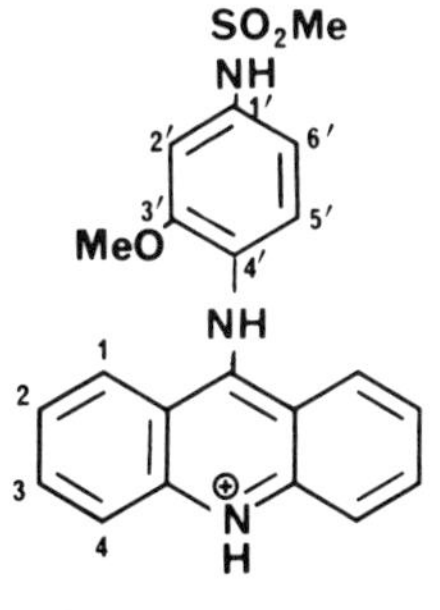

Figure 6.8. The structure of amsacrine

fully explored by molecular mechanics optimizations of local minima. A preliminary analysis of this question (Neidle *et al.*, 1987a) has shown that the lowest-energy conformer of the isolated drug is not of such low energy when the compound becomes intercalated. Even with this approximation, it was possible (Neidle *et al.*, 1987a) to shown that the low-energy intercalated amsacrine complexes found (Figure 6.9), did provide useful structure–activity data on the effects of substituents at various positions on the drug molecule. A more recent study with an intercalated hexamer duplex model has concurred with this general conclusion (Abraham *et al.*, 1988). In general, the findings support the overall hypothesis that the extent of calculated DNA intercalative binding correlates with *in vitro* cytotoxicity though less well in detail with *in vivo* anti-tumour effects, in part undoubtedly on account of DNA topoisomerase II being a factor in the latter situations, and in part due to pharmacokinetic factors. It is of importance to note that measurements of equilibrium binding constants with natural sequence DNA tend to average over all possible binding sites, as well as both major and minor groove insertion, whereas topoisomerase action is believed to be very specific, arising only at certain sequences as well as presumably being confined to one groove, in common with many other DNA-binding proteins. The molecular modelling has likewise focused on single binding sites. It is thus surprising that results in the amsacrine series tend to correlate rather better with the biological data than with the solution DNA-binding measurements. Somewhat more consistent correlations have been found for other series of intercalative drugs such as the anthracyclines (Arcamone, 1981), although this is by no means always true. The newly-developed 4-carboxamide acridine derivatives (Figure 6.10) have high *in vitro* and *in vivo* activity (Denny *et al.*, 1986; Rewcastle *et al.*, 1986). Examination of the solid-state structure and conformational flexibility of the drug itself (Hudson *et al.*, 1987) has shown that the side-chain can rotate about its minimum energy position with respect to the acridine ring, with little energy cost. A model for intercalative interaction with a DNA site has been proposed on the basis of this study, with the side chain residing in the major groove and intermolecular

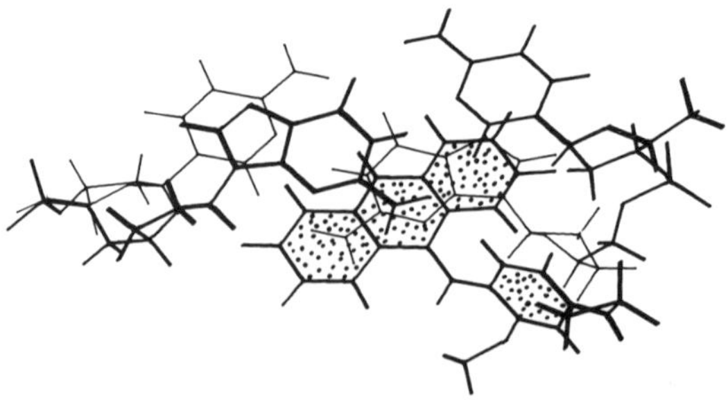

Figure 6.9. A view of a low-energy position for amsacrine in the minor groove of an intercalated helix

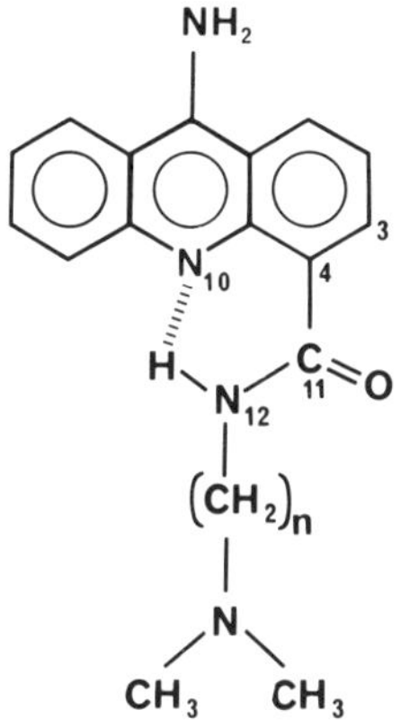

Figure 6.10. The structure of amsacrine 4-carboxamides

hydrogen bonding between an O6 atom of a guanine and the two NH groups on the side chain. A similar conclusion of major groove preference at GC sites has been reached by calculation of sequence preferences at two tetranucleotide sequences (Chen *et al.*, 1987b), although the pattern of intermolecular interactions differ in detail. No conformational features of these models have been published, so it is not possible to compare results with the model of Hudson *et al.*

A study of phenylquinoline carboxamides with the position of phenyl ring substitution being systematically varied, and using an intercalated hexamer duplex model, has found an overall high correlation between calculated binding energy and cytotoxicity (McKenna *et al.*, 1989).

6.7.2 DAUNOMYCIN, ANTHRACYCLINES AND ANTHRAQUINONES

The anthracycline compound doxorubicin has already been mentioned as being the single most widely used anti-cancer drug in current clinical use. Its close analogue daunomycin (Figure 6.3) has a narrower spectrum of activity, being chiefly of use in the treatment of acute leukaemia. Both compounds bind to DNA by intercalation, with very similar binding constants and thermodynamics of binding (reviewed *inter alia*, by Neidle, 1979; Arcamone, 1981). Thus, differences in their biological effects are most likely due to differences in pharmacokinetics and metabolism, factors that are beyond the scope of this chapter.

Early studies of anthracyclines by molecular modelling concentrated on the conformational properties of daunomycin. Crystal-structure analyses of the drug in several crystallographic environments showed a single conformation around the ether linkage connecting the chromophore and the daunosamine sugar group (Neidle and Taylor, 1977 and references therein). This was subsequently confirmed by energy calculations (Brown *et al.*, 1982; Islam and Neidle, 1983), which also indicated that the puckering of the cyclohexene ring A of the chromophore was an important factor. Molecular modelling using physical models and very limited qualitative data from fibre diffraction studies resulted in the first model for a daunomycin–DNA complex (Pigram *et al.*, 1972). This had the drug intercalated from the major groove direction with the chromophore aligned approximately parallel with the long axis of base pairs. The amino-sugar group was aligned in the major groove with potential electrostatic interaction between the amino group and a phosphate oxygen atom. This interaction was apparently in accord with solution DNA-binding and activity data which demonstrated the importance of the presence of the amino group, which is protonated at physiological pH (Zunino *et al.*, 1972, 1976). NMR studies of daunomycin with various oligonucleotides have provided important information regarding the geometry of these intercalation complexes (Patel and Canuel, 1978; Phillips and Roberts, 1980), even though the lack of non-exchangeable protons on the chromophore other than at its extremities has produced some ambiguity in probing the geometry of the complexes by chemical shift methods. The exchange rate of the phenolic protons was reduced on intercalation, whereas the chemical shift of ring D protons was barely changed. These results are not consistent with the major groove, parallel-binding model, but are in excellent agreement with the X-ray structure of the daunomycin–hexanucleotide complex (Quigley *et al.*, 1980; Wang *et al.*, 1987). A pioneering molecular-modelling study of the interaction of doxorubicin with dinucleoside duplexes (Nakata and Hopfinger, 1980) independently concluded on the basis of intermolecular energy calculations, that the minor groove is preferred, at least for some sequences.

The question of the sequence selectivity of the anthracyclines has been addressed by several authors, not least because solution studies on bulk DNA and polynucleotides have not provided a consistent set of conclusions (surveyed by Neidle and Abraham, 1984) . A study of daunomycin intercalation into tetranucleotide duplexes, with a simplified intermolecular force field (Newlin *et al.*, 1984) found a preference for TA base pairs at the binding site. This is a surprising result for any intercalator on account of the known greater polarizability of GC base pairs, but could conceivably be rationalized on the basis of the particular overlap geometries at the binding site. Thus, use of a force field that incorporates full geometry optimization would provide further insight into the question. The perennial problem of definition of oligonucleotide intercalation geometry in relation to sequence specificity has been circumvented by an analysis that has taken the crystal structure results for the d(CGTACG) hexanucleotide duplex as a starting point to produce a number of hexamer sequences, all with the intercalation site at the ends (Chen *et al.*, 1985, 1986a).

The force field (equation 2) outlined in Section 6.3, which does not incorporate optimization of oligonucleotide conformation, was used with both daunomycin and doxorubicin as intercalants, although stronger binding was found for the alternating sequence $(AT)_3$ than for $(GC)_3$, in accord with some solution data on synthetic polynucleotide complexes (Graves and Krugh, 1983). However, and more significantly, the most stable complexes were computed to be mixed-sequence ones with the drugs intercalated between GC sites, in accord with the preference found in the crystal. These preferences are unsurprisingly a result of a number of intermolecular factors in addition to chromophore–base-pair interactions, as suggested by the crystallographic results themselves. The modelling analysis furthermore predicted that an adenosine residue 5′ to the intercalation site would be involved in the sequence specificity i.e. that the triplet sequence ACG would be the most preferred for these drugs—the recent footprinting analyses of daunomycin (Chaires *et al.*, 1987, 1990) have fully confirmed these predictions. This study has been taken a stage further by the analysis of relative binding energies for daunomycin and several synthetic analogues (Chen *et al.*, 1986a, b) using the two CG intercalation sites in the hexamer sequence CGTACG. Experimental equilibrium binding constants and *in vitro* activities have been previously determined by other workers for some of these. This study is therefore an interesting test of the extent to which empirical force-field calculations can be of use in intercalative drug design. Table 6.2 details the results from this study. The binding constants for the known compounds have been measured with calf thymus random-sequence DNA, so comparison with calculated energies solely at CG sites is strictly not formally valid. Although daunomycin itself has a CG site preference, no sequence preference data are available for the derivatives and one must assume that the binding constants represent averages over all sites. The energies in Table 6.2 are relative to that for daunomycin, which has been set at 0.0. The table shows a very wide variation in energies, with the extreme being a difference of over 60 kcal/mol. Such a difference is physically unreasonable, and illustrates the lack of quantitative compatibility between such calculated energies, and experimentally derived thermodynamic parameters, largely because of solvent and ionic shielding

Table 6.2. Comparison of relative energies (as the sum of intermolecular and conformational contributions) with experimental binding affinities

	DM	**1**	**2**	**3**	**4**	**5**	**6**	**7**
E(kcal/mol)	0.0	−1.5	25.9	48.9	60.9	20.8	−18.4	−5.3
$K_A(\mathrm{M}^{-1})$		3.3×10^6	4.4×10^6			7.1×10^3		
$IC_{50}(\mu\mathrm{M})$		18	18	38	50	≥ 100		

Key: DM, daunomycin; **1**, 4-demethoxy-DM; **2**, β-anomer of 4-demethoxy-DM; **3**, 4-demethoxy-7,9-bis-epi-DM; **4**, β-anomer of 4-demethoxy-7,9-bis-epi-DM; **5**, glucosamine-DM; **6**, methylamine-DM; **7**, ethylamine-DM.

factors. Nonetheless, it is encouraging that the *trends* in calculated energies are paralleled by the experimental data, and especially by the *in vitro* cytotoxicity results. These provide further evidence of the relationship between DNA binding and biological activity, and explain why stereochemical changes such as epimerization at various asymmetric centres of daunomycin results in dramatic decreases in activity. The slightly enhanced interaction energy of the 4-demethoxy derivative **1** compared to daunomycin itself is primarily due to the enhancement of stacking interactions possible in the absence of the methoxy group. Compounds **6** and **7** are predicted rather than actual analogues. The suggestion is that the presence of an extra one- or two-carbon link between the daunosamine sugar ring and the charged amine group would confer additional stabilization by means of new hydrogen bonds and electrostatic interactions compared to daunomycin itself. A similar derivative was suggested some years previous to this, albeit on the basis of the binding model with the sugar group being in the major groove (Neidle, 1977).

There has been a continuing search for new anthracycline analogues with equal or superior efficacy to daunomycin and doxorubicin, yet with reduced cardiotoxicity—this toxic effect is particular to these compounds and may be related to their redox potentials (Brown and Iman, 1984). The realization that structurally simpler anthraquinones similarly show DNA-intercalative properties has led to the development (Murdock *et al.*, 1979) of the compound mitozantrone (Figure 6.11), which has promising activity against, in particular, breast tumours (Stuart-Harris and Smith 1982). As yet, there is no X-ray structure of a mitozantrone–oligonucleotide complex. NMR data supports an intercalation model for the drug (Lown and Hanstock, 1985; Kotovych *et al.*, 1986) with the side chains situated in the major groove. A molecular modelling study of four anthraquinones (Figure 6.12) with alkylaminoalkyl side chains at different positions on the chromophore (Islam *et al.*, 1983) has shown that the 1,4-disubstituted compound, a close analogue of mitozantrone, has a low-energy conformation with major groove binding. This study used interactive computer graphics to locate low-energy conformations for the side chains, followed by use of a simplified force field to rank the interaction energies of the different anthraquinones (Islam *et al.*, 1983). Parallel solution DNA-binding studies and *in vitro* cytotoxicity assays have given results in overall agreement with the

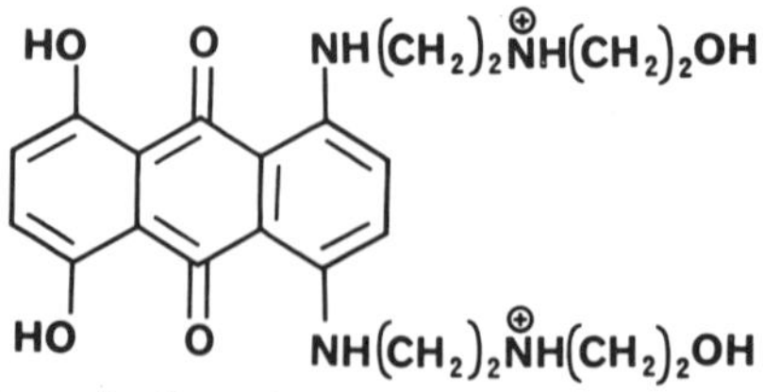

Figure 6.11. The structure of mitozantrone

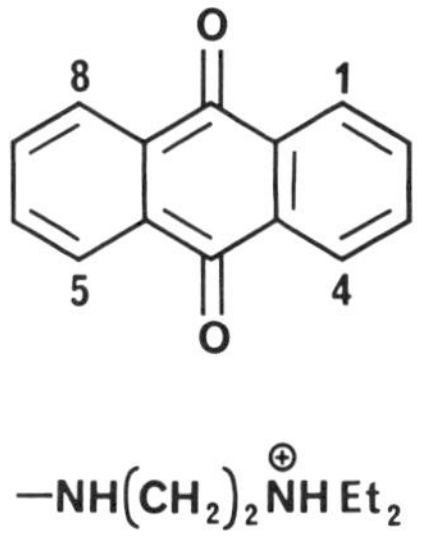

Figure 6.12. The structures of the substituted anthraquinones. The aminoalkylamino group is shown. This is variously substituted or disubstituted in the four compounds

calculated ranking order of interaction energies. A detailed analysis of the relative roles of side chain and chromophore has been made (Islam *et al.*, 1983). This shows that the role of the former is to provide attractive non-bonded interactions in a non-specific sense with the sides of the DNA base pairs that face into the grooves. These interactions are enhanced when the side chain carries a positive charge. Furthermore, the nature, length and substitution pattern of this side chain are constitutive factors in the overall enhancement of DNA-binding ability (Brown and Neidle, 1989). The modelling showed that these compounds do not invariably have a single mode of intercalative binding with, for example, both major and minor grooves being accessible to the 1-substituted compound (Table 6.3, Figure 6.13).

The modelling showed that the possession of the group —NH—(CH_2)N— is critical for optimal attractive non-bonded and electrostatic interactions with the sides of the base pairs that form the walls of both major and minor grooves. The distinct geometries predicted for the complexes are controlled by the positions of the side chains. Thus, the 1,8 compound can only intercalate from the major groove direction, since steric hindrance excludes the minor groove. The 1,5

Table 6.3. Calculated binding energies and binding modes for some substituted anthraquinones

Substitution	E_{total}	Binding mode
1	−310.2	M
1	−312.1	m
1	−305.0	P
1,4	−345.5	P
1,4	−373.4	S
1,5	−353.6	S
1,8	−338.2	M

M, major groove; P, perpendicular mode; m, minor groove; S, straddle mode.

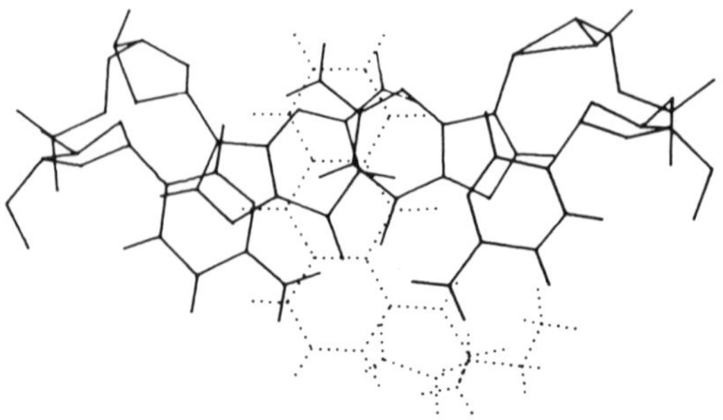

Figure 6.13. A binding mode for the 1-substitued anthraquinone

compound can only form an intercalation complex with one side chain situated in each groove (Figure 6.14). This non-classical, straddling intercalation mode implies that the drug cannot be straightforwardly inserted into the intercalation site, since the minimal van der Waals thickness of the side chains is greater than the maximum possible base-pair opening, even with staggering of backbone conformations. Since both the 1,5 and the 1,4 compound (which can also conceivably access this binding mode) have been found to intercalate into DNA in solution, this 'straddle' binding mode can only occur by a non-classical mechanism that involves binding to an initially non-base-paired sequence. Such a model predicts a preference for AT-rich sequences as well as slow dissociation kinetics from polynucleotides. This latter prediction has been verified experi-

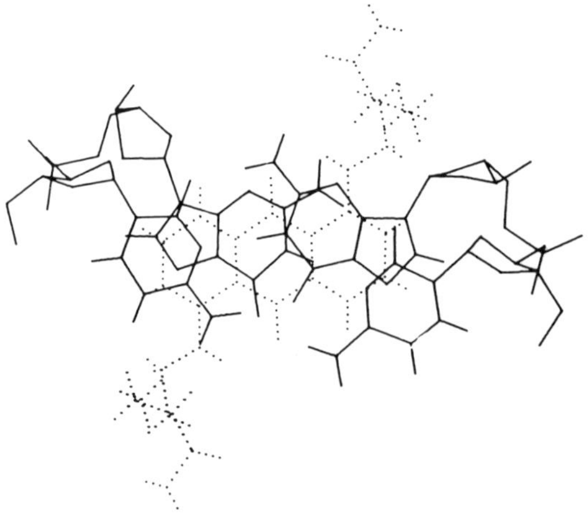

Figure 6.14. The straddle intercalation mode of the 1,5-anthraquinone

mentally (Gandecha *et al.*, 1985). A series of 1- and 1,4-substituted amidoanthraquinones has been developed with the aim of further evaluating the relationships between predicted (by molecular graphics and approximate force-field calculations) and observed relative DNA affinities, and *in vitro* and *in vivo* anti-tumour properties (Collier and Neidle, 1988). For this series at least, where DNA affinity was controlled by specific hydrogen bonds between hydroxyl and amine groups at the termini of side chains and groups on DNA, consistently significant correlations were found, after due allowance was made for factors such as cell wall transport and possible differences in metabolism. A further series of 2,6-disubstituted aminoanthraquinones has been designed which also adopts the straddling mode of intercalation (Agbandje *et al.*, 1992). Molecular modelling studies on the anthracycline antibiotic nogalamycin (Figure 6.15) have been reported (Collier *et al.*, 1984). This drug has two bulky sugar groups, one at each end of the molecule. Its crystal structure shows that it adopts a rigid saucer-like overall shape (Arora, 1983). It therefore resembles the 1,5-disubstituted anthraquinones described above in not being able to straightforwardly intercalate into DNA by virtue of the size of its substituents. The molecular models conceived show (Figure 6.16, colour section) one sugar in each groove; on the basis of crude energy calculations there does not appear to be a significant difference between 180°-rotated positions for the bound drug. The orientation of the nogalamycin chromophore with respect to the base pairs is similar to that found in the daunomycin-hexanucleotide crystal structure (Quigley *et al.*, 1980). This model for nogalamycin intercalation, with its implications for mechanism and kinetics, is in accord with subsequent kinetic and footprinting studies (Fox and Waring, 1984, 1986a, b), and most recently, crystal-structure analyses of complexes with hexanucleoside duplexes (Liaw *et al.*, 1989; Williams *et al.*, 1990a; Egli *et al.*, 1991).

The intercalative binding of the anthraquinone mitozantrone (Figure 6.11) has itself been the subject of several modelling studies. The most extensive (Balaji *et al.*, 1985) is from Lederle Laboratories, where the drug was discovered. This used a tetranucleotide duplex as an intercalation model, with a full geometry force field for energy minimization. Although the full conformational

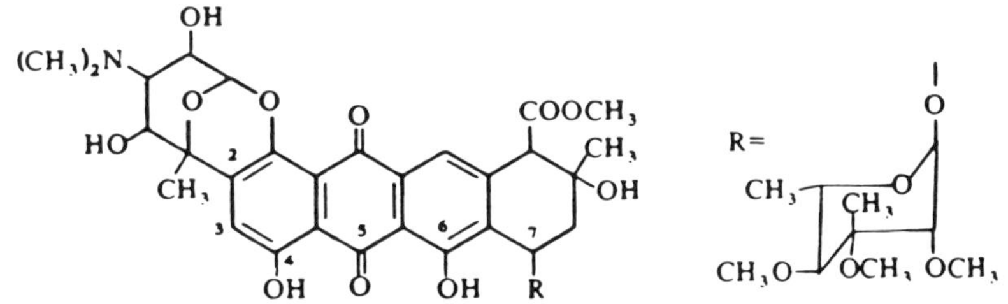

Figure 6.15. The structure of nogalamycin

details of the models have not been published, they appear to have stereochemical plausibility. Different sequences were examined by means of appropriate base substitutions, although subsequent minimizations were not performed —thus, relative energies of the various altered sequences may not be reliable. The problem of tackling the large number of conformational possibilities for the flexible side chains was circumvented by using computer graphics, including representations of electrostatic potential, as filtering procedures. Almost all of the low-energy structures that were found have minor-groove intercalation of the drug, closely analogous to that found for daunomycin in the crystal structure of the hexamer complex. This does not agree with the modelling study of the closely-analogous 1,4-disubstituted anthraquinones (Islam *et al.*, 1985) or with a subsequent analysis of mitozantrone intercalated into tetranucleotide sequences (Chen *et al.*, 1986c). It may be that the precise conformational characteristics of the various intercalation models used in these three investigations are actually of importance in defining relative stabilities of bound complexes; this further underlines the importance and necessity of obtaining experimental structural data so as to unequivocally define global energy mimima for oligonucleotides. This study by Chen *et al.* has also shown that mitozantrone intercalation is preferred at C-3′,5′-G sites, in accord with a footprinting analysis (Fox *et al.*, 1986) of mitozantrone and other substituted anthraquinones; the major-groove positional preferences agree with NMR assignments (Lown and Hanstock, 1985; Kotovych *et al.*, 1986).

6.8 CONCLUDING REMARKS

The overall picture that emerges from the studies of drug-DNA intercalation modelling is that theory can accord very well with *in vitro* experiment in terms of describing the binding behaviour of these drugs, especially when considering variations within a series of components. Relationships with *in vivo* activity are less clearly defined. Further advances in modelling will come from continuing improvements in force field parameterizations and of methodology for computationally cheap predictions of solvent arrangements, which will result in greater use of molecular dynamics methods both to produce thermodynamically relevant free energies and entropies, and to access conformationally complex problems. Advances in our understanding of the basic molecular mechanisms whereby intercalating drugs exert their action, beyond the simplistic picture used until very recently, will undoubtedly provide further layers of complexity for molecular modelling to tackle. It is to be hoped that relevant X-ray crystallographic results will continue to be obtained, as they provide the single richest source of raw data as starting points for molecular modelling in this field.

REFERENCES

Abraham, Z. H. L., Agbandje, M., Neidle, S. and Acheson, R. M. (1988). *J. Biomol. Struct. Dyn.*, **6**, 471-88.
Abraham, Z. H. L., Cutbush, S. D., Kuroda, R., Neidle, S., Acheson, R. M. and Taylor, G. N. (1985). *J. Chem. Soc., Perkin Trans 2*, 461-5.
Agbandje, M., Jenkins, T. C., McKenna, R., Reszka, A. P. and Neidle, S. (1992). *J. Med. Chem.*, **35**, 1418-1429.
Aggarwal, A., and Neidle, S. (1985). *Nucleic Acids Res.*, **13**, 5671-84.
Alden, C. J. and Arnott, S. (1975). *Nucleic Acids Res.*, **2**, 1701-18.
Alden, C. J., and Arnott, S. (1977). *Nucleic Acids Res.*, **4**, 3855-61.
Arcamone, F. (1981). *Doxorubicin Anti Cancer Antibiotics*, Academic Press, New York.
Arnott, S. (1981). In *Topics in Nucleic Acid Structure* (Ed. S. Neidle, Macmillan Press, London, pp. 65-82.
Arnott, S., Chadrasekaran, R., Hall, I. H., Puigjaner, L. C., Walker, J. K. and Want, M. (1982). *Cold Spring Harbour Symp. Quant. Biol.*, **47**, 53-65.
Arnott, S., Chandrasekaran, R., Banerjee, A. K., He, R. and Walker, J. K. (1983a). *J. Biomol. Struct. Dyn.*, **1**, 437-52.
Arnott, S., Chandrasekaran, R., Hall, I. H. and Puigjaner, L. C. (1983b). *Nucleic Acids Res.*, **11**, 4141-55.
Arora, S. K. (1983). *J. Am. Chem. Soc.*, **105**, 1328-32.
Baguley, B. C. (1991). *Anti-Cancer Drug Design*, **6**, 1-35.
Baguley, B. C., Denny, W. A., Atwell, G. J. and Cain, B. F. (1981). *J. Med. Chem.*, **24**, 170-7.
Balaji, V. N., Dixon, J. S., Smith, D. H., Venkataraghavan, R. and Murdock, K. C. (1985). *Ann. New York Acad. Sci.*, **439**, 140-61.
Barbacid, M. (1986). *Important Avenues in Oncology 1986*, (Eds V. DeVita, K. Hellman and M. Rosenberg), J. B. Lippincott, Philadelphia.
Bash, P. A., Singh, U. C., Langridge, R., Kollman, P. A. (1987). *Science*, **236**, 564-8.
Berman, H. M. and Neidle, S. (1979). In *Nucleic Acid Geometry and Dynamics* (Ed. R. H. Sarma), Pergamon Press, New York.
Berman, H. M. and Young, P. R. (1981). *Ann. Rev. Biophys.*, **10**, 87-114.
Berman, H. M., Neidle, S. and Stodola, R. K. (1978). *Proc. Natl. Acad. Sci. USA*, **75**, 828-32.
Bird, A. P. (1986). *Nature*, **321**, 209-13.
Boehncke, K., Nonella, M., Schulten, K. and Wang, A. H.-J. (1991). *Biochemistry*, **30**, 5465-75.
Brookes, B. R., Bruccoleri, R. E. Olafson, B. D., States, D. J., Swaminathan, S. and Karplus, M. (1983). *J. Comp. Chem.*, **4**, 187-217.
Brown, J. R. and Iman, S. H. (1984). *Prog. Med. Chem.*, **21**, 169-236.
Brown, J. R. and Neidle, S. (1989). In *Anthracycline and Anthracene-dione-based Anticancer Agents* (Ed. J. W. Lown), Elsevier Press, Amsterdam.
Brown, S. C., Kollman, P. A., and Weiner, P. K. (1982). *Biochem. Biophys. Acta*, **717**, 49-55.
Broyde, S. and Hingerty, B. (1979). *Biopolymers*, **25**, 249-66.
Calladine, C. R. (1982). *J. Mol. Biol.*, **161**, 343-52.
Chaires, J. B., Herrera, J. E. and Waring, M. J. (1990). *Biochemistry*, **29**, 6145-53.
Chaires, J. B., Fox, K. R., Herrera, J. E., Britt, M. and Waring, M. J. (1987). *Biochemistry*, **26**, 8227-36.
Chandrasekaran, R. and Arnott, S. (1989). In *Landolt-Börnstein*, Group II, Vol. 1b, Springer-Verlag, Berlin, pp. 31-170.

Chen, K. X., Gresh, N. and Pullman, B. (1985). *J. Biomol. Struct. Dyn.*, **3**, 445-65.
Chen, K. X., Gresh, N. and Pullman, B. (1986a). *Mol. Pharmacol.*, **30**, 279-86.
Chen, K. X., Gresh, N. and Pullman, B. (1986b). *Nucleic Acids Res.*, **14**, 2251-67.
Chen, K. X., Gresh, N. and Pullman, B. (1986c). *Nucleic Acids Res.*, **14**, 3799-813.
Chen, K. X., Gresh, N. and Pullman, B. (1987a). *Biopolymers*, **26**, 831-48.
Chen, K. X., Gresh, N. and Pullman, B. (1987b). *FEBS Lett.*, **224**, 361-4.
Cieplak, P., Rao, S. N., Grootenhuis, P. D. J. and Kollman, P. A. (1990). *Bipolymers*, **29**, 717-27.
Clark, G. R., Brown, D. G., Sanderson, M. R., Chwalinski, T., Neidle, S., Veal, J. M., Jones, R. L., Wilson, W. D., Zon, G., Garman, E. and Stuart, D. I. (1990). *Nucleic Acids Res.*, **18**, 5521-28.
Collier, D. A. and Neidle, S. (1988). *J. Med. Chem.*, **31**, 847-57.
Collier, D. A., Neidle, S. and Brown, J. R. (1984). *Biochem. Pharmacol.*, **33**, 2877-80.
Crick, F. H. C. and Watson, J. D. (1954). *Proc. R. Soc.*, **A223**, 80-96.
Crothers, D. M. (1986). *Biopolymers*, **6**, 575-84.
Dean, P. M. and Wakelin, L. P. G. (1979). *Phil. Trans. R. Soc. Lond.*, **287**, 571-607.
Dean, P. M., and Wakelin, L. P. G. (1980a). *Proc. R. Soc. Lond.*, **209**, 453-71.
Dean, P. M., and Wakelin, L. P. G. (1980b). *Proc. R. Soc. Lond.*, **209**, 473-87.
Dearing, A., Weiner, P. and Kollman, P. A. (1981). *Nucleic Acids Res.*, **9**, 1483-95.
Denny, W. A., Baguley, B. C., Cain, B. F. and Waring, M. J. (1983). In *Molecular Aspects of Anticancer Drug Action* (Eds S. Neidle, and M. J. Waring), Macmillan Press Ltd, London.
Denny, W. A., Twigden, S. J. and Baguley, B. C. (1986). *Anti-Cancer Drug Design*, **1**, 125-32.
Dervan, P. B. (1986). *Science*, **232**, 464-71.
Dickerson, R. E. and Drew, H. R. (1981a). *J. Mol. Biol.*, **149**, 761-86.
Dickerson, R. E. and Drew, H. R. (1981b). *Proc. Natl. Acad. Sci. USA*, **78**, 7318-22.
Drew, H. R. and Dickerson, R. E. (1981). *J. Mol. Biol.*, **151**, 535-56.
Drew, H. R. and Travers, A. A. (1985). *Nucleic Acids Res.*, **13**, 4445-67.
Egli, M., Williams, L. D., Frederick, C. A. and Rich, A. (1991). *Biochemistry*, **30**, 1364-72.
Feigon, J., Denny, W. A., Leupin, W. and Kearns, D R. (1984). *J. Med. Chem.*, **27**, 450-65.
Fox, K. R. and Waring, M. J. (1984). *Biochim. Biophys. Acta*, **802**, 162-8.
Fox, K. R. and Waring, M. J. (1986a). *Biochemistry*, **25**, 4349-56.
Fox, K. R. and Waring, M. J. (1986b). *Nucleic Acids Res.*, **14**, 2001-14.
Fox, K. R., Waring, M. J., Brown, J. R. and Neidle, S. (1986). *FEBS Lett.*, **202**, 289-94.
Frederick, C. A., Williams, L. D., Ughetto, G., van der Marel, G. A., van Boom, J. H., Rich, A. and Wang, A. H.-J. (1990). *Biochemistry*, **29**, 2538-49.
Fuller, W. and Waring, M. J. (1964). *Ber. Bunsenges. Physik. Chem.*, **68**, 805-8.
Gago, F., Reynolds, C. A. and Richards, W. G. (1989). *Mol. Pharmacol.*, **35**, 232-41.
Gale, E. F., Cundliffe, E., Reynolds, P. D., Richmond, M. H. and Waring, M. J. (1981). *The Molecular Basis of Antibiotic Action*, 2nd edn., John Wiley, London.
Gandecha, B. M., Brown, J. R. and Crampton, M. R. (1985). *Biochem. Pharmacol.*, **34**, 733-36.
Gao, X. and Patel, D. J. (1988). *Biochemistry*, **27**, 1744-51.
Gao, Y-G., Liaw, Y. C., Robinson, H. and Wang, A. H.-J. (1990). *Biochemistry*, **29**, 10307-316.
Gottesfeld, J. M., Bianco, J. and Tennant, L. L. (1987). *Nature*, **329**, 460-5.
Graves, D. E. and Krugh, T. R. (1983). *Biochemistry*, **22**, 3941-7.
Heinemann, U., Lauble, H., Frank, R. and Blocker, H. (1987). *Nucleic Acids Res.*, **15**, 9531-51.

Herzyk, P., Goodfellow, J. M. and Neidle, S. (1991). *J. Biomol. Struct. Dyn.*, **9**, 363-386.
Herzyk, P., Neidle, S. and Goodfellow, J. M. (1992). *J. Biomol. Struct. Dyn.*, in press.
Holbrook, S R. and Kim, S. H. (1979). *Int. J. Biolog. Macromol.*, **1**, 233-40.
Hudson, B. D., Kuroda, R., Denny, W. A. and Neidle, S. (1987). *J. Biomol. Struct. Dyn.*, **5**, 145-58.
Islam, S. A. and Neidle, S. (1983). *Acta. Crystallogr.*, **B39**, 114-19.
Islam, S. A. and Neidle, S. (1984). *Acta. Crystallogr.*, **B40**, 424-9.
Islam, S. A., Neidle, S., Gandecha, M. and Brown, J. R. (1983). *Biochem. Pharmacol.*, **32**, 2801-8.
Islam, S. A., Neidle, S., Gandecha, M., Partridge, M., Patterson, H. and Brown, J. R. (1985). *J. Med. Chem.*, **28**, 857-64.
Jain, S. C., Tsai, C. C. and Sobell, H. M. (1977). *J. Mol. Biol.*, **114**, 317-31.
Kohn, K. W., Hartley, J. A. and Mattes, W. B. (1987a). *Nucleic Acids Res.*, **15**, 10531-49.
Kohn, K. W., Pommier, Y., Kerrigan, D., Markovits, J. and Covey, J. M. (1987b). *NCI Monogr.*, **4**, 61-71.
Koo, H. S., Wu, H.-M. and Crothers, D. M. (1986). *Nature*, **320**, 501-6.
Kotovych, G., Lown, J. W. and Tong, J. P. K. (1986). *J. Biomol. Struct. Dyn.*, **4**, 111-24.
Lerman, L. S. (1961). *J. Mol. Biol.*, **3**, 18-30.
Lerman, L. S. (1963). *Proc. Natl. Acad. Sci. USA*, **49**, 94-102.
Levitt, M. (1983). *Cold Spring Harbor Symp. Quant. Biol.*, **347**, 271-5.
Liaw, Y.-C., Gao, Y.-G., Robinson, H., van der Marel, G. A., van Boom, J. H. and Wang, A. H.-J. (1989). *Biochemistry*, **28**, 9913-18.
Lock, R. B. and Ross, W. E. (1987). *Anti-Cancer Drug Design*, **2**, 151-64.
Low, C. M. L., Drew, H. R. and Waring, M. J. (1984). *Nucleic Acids Res.*, **12**, 4865-79.
Lown, J. W. and Hanstock, C. C. (1985). *J. Biomol. Struct. Dyn.*, **2**, 1097-106.
Lybrand, T. and Kollman, P. (1985). *Biopolymers*, **24**, 1863-79.
Lybrand, T. P., Brown, S. C. Creighton, S., Shafer, R. H. and Kollman, P. A. (1986). *J. Mol. Biol.*, **191**, 495-507.
Mahendrasingam, A., Forsyth, V. T., Hussain, R., Greenall, R. J. Pigram, W. J. and Fuller, W. (1986). *Science*, **233**, 195-7.
Malhotra, D. and Hopfinger A. J. (1980). *Nucleic Acids Res.*, **8**, 520-30.
McCammon, J. A. (1987). *Science*, **238**, 486-91.
McCammon, J. A. and Harvey, S. C. (1987). *Dynamics of Proteins and Nucleic Acids*. Cambridge University Press, Cambridge, UK.
McKenna, R. Beveridge, A. J., Jenkins, T. C., Neidle, S. and Denny, W. A. (1989). *Mol. Pharmacol.*, **35**, 720-8.
McLean, M. J. Seela, F. and Waring, M. J. (1989). *Proc. Natl. Acad. Sci. USA*, **86**, 9687-91.
Mezei, M., Beveridge, D. L., Berman, H. M., Goodfellow, J. M., Finney, J. L. and Neidle, S. (1983). *J. Biomol. Struct. Dyn.*, **1**, 287-97.
Miller, K. J. (1981). *Convers. Biomol. Stereodyn.*, **2**, 469.
Miller, K. J. and Pycior, J. F. (1979). *Biopolymers*, **18**, 2683-19.
Minford, J., Pommier, Y., Filipiski, J., Kohn, K. W., Kerrigan, D., Mattern, M., Michaels, S., Schwartz, R. and Zwelling, L. A. (1986). *Biochemistry*, **25**, 9-16.
Moore, M. H., Hunter, W. N., Langlois d'Estaintot, B. and Kennard, O. (1989). *J. Mol. Biol.*, **206**, 693-705.
Murdock, K. C., Child, R. G., Fabio, P. F., Augier, P. F., Wallace, R. E., Durr, F. E. and Citerella, R. V. (1979). *J. Med. Chem.*, **22**, 1024-30.
Nakata, Y. and Hopfinger, A. J. (1980). *Biochem. Biophys. Res. Commun.*, **95**, 583-8.
Neidle, S. (1977). *Cancer Treat. Rep.*, **61**, 928-30.

Neidle, S. (1979). *Prog. Med. Chem.*, **16**, 151–221.
Neidle, S. (1989). In *Landolt-Bornstein*, Group VII, Vol. 1b (Ed. W. Saenger), Springer-Verlag, Berlin, pp. 247–76.
Neidle, S. and Abraham, Z. H. L. (1984). *CRC Crit. Rev. Biochem.*, **17**, 73–121.
Neidle, S. and Jenkins, T. C. (1991). *Methods in Enzymology.*, **203**, 433–58.
Neidle, S. and Taylor, G. (1977). *Biochem. Biophys. Acta*, **479**, 450–9.
Neidle, S., Achari, A., Taylor, G. L., Berman, H. M., Carrell, H. L., Glusker, J. P. and Stallings, W. C. (1977). *Nature*, **269**, 304–7.
Neidle, S., Berman, H. M. and Shieh, H.-S. (1980). *Nature*, **285**, 129–34.
Neidle, S. and Waring, M. J. (eds) (1983). *The Molecular Basis of Antibiotic Action*, Macmillan Press, London.
Neidle, S., Abraham, Z. H. L., Collier, D. A. and Islam, S. A. (1987a). In *New Avenues in Developmental Cancer Chemotherapy* (Eds K. R. Harrap and T. A. Connors), Academic Press, Orlando, Florida, pp. 83–109.
Neidle, S., Pearl, L. H. and Herzyk, P. (1988). *Nucleic Acids Res.*, **16**, 8999-9016.
Neidle, S., Pearl, L. H. and Skelly, J. V. (1987b). *Biochem. J.*, **243**, 1–13.
Nelson, E. M., Tewey, K.M., Liu, L. F. (1984). *Proc. Natl. Acad. Sci. USA*, **81**, 1361–5.
Nelson, H. C. M., Finch, J. T., Luisi, B. F. and Klug, A. (1987). *Nature*, **330**, 221–3.
Newlin, D. D., Miller, K. J. and Pilch, D. F. (1984). *Biopolymers*, **23**, 139–58.
Nuss, M. E., Marsh, F. J. and Kollman, P A. (1979). *J. Am. Chem. Soc.*, **101**, 825–33.
Ohlendorf, D. H., Anderson, W. F., Fisher, R. G., Takeda, Y. and Matthews, B. W. (1982). *Nature*, **298**, 718–23.
Olson, W. K. (1982). In *Topics in Nucleic Acid Structure*, **2**, (Eds. S. Neidle), Macmillan Press, London.
Orozco, M., Laughton, C. A., Herzyk, P. and Neidle, S. (1990). *J. Biomol. Struct. Dyn.*, **8**, 359–73.
Ornstein, R. L. and Rein, R. (1979). *Biopolymers*, **18**, 1277–91.
Pai, E. F. (1991). *Current Opinion in Structural Biology*, **1**, 941–45.
Park, H.-S, Arnott, S., Chandrasekaran, R., Millane, R. P. and Campagnari, F. (1987). *J. Mol. Biol.*, **197**, 513–23.
Patel, D. J. (1979). *Acc. Chem. Res.*, **12**, 118–25.
Patel, D. J. and Canuel, L. L. (1977). *Proc. Natl. Acad. Sci. USA*, **74**, 2624–8.
Patel, D. J. and Canuel, L. L. (1978). *Eur. J. Biochem.*, **90**, 247–54.
Patel, D. J., Kozlowski, S. and Rice, J. (1981). *Proc. Natl. Acad. Sci. USA*, **78**, 3333–7.
Pearl, L. and Neidle, S. (1986). *FEBS Lett.*, **209**, 269–76.
Phillips, D. R. and Roberts, G. C. K. (1980). *Biochemistry*, **19**, 4795–801.
Pigram, W. J., Fuller, W. and Hamilton, L. H. (1972). *Nature*, **235**, 17–19.
Prabhakaran, M. and Harvey, S. C. (1985). *J. Phys. Chem.*, **89**, 5767–9.
Prabhakaran, M. and Harvey, S. C. (1988). *Biopolymers*, **27**, 1239–48.
Prive, G. G., Heinemann, U., Chandrasegaran, S., Kan, L.-S., Kopka, M. L. and Dickerson, R. E. (1987). *Science*, **238**, 498–504.
Pullman, A., and Pullman, B. (1981). *Q. Rev. Biophys.*, **14**, 289–380.
Pullman, B. (1989). *Adv. Drug Res.*, **18**, 1–113.
Quigley, G. J., Wang, A. H.-J., Ughetto, G., van der Marel, G., van Boom, J. H. and Rich, A. (1980). *Proc. Natl. Acad. Sci. USA*, **77**, 7204–8.
Quigley, G. J., Ughetto, G., Marel, J. H., van Boom, A. H.-J. and Rich, A. (1986). *Science*, **232**, 1255–58.
Rao, N. and Kollman, P. A. (1987). *Proc. Natl. Acad. Sci. USA*, **84**, 5735–9.
Rao, N. S., Singh, C. U. and Kollman, P. A. (1986). *J. Med. Chem.*, **29**, 2484–92.
Reinhardt, C. G. and Krugh, T. R. (1978). *Biochemistry*, **17**, 4845–54.

Remers, W. A., Mabilia, M. and Hopfinger, A. J. (1986). *J. Med. Chem.*, **29**, 2492–503.
Rewcastle, G. W., Atwell, G. J., Chambers, D., Baguley, B. C. and Denny, W. A. (1986). *J. Med. Chem.*, **29**, 472–7.
Saenger, W. (1984). *Principles of Nucleic Acid Structure*, Springer-Verlag, Berlin.
Shakked, Z., Rabinovich, D., Kennard, O., Cruse, W. B. T., Salisbury, S. A. and Viswamitra, M. A. (1983). *J. Mol. Biol.*, **166**, 183–201.
Shieh, H.-S., Berman, H.-M., Dabrow, M. and Neidle, S. (1980). *Nucleic Acids Res*, **8**, 85–97.
Sigal, I. S., Smith, G. M., Jumak, F., Marsico-Ahern, J. D., Dalonzo, J. S., Scolnick, E. M. and Gibbs, J. B. (1987). *Anti-Cancer Drug Design*, **2**, 107–15.
Singleton, C. K. (1987). In *Topics in Nucleic Acid Structure*, **3** (Ed. S. Neidle), Macmillan Press, London, pp. 163–91.
Sobell, H. M., Tsai, C.-C., Jain, S. C., and Gilbert, S. G. (1977). *J. Mol. Biol.*, **114**, 333–65.
Srinivasan, A. R., Torres, R., Clark, W. and Olson, W. K. (1987). *J. Biomol. Struct. Dyn.*, **5**, 459–96.
Stuart-Harris, R. C. and Smith, I. E. (1982). *Cancer Chemother. Pharmacol.*, **8**, 179–82.
Swaminathan, S., Beveridge, D. L. and Berman, H. M. (1990). *J. Phys. Chem.*, **94**, 4660–665.
Swaminathan, S., Ravishanker, G. and Beveridge, D. L. (1991). *J. Am. Chem. Soc.*, **113**, 5027–40.
Takusagawa, F. Dabrow, M., Neidle, S. and Berman, H. M. (1982). *Nature*, **296**, 466–9.
Tan, K. B., Mattern, M. R., Boyce, R. A. and Schein, P. S. (1987). *Proc. Natl. Acad. Sci. USA*, **84**, 7668–71.
Taylor, E. R. and Olson, W. K. (1983). *Biopolymers*, **22**, 2667–702.
Tilton, R. F., Weiner, P. K. and Kollman, P. A. (1983). *Biopolymers*, **22**, 969–1002.
Tsai, C. C., Jain, S. C. and Sobell, H. M. (1977). *J. Mol. Biol.*, **144**, 317–31.
Ughetto, G., Wang, A. H.-J., Quigley, G. J., van der Marel, G. A., Boom, J. H. and Rich, A. (1985). *Nucleic Acids Res.*, **13**, 2307–20.
Van Dyke, M. M. and Dervan, P. B. (1984). *Science*, **225**, 1122–7.
van Gunsteren, W. F. and Berendsen, H. J. C. (1986). *GROMOS86*: *Groningen Molecular Simulation System*, University of Groningen, The Netherlands.
Wakelin, L. P. G. (1986). *Med. Chem. Res. Rev.*, **6**, 275–340.
Wang, A. H.-J., Quigley, G. J., Kolpack, F. J., Crawford, J. L., van Boom, J. H., van der Marel, G. and Rich, A. (1979). *Nature*, **282**, 680–6.
Wang, A. H.-J., Ughetto, G., Quigley, G. J., Hakoshima, T., van der Marel, A., van Boom, J. H. and Rich, A. (1984). *Science*, **225**, 1115–20.
Wang, A. H.-J., Ughetto, G., Quigley, G. J. and Rich, A. (1987). *Biochemistry*, **26**, 1152–63.
Waring, M.J. (1981). *A. Rev. Biochem.*, **50**, 159–92.
Waring, M. J. and Fox, K. R. (1983). In *Molecular Aspects of Anti-Cancer Drug Action* (Eds S. Neidle and M. J. Waring), Macmillan Press, London, pp. 127–56.
Warpehoski M. A., McGovren, P., Mitchell, M. A. and Hurley, L. H. (1990). In *Molecular Basis of Specificity in Nucleic Acid-Drug Interactions* (Eds. B. Pullman and J. Jortner), Kluwer Academic Publishers, Dordrecht, pp. 531–50.
Webster, G. D., Sanderson, M. R., Skelly, J. V., Neidle, S., Swann, P. F., Li, B. F. and Tickle, I. J. (1990). *Proc. Natl. Acad. Sci. USA*, **87**, 6693–97.
Weiner, P. (1984). *J. Am. Chem. Soc.*, **106**, 765–84.
Weiner, S. J. and Kollman, P. A. (1986). *J. Comput. Chem.*, **7**, 230–52.
Weiner, S. J., Kollman, P. A., Case, D. A., Singh, U. C., Ghio, C. Alagona, G., Profeta, S. and Weiner, P. (1984). *J. Am. Chem. Soc.*, **106**, 765–84.

Weiner, P. K., Langridge, R., Blaney, J. M., Schaefer, R. and Kollman, P. A. (1982). *Proc. Natl. Acad. Sci. USA*, **79**, 3754–58.

Williams, L. D., Egli, M., Gao, Q., Bash, P., van der Marel, G. A., van Boom, J. H., Rich, A. and Frederick, C. A. (1990a). *Proc. Natl. Acad. Sci. USA*, **87**, 2225–9.

Williams, L. D., Frederick, C. A., Ughetto, G. and Rich, A. (1990b). *Nucleic Acids Res.*, **18**, 5533–41.

Wilson, W. D. and Jones, R. L. (1981). *Adv. Pharmacol. Chemother.*, **18**, 177–222.

Wilson, W. R., Baguley, B. C., Wakelin, L. P. G. and Waring, M. J. (1981). *Mol. Pharmacol.*, **20**, 404–14.

Wing, R. M., Drew, H. R., Takano, T., Broka, C., Tanaka, S., Itakura, K. and Dickerson, R. E. (1980). *Nature*, **287**, 755–8.

Zhang, R.-G., Joachimiak, A., Lawson, C. L., Schevitz, R. W., Otwinowski, Z. and Sigler, P. B. (1987). *Nature*, **327**, 591–5.

Zunino, F., Gambetta, R., DiMarco, A. and Zaccara, A. (1972). *Biochem. Biophys. Acta*, **277**, 489–98.

Zunino, F., Gambetta, R., DiMarco, A., Luoni, G. and Zaccara, A. (1976). *Biochem. Biophys. Res. Commun.*, **69**, 744–50.

7 Crystallographic Investigations of Glycolytic Enzymes from *Trypanosoma brucei*: Potential Starting Points for the Design of New Sleeping Sickness Drugs

WIM G. J. HOL
BIOSON Research Institute, University of Groningen, The Netherlands
AND RIK K. WIERENGA
EMBL, Heidelberg, Germany

This paper is dedicated to the memory of Alan S. Horn

The Design of Drugs to Macromolecular Targets. Edited by C. R. Beddell
© 1992 John Wiley & Sons Ltd

7.1 INTRODUCTION

Living organisms owe their unique characteristics to the properties of a large number of complex biomacromolecules. Among these, proteins occupy a special position because they carry out an incredible range of different and often crucial functions. Their importance is shown, for instance, by the fact that several diseases can be traced back to minute alterations in a single protein molecule. An example is the point mutation in sickle cell haemoglobin which leads to sickle cell anaemia (Perutz, 1981).

In addition to inherited defects, errors in the amino acid sequence of a protein can be caused by damage at the DNA level which occurs during the life of a human being. An example is the point mutation of the *ras* gene, leading to mutated p21 protein, which is closely associated with some forms of cancer (Reddy *et al.*, 1982; Tabin *et al.*, 1982). Human disease is thus often caused by errors in proteins themselves, either inherited or acquired. If drugs are to be useful in such cases they should counteract the effects of erroneous protein structures.

Many other diseases are, of course, not due to deficiencies of human proteins, but to invasions by harmful organisms. In such cases drugs have to interfere with crucial pathways of the pathogenic organism without, ideally, affecting the human host. Sufficient selectivity of the compounds used is the goal to be achieved (Albert, 1985).

This goal has been reached in many cases by approaches which consisted mainly of trial-and-error procedures no matter how sophisticated the methods employed were. The advent of a number of new techniques—ranging from recombinant DNA technology, rapid gene sequencing techniques, modern protein purification and protein microsequencing methods, oligonucleotide synthesis, X-ray crystallography, computer graphics, molecular modelling to molecular dynamics calculations—are making it now possible to try to design new drugs based upon three-dimensional structures of target proteins elucidated at the atomic level (see, e.g. Hol, 1986).

Of course, such a rational drug design approach has still several trial-and-error components because of:

(a) our incomplete understanding of intermolecular interactions which may, for example, result in unexpected modes of binding of newly synthesized compounds to the target protein;

(b) the potential drug may not reach the target protein *in vivo*;

(c) metabolic processes may transform the potential drug into an ineffective compound;

(d) the newly synthesized compound may be toxic, teratogenic, mutagenic or carcinogenic.

Therefore, it is useful to consider a rational drug design *cycle* (Figure 7.1) which leads to a gradual increase of understanding of the mode of binding of a range of compounds to the potential target protein. Once such a database of three-dimensional structures of potential drugs bound to target proteins, plus associated binding constants, is obtained, the efficiency of the drug design procedure should be significantly increased.

In this contribution we describe the status (as of mid-1990) of a collaborative project aimed at development of therapeutically useful compounds for the treatment of African trypanosomiasis, or sleeping sickness, on the basis of three-dimensional structures of key parasite enzymes.

7.2 SLEEPING SICKNESS, OR AFRICAN TRYPANOSOMIASIS

Sleeping sickness is a serious threat to the health of the people living in large parts of Africa. If untreated it is virtually always fatal. Treatment requires hospitalization, because the drugs currently in use are very toxic (Molyneux and Ashford, 1983). About 50 million human beings in 34 African countries are at risk of developing the disease (WHO, 1985), which is caused by a unicellular protozoan haemoflagellate, the trypanosome. Two subspecies of *Trypanosoma*

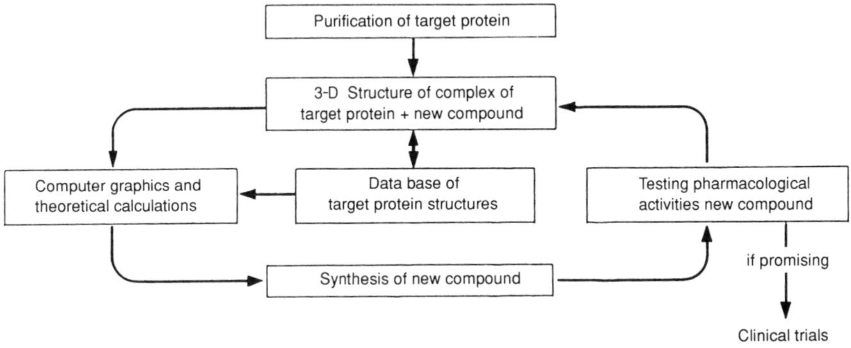

Figure 7.1. Schematic drawing of the 'rational drug design cycle' (Hol, 1986)

brucei are the causative agents: *T. b. gambiense* and *T. b. rhodesiense*. Trypanosomes also cause an infectious cattle disease ('nagana'). Of greatest importance in this respect are *T. vivax* and *T. congolense*. The various trypanosomal species belong to the family Trypanosomatidae, which is a subdivision of the Kinetoplastida.

As trypanosomes affect both the physical as well as the economical well-being of the inhabitants of tropical Africa, the World Health Organization (WHO) has included this disease in its special Program for Research and Training in Tropical Diseases (WHO, 1985).

Trypanosomes are transmitted by a limited number of *Glossina* species, insects commonly known as the tsetse fly. After a bite by an infected fly, the parasites move to the bloodstream of the host. *T. brucei* cells, quite remarkably, do not hide in any of the host blood cells. They are in this respect quite different from, for example, the malaria parasite which is found in the erythrocyte. The trypanosomes multiply in the blood by binary fission. Eventually, other organs become infected whereby, surprisingly, the organism is able to cross the blood–brain barrier. The invasion of the central nervous system causes the final symptoms of sleeping sickness: insomnia, epileptic attacks, coma and eventually death (Molyneux and Ashford, 1983).

In view of the shortcomings of the available drugs (Gutteridge, 1985), we have embarked on a project which aims at obtaining new drugs by exploiting peculiarities of the trypanosomal biochemical machinery.

7.3 TRYPANOSOMES

Trypanosomes have a quite complicated life cycle, both in the tsetse fly and in the host (Molyneux and Ashford, 1983). Major morphological changes occur during this cycle. In this chapter we will focus on the bloodstream form which is schematically depicted in Figure 7.2. This form has several unusual features which have recently been reviewed by Opperdoes (1985) and some of which will be briefly discussed in the next few sections.

7.3.1 KINETOPLAST DNA

This is a network of small and large circular DNA molecules in mitochondria. Each mitochondrion has one such network, which represents between 5 and 30% of total cellular DNA (Opperdoes, 1985). The 'mini-circles', $\sim 0.3\ \mu m$ in length, make up 90–95% of the network mass. The 'maxi-circles', $\sim 16\ \mu m$ in length, code for mitochondrial proteins. Some species, such as *T. brucei evansi* and *T. brucei equiperdum* have lost their maxi-circles, therefore they cannot develop a functional mitochondrial respiratory chain. Consequently they have lost the ability to develop in the insect host.

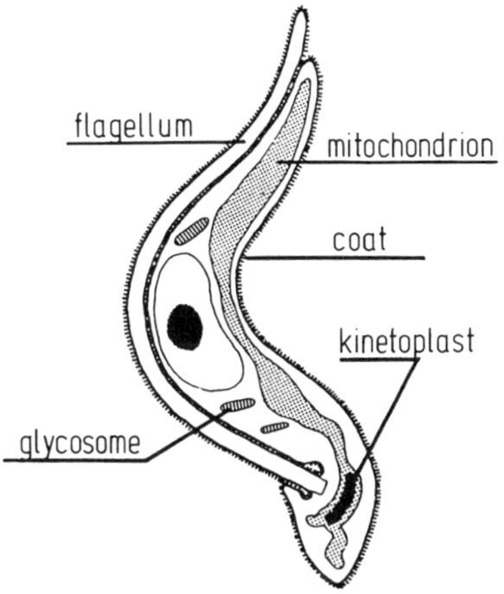

Figure 7.2. Schematic drawing of the cross-section of bloodstream form of the trypanosome which is approximately 10 μm long. Due to the flagellum, trypanosomes in the blood move continuously and rapidly. Three peculiar features of trypanosomes—the surface coat, the kinetoplast and the glycosome—are discussed more extensively in the text. The black dot represents nuclear DNA. The grey organelle running along the entire length of the trypanosome is the mitochrondrion, which is hardly developed in the bloodstream form of *Trypanosoma brucei.* Drawing kindly provided by Dr F. R. Opperdoes

Exciting recent discoveries have led to deeper insight as to the function of these circular DNA molecules. Some mitochondrial proteins of trypanosomes appear to be encoded on the DNA in a most unusual fashion. For example, thc gene for a cytochrome *c* oxidase subunit is missing numerous thymines—up to 50% (!)—when compared with the eventual amino acid sequence (Benne *et al.*, 1986; Feagin *et al.*, 1988; Shaw *et al.*, 1988). The gene is therefore called a 'cryptogene'. Extensive RNA editing occurs in the trypanosome which appears to involve the use of numerous short stretches of 'guide RNAs', encoded elsewhere on the DNA (Simpson and Shaw, 1989; Weiner and Maizels, 1990). In a fascinating editing process the missing bases are somehow inserted, and a few deleted, so that a proper mature RNA eventually arises.

Clearly, this process is distinctly different from that of mammalian protein biosynthesis. Since the mitochondrion is poorly developed in the bloodstream form of the parasite (Opperdoes, 1985), RNA-editing may occur for only a small number of proteins in the bloodstream form of the parasite. However, if such proteins turn out to be essential for the parasite's metabolism, these unique large scale editing processes would be intriguing targets for drug design. So far, the proteins and nucleoprotein complexes involved have not yet been characterized at the atomic level and, hence, will not receive further attention in this overview.

7.3.2 TRYPANOTHIONE AND POLYAMINE METABOLISM

In trypanosomes, the ratio of intracellular levels of reduced and oxidized thiols is maintained by a 'novel' enzyme, trypanothione reductase (Fairlamb *et al.*, 1985; Shames *et al.*, 1986). Trypanothione is a glutathione-spermidine conjugate (Figure 7.3), whose structure has been elucidated by Fairlamb *et al.* (1985). Trypanothione has been found in several parasitic Kinetoplastida, such as *Trypanosoma brucei*, *Trypanosoma cruzi*, *Crithidia fasciculata* and *Leishmania mexicana*. Trypanothione reductase of both *T. cruzi* (Krauth-Siegel *et al.*, 1987) and *Crithidia fasciculata* (Shames *et al.*, 1986) has been purified. It is a flavoenzyme which is homologous to glutathione reductase. The molecular weight of one subunit of trypanothione reductase, a dimeric protein, is approximately 50 000 daltons. It catalyses the NADPH-dependent reduction of oxidized trypanothione (Figure 7.3). The active site peptide containing the catalytically important cysteine residues is identical to the analogous peptide of human glutathione reductase. Inhibition of this unique trypanosomal enzyme will affect the ratio of the intracellular concentrations of reduced and oxidized thiols, which is expected to be deleterious to the trypanosome (Fairlamb *et al.*, 1985), because it will increase the sensitivity to reactive oxygen-containing molecules.

Recent developments are that this important trypanosomal protein from *Crithidia fasciculata*, an insect trypanosome without bloodstream form, has been overexpressed and crystallized. The structure determination of this enzyme is now underway (W. Hunter *et al.*, personal communication; J. Kuriyan, personal communication). Once high-resolution structures are available this enzyme will provide an exciting starting point for drug design, after molecular modelling procedures have created an approximate model of the true trypanothione reductase of *Trypanosoma brucei.*

The spermidine moiety of trypanothione is a polyamine. It is therefore

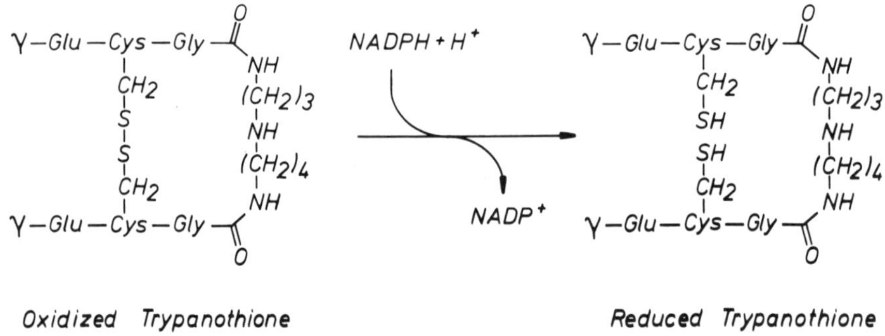

Figure 7.3. Reaction catalysed by trypanothione reductase. The oxidized form of trypanothione contains a 24-atom ring, which is formed by spermidine and by atoms of the glutathione moities

interesting that a compound which interferes with the polyamine metabolism turns out to be a new drug against sleeping sickness. This compound is DFMO, difluoromethyl ornithine, a suicidal inhibitor of ornithine decarboxylase (Bitonti *et al.*, 1985). This enzyme decarboxylates ornithine into putrescine ($H_2N—(CH_2)_4—NH_2$). Putrescine can subsequently be converted (Bacchi, 1981) by another enzyme into spermidine ($H_2N—(CH_2)_3—NH—(CH_2)_4—NH_2$) which is an essential component of trypanothione. DFMO is now in use (Sjoerdsma and Schechter, 1984) as a chemotherapeutic agent against sleeping sickness. Some disadvantages associated with the general use of this compound as a therapeutic agent are the large quantities needed to be effective, the occurrence of resistance and the relative insensitivity of *T. brucei rhodesiense*, the most virulent form of the parasite for DFMO (WHO, 1987).

7.3.3 VARIABLE SURFACE GLYCOPROTEINS

The bloodstream form of the trypanosome is continuously exposed to the immune system of the host. The surface of these trypanosomes is a dense coat of protein molecules that is 120–150 Å thick. It is a monolayer of identical copies of a particular coat protein, the variable surface glycoprotein (VSG), which has antigenic properties (Vickerman, 1987; Borst, 1986). The immune system successfully produces antibodies against this coat protein, but nevertheless the trypanosomal infection persists. This is due to the phenomenon of antigenic variation: regularly a new VSG is synthesized which eventually completely replaces the previous coat protein. Altogether there are perhaps more than 1000 different genes coding for VSGs. Only one VSG gene is being expressed at the same time, although at the time of switching two different VSGs, distributed uniformly over the cell surface, can be detected by monoclonal antibodies (Esser and Schoenbehler, 1985; Baltz *et al.*, 1986). The molecular mechanism of the expression of VSG genes has been described in a recent review (Pays and Steinert, 1988).

VSGs consist of two domains. Upon treatment by proteases the N-terminal domain is cleaved off and a soluble VSG fragment can be purified. These VSG fragments form dimers with subunits of 40 000 daltons each. N-terminal domains of two different VSGs have been crystallized. Despite lack of sequence homology these domains have very similar structures as can be deduced from low resolution electron density maps (Freymann *et al.*, 1984; Metcalf *et al.*, 1987). Prominent features in these structures are two long α helices (80 Å) per subunit. Disruption of the folding of VSGs in such a manner that more conserved regions become exposed to circulating antibodies may be a potential, but rather speculative, approach to developing new anti-trypanosomal drugs. Other potential targets are interference with the machinery which ensures copying of a new VSG gene to the expression site, and blocking the attachment of VSGs to the membrane. So far, however, the proteins involved have not been characterized in detail.

An important consequence of the effective manner by which the trypanosome avoids the immune system is that development of a protective vaccine is very difficult indeed. This is a major reason to try to develop new drugs on the basis of a vital metabolic pathway described in the next section.

7.3.4 GLYCOLYSIS AND GLYCOSOMES

Glucose uptake and utilization by the bloodstream form of the trypanosomes is very fast. Per mg cellular protein the glucose consumption is $\sim 85\ \text{nmol}\cdot\text{min}^{-1}$ (mg protein)$^{-1}$. This is 50 times faster than the glucose consumption of mammalian cells (Wang, 1984). From this turnover rate, and assuming a protein content of 10% (w/w) (Misset *et al.*, 1986), it can be estimated that the amount of glucose consumed by trypanosomes in a period of 7 hours, is roughly equal to the total cellular weight.

In the bloodstream form of the trypanosome the single mitochondrion has been reduced to a peripheral canal without a functional Krebs' cycle. Moreover, the trypanosomes lack storage forms of metabolic energy, such as carbohydrates or 'high energy phosphate' molecules, which further increases their vulnerability to compounds interfering with glycolysis (Opperdoes, 1983). Not surprisingly, therefore, a mixture of salicylhydroxamic acid and glycerol, which is expected to block glycolysis (Fairlamb *et al.*, 1977), kills the trypanosomes *in vitro* (Clarkson and Brohn, 1976) as well as *in vivo* (Opperdoes and Borst, 1977). For pharmacological reasons, such as toxicity for the host, the combination of these two compounds is obviously not an acceptable drug.

Trypanosomes have a specialized organelle, called a 'glycosome' by the discoverers Opperdoes and Borst (1977), involved in efficient glycolysis (Opperdoes, 1987). Glycosomes are microbodies with a diameter of $\sim 0.3\ \mu$m which are surrounded by a single membrane. The lipids occurring in this membrane are the common phospholipids phosphatidylcholine and phosphatidyl ethanolamine (Opperdoes *et al.*, 1984). The major proteins sequestered in the glycosomes are nine enzymes involved in glucose and glycerol metabolism, catalysing the metabolic path shown in Figure 7.4. Other glycosomal enzymes are involved in carbon dioxide fixation (Fairlamb and Opperdoes, 1986), as well as in etherlipid and pyrimidine synthesis (Opperdoes, 1987). This indicates common features, and hence a possible common origin, with other organelles such as peroxisomes. It should be pointed out that the glycosomes have not only been found in *T. brucei*, but have also been identified in several other members of the order Kinetoplastida, such as *T. cruzi*, *Crithidia* spp. and in four representatives of *Leishmania* spp. (see, e.g., Fairlamb and Opperdoes, 1986, p. 199).

Given the importance of glycosomes for trypanosomes, it was decided to focus on glycosomal enzymes as potential targets for drug design.

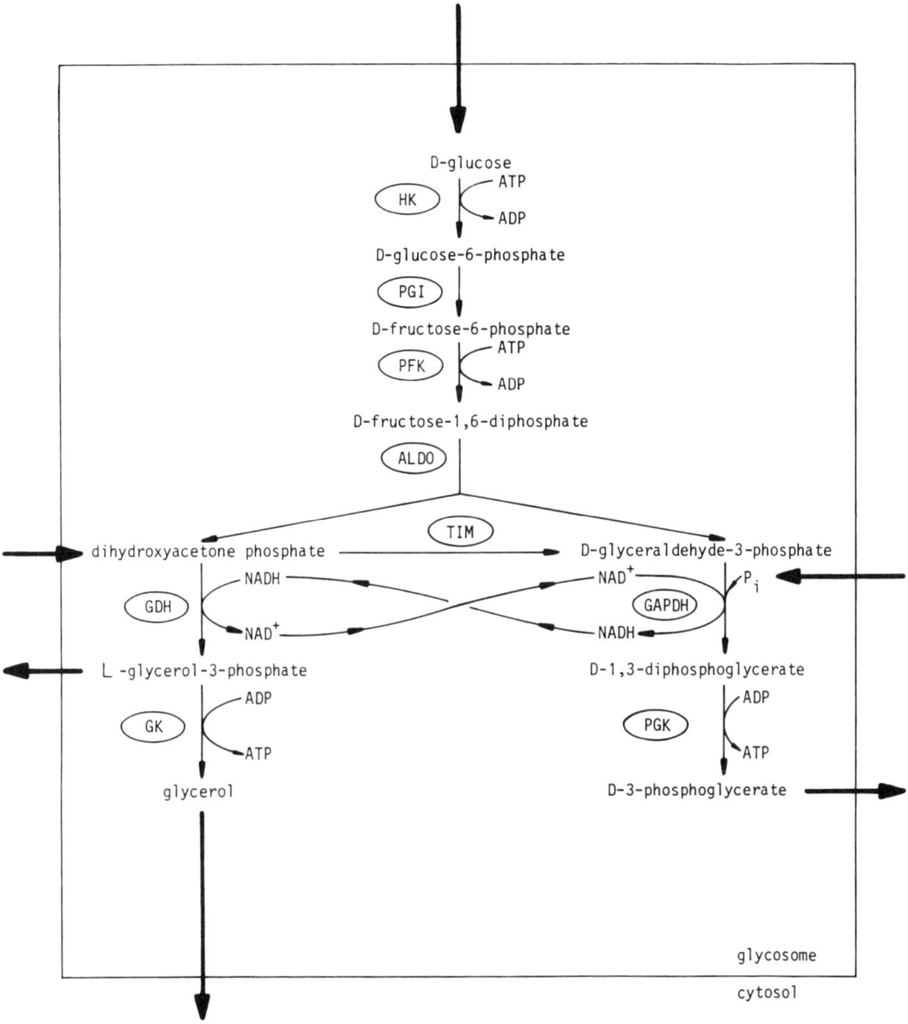

Figure 7.4. The glucose and glycerol metabolism in the glycosome (Fairlamb and Opperdoes, 1986). The abbreviations (Table 7.1) of the nine enzymes are encircled. Six compounds can be exchanged between the glycosomal matrix and the cytosol (indicated by arrows): D-glucose, dihydroxyacetone phosphate, glycerolphosphate, glycerol, inorganic phosphate (P_i), and D-3-phosphoglycerate. Under aerobic conditions one molecule of D-glucose is converted into two molecules of D-3-phosphoglycerate. The net ATP production occurs in the cytosol

7.4 TRYPANOSOMAL GLYCOLYSIS AND RATIONAL DRUG DESIGN

The unique features and crucial importance of glycolysis in trypanosomes offer several opportunities for designing new therapeutic compounds following the 'rational drug design cycle' (Hol, 1986) depicted in Figure 7.1. Various possible approaches are:

(a) exploiting the fact that glycosomal enzymes have to be translocated into the glycosome. Interference with this membrane passage process into an organelle which has no equivalent in the human host is a potentially promising approach, although our understanding of the import mechanism is still quite limited at present;
(b) exploiting differences in the catalytic centres of glycosomal enzymes and their human analogues. With knowledge of kinetic properties, amino acid sequences and three-dimensional structures, it might be feasible to arrive in a rational way at compounds which bind well to glycosomal enzymes and inhibit their activity while leaving the human analogous enzymes largely unaffected;
(c) interfering with the assembly of glycosomal enzymes. All of these enzymes are multimers, except phosphoglycerate kinase. For several, if not all, of these proteins proper assembly is required for activity. Compounds interfering with correct association of glycosomal, and not of human, multi-subunit enzymes are potentially useful as drugs;
(d) the monomeric enzyme phosphoglycerate kinase (PGK) has to make a considerable hinge bending motion during its catalytic cycle. Compounds which inhibit this motion in trypanosomal PGK may be useful as drugs as long as interference with human PGK is limited.

With these strategies in mind we will now look at glycosomal enzymes from *Trypanosoma brucei*. Ideally we would like to have at our disposal amino acid sequences, full kinetic data and three-dimensional X-ray structures of all nine enzymes complexed with a wide variety of ligands. This is not achieved yet, but the next sections describe the progress being made towards this goal.

7.5 GLYCOSOMAL ENZYMES

Essential steps to be taken in the first phase of the project were the purification of the target enzymes and the determination of the amino acid sequences of glycosomal proteins. Next physicochemical properties had to be determined, and single crystals to be grown in order to unravel three-dimensional structures by X-ray crystallography.

An elegant procedure has been developed at the Institute of Cellular and

Table 7.1. The nine enzymes occurring in the glycosome with number of subunits per oligomer, molecular weight per subunit and isoelecric point, p*I* (Misset *et al.*, 1986)

Enzyme	Abbreviation	Number of subunits	Mol. wt (subunit)	p*I*
Hexokinase	HK	6	50 000	10.0
Phosphoglucose isomerase	PGI	2	62 000	7.5
6-Phosphofructokinase	PFK	4	50 000	8.7
Fructose-bisphosphate aldolase	ALDO	4	41 000	9.1
Glycerol-3-phosphate dehydrogenase	GDH	2	37 000	10.0
Glycerol kinase	GK	2	53 000	9.0
Triosephosphate isomerase	TIM	2	27 000	9.8
Glyceraldehyde-3-phosphate dehydrogenase	GAPDH	4	39 000	9.3
3-Phosphoglycerate kinase	PGK	1	47 000	9.4

Molecular Pathology (ICP) in Brussels which allows rapid purification of all nine enzymes involved (Misset and Opperdoes, 1987). This was followed by investigations of physicochemical properties, the results of which are summarized in Table 7.1. Kinetic studies which are currently being carried out in Brussels by Dr Fred Opperdoes and co-workers are revealing interesting differences between glycosomal enzymes and their mammalian counterparts.

Some of the major differences observed so far are the following:

(a) Hexokinase and phosphofructokinase have completely different molecular weights compared with their mammalian homologues (Misset *et al.*, 1986);
(b) All enzymes, except phosphoglucose isomerase, have a very high isoelectric point (Misset *et al.*, 1986);
(c) The nucleotide specificity of glycosomal PGK is different from other PGKs (Misset and Opperdoes, 1987);
(d) At low ionic strength and pH above 7 the activity of triosephosphate isomerase (TIM) is considerably less than that of TIMs from other organisms (Lambeir *et al.*, 1987);
(e) The affinity of GAPDH for the cofactor NAD is 5–10 times less in the trypanosome enzyme than in the human isozyme (Callens, M., Kuntz, D. and Opperdoes, F. R., submitted for publication).

Simultaneously with physicochemical and kinetic characterization of glycosomal enzymes, their genes have been the subject of intense investigations. As a result *T. brucei* gene sequences are now known for glycosomal TIM (Swinkels *et al.*, 1986), cytosolic and glycosomal PGK (Osinga *et al.*, 1985), glycosomal GAPDH (Michels *et al.*, 1986), cytosolic GAPDH (Lambeir *et al.*, 1990), glycosomal aldolase (Clayton, 1985; Marchand *et al.*, 1988) and glycosomal PGI (Marchand *et al.*, 1989).

7.6 IMPORT OF ENZYMES INTO GLYCOSOMES

Glycosomal enzymes are synthesized in the cytosol on free polysomes. The half-life of the mature enzyme in the cytosol is approximately 3 minutes (Clayton, 1987; Hart *et al.*, 1987). Subsequently they are taken up by the glycosome without post-translational modification (Hart *et al.*, 1987). The mechanism of import into glycosomes is unknown as yet. From a comparative study of amino acid sequences and inspection of approximate models of three glycosomal enzymes, a hypothesis was proposed by Wierenga *et al.* (1987b). The basic idea was that two clusters of positive charges about 40 Å apart on the surface of glycosomal enzymes might constitute an important component of the import signal for glycosomes.

This hypothesis does not appear to hold up well in the light of new data which has become available. Recent gene sequence data from Swinkels *et al.* (1988) on glycosomal and cytosolic PGKs from the insect trypanosome *Crithidia fasciculata*, provided no evidence at all for positive clusters of charges being crucial differences between these two enzymes. Virtually the only difference in sequence of these PGKs is the C-terminal extension of glycosomal PGK. This suggests that the import signal resides in the C-terminal extension of glycosomal PGK. Not all glycosomal enzymes have a C-terminal extension, however (Clayton, 1985; Michels *et al.*, 1986; Wierenga *et al.*, 1987b; Marchand *et al.*, 1988, 1989). Very recently, the gene sequence of glycosomal GAPDH from *Trypanosoma cruzi* (Kendall *et al.*, 1990) shows that one of the positive clusters suggested to be important for glycosomal import was no longer intact. Hence, the excess of positive charge which so prominently occurs in glycosomal enzymes from *Trypanosoma brucei* (Wierenga *et al.*, 1987b) may have other functions. Opperdoes *et al.* (1990) and Michels (1988) suggested neutralization of negatively charged metabolites which are present in high concentrations in the glycosomes. Clearly, further research is required to establish the glycosomal import signal.

One final remark concerns potential possibilities to block import which would not depend on precise knowledge of import signals. In the case of import into mitochondria, studies of Schatz *et al.* (Eilers and Schatz, 1986) indicate that compounds with high affinity for the protein to be translocated prevent membrane passage. This is interpreted as evidence for at least partial unfolding during the import process. If this unfolding were to be essential for import into glycosomes, then every low molecular weight compound which binds very tightly in any of the clefts of a glycosomal enzyme might prevent uptake into the glycosome. If this were true then we would not be limited to compounds binding selectively to active sites or subunit interfaces, but could explore the entire surface of glycosomal proteins. It will obviously be of great interest to see if unfolding is indeed a feature of the glycosome import process. As this has not yet been established, we refrain from further discussions of glycosome targeting and turn to two glycosomal enzymes which have been studied with respect to their three-dimensional structures.

7.7 GLYCOSOMAL GLYCERALDEHYDE PHOSPHATE DEHYDROGENASE

Glyceraldehyde phosphate dehydrogenase (GAPDH) is a tetramer with subunits of approximately 37 000 daltons. The enzyme catalyses the following reaction:

$$\text{glyceraldehyde-3-phosphate} + NAD^+ + P_i \rightleftharpoons \text{1,3-diphosphoglycerate} + NADH + H^+.$$

The enzyme undergoes considerable conformational changes upon binding of NAD, as has been revealed by crystallographic studies on holo- and apo-GAPDH from lobster and *Bacillus stearothermophilus* (Moras *et al.*, 1975; Leslie and Wonacott, 1984).

The degree of amino acid sequence identity of lobster, *B. stearothermophilus* and human enzyme with each other and with the glycosomal *T. brucei* enzyme (Michels *et al.*, 1986) is shown in Table 7.2. It is obvious that the *T. brucei* enzyme differs significantly from the two enzymes with known three-dimensional structure (Moras *et al.*, 1975; Leslie and Wonacott, 1984; Skarzynski *et al.*, 1987) so that an accurate structure determination of *T. brucei* GAPDH is required for our drug design purpose. A low resolution structure of human GAPDH (Mercer *et al.*, 1976) was available at the beginning of our studies with obvious limitations as to its accuracy. Hence we needed not only to determine the atomic structure of *T. brucei* GAPDH, but also to improve the accuracy of the human structure.

We are most grateful to Prof. H. Watson from Bristol, UK, for providing us

Table 7.2. Percentage of sequence identity for lobster *B. stearothermophilus*, human and *T. brucei* glyceraldehyde-3-phosphate dehydrogenase (GAPDH) (Michels *et al.*, 1986)

	Lobster	*B. stearothermophilus*	Human	*T. brucei* (glycosomal)
Lobster	100	53	70	57
B. stearothermophilus	53	100	52	54
Human	70	52	100	55
T. brucei (glycosomal)	57	54	55	100

with enzyme, crystals, data and initial coordinates which allowed us to obtain a quite accurate structure of the human enzyme. Using one crystal, an 84% complete data set out to 2.4 Å resolution was collected on the FAST television detector diffractometer in Groningen (R. J. Read *et al.*, unpublished results). Refinement of the starting model, which has half a tetramer, or 80 000 daltons, in the asymmetric unit, has resulted in a structure which has a quite reasonable R-factor of 22%, with good geometry and without water molecules. In particular, near the substrate and the coenzyme binding sites the currently available model provides a reliable reference point for assistance in the design of compounds which would block the functioning of glycosomal GAPDH (gGAPDH) but should leave the human enzyme largely unaffected.

The structure determination of glycosomal GAPDH is making good progress as well. Purified enzyme gave two crystal forms (Read *et al.*, 1987) both with quite large cell dimensions, rather small size and considerable radiation sensitivity. Only with the use of synchrotron radiation could an 85% complete 2.8 Å resolution data set of form II crystals be recorded on film. For the data collection a total of 16 crystals were used, and altogether 134 films were processed. After merging and scaling, a dataset of 57 050 unique reflections was obtained. Next, the molecular replacement method (Rossmann, 1972), starting from the *B. stearothermophilus* GAPDH structure (Figure 7.5), allowed a fairly straightforward initial solution of the *T. brucei* GAPDH form II structure, in spite of the considerable complexity of the subunit arrangement in these crystals. It appeared that the crystals contain one complete and one half tetramer in the asymmetric unit, which means that we are dealing with six subunits of 37 000 daltons each, or a total of ~ 220 000 daltons. Refinement of this crystal structure is still underway. The current model has an R-factor of 44% but still needs careful rebuilding at several positions.

Due to scarcity of protein material too few crystals were left to do either soaking studies or to determine the structure of crystal form I using the 'classical' rotation method. Since the morphology of the crystals did not allow distinction of forms I or II we decided to see if the application of the 'Laue method' would yield structures of sufficient accuracy for our design purposes, while keeping the need for crystals to a minimum. It turned out that this led to a quite unconventional structure determination of crystal form I of glycosomal GAPDH from *T. brucei.*

Only two crystals were required to collect four Laue film packs with a total exposure time of 12.5 seconds. The film packs of six films each were processed using the CCP4 Laue film processing software. This eventually led to a dataset which was 37% complete between 7 and 3.2 Å resolution, containing 22 022 unique reflections.

Cell dimensions of crystal form I are $a = 135.5$ Å, $b = 256.3$ Å, $c = 114.9$ Å, space group $P2_12_12$, which allows for a large number of subunits in the cell. Possibilities range from 1.5 to 3 tetramers, or 220 000 to 440 000 daltons per asymmetric unit. Starting from the well refined *B. stearothermophilus* GAPDH structure (Leslie and Wonacott, 1984; Skarzynski *et al.*, 1987) it was possible to solve this molecular replacement problem, resulting in an initial model of 1.5 tetramers, or six subunits, per asymmetric unit. The sixfold redundancy was subsequently used for: (a) density averaging; (b) introduction of calculated phases and amplitudes for unobserved reflections; and (c) constrained refinement imposing sixfold symmetry on the atomic coordinates using simulated annealing procedures. The resultant model needs still further rebuilding and refinement, but a reasonably accurate model of also this crystal form of gGAPDH has now been obtained (F. M. D. Vellieux, J. Hajdu, C. L. M. J. Verlinde, R. J. Read, T. Greenough, J. Littlechild, H. Watson and W. G. J. Hol, in preparation). So, for future experiments structures of both crystal forms of the glycosomal enzyme, as well of the human enzyme, are now available.

Exploitation of this structural information is only just starting. Fortunately, the sequence of glycosomal GAPDH has been determined within the framework of this collaborative project by Michels *et al.* (1986). Inspection of sequences and structural models revealed most interesting differences near the adenine binding pocket of the cofactor NAD (Table 7.3; Figure 7.6). For instance, in the 'back' of this pocket the asparagines 33 and 200 appear to be hydrogen bonded to each other, both in the human as well as in the *B. stearothermophilus* enzymes. In glycosomal GAPDH, however, Asn 33 is replaced by a valine. Hence, the characteristics of the binding pocket change significantly. Initial and preliminary suggestions are that modified adenines, adenosines, AMP or ADP or molecules with appropriate 'variants' of the adenine ring might be better inhibitors of *T. brucei* glycosomal GAPDH than of the human enzyme. Other modifications, involving for instance the 2′ and 3′ hydroxyls of the adenine ribose, might also lead to selective inhibition of the trypanosomal enzyme.

While these structure determinations and modelling efforts were proceeding, the good news reached us that *T. brucei* glycosomal GAPDH was successfully overexpressed in *E. coli* in Brussels by Dr P. A. Michels and co-workers. This is a dramatic change: the target enzyme was so far only available in small amounts,

Table 7.3. Residues near the adenine moiety of NAD differing between human and glycosomal GAPDH of *Trypanosoma brucei*

T. brucei	Human
Asn 7	Asp 7
Val 36	Asn 33
Asn 39	Phe 36
Gln 90	Glu 78
Asn 92	Asp 80
Leu 112	Val 100

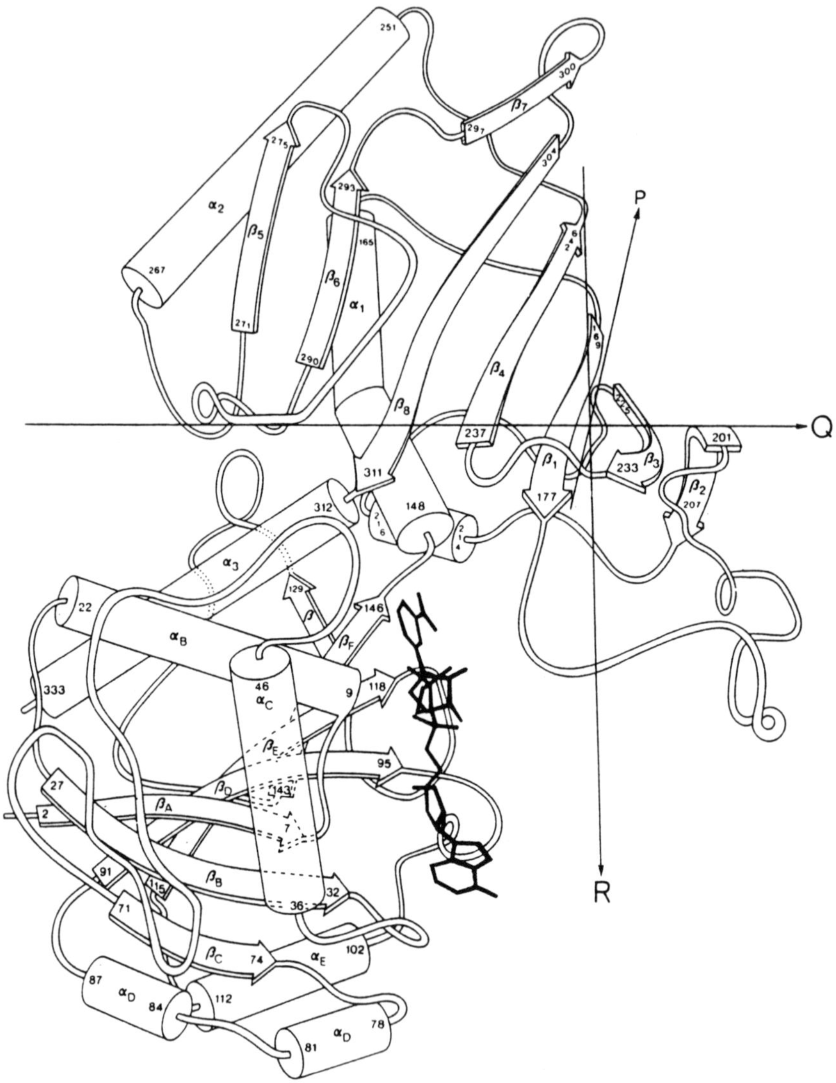

Figure 7.5. Schematic drawing of *B. stearothermophilus* glyceraldehyde phosphate dehydrogenase. Cylinders indicate α helices, arrows β strands. The NAD is shown in heavy lines. The twofold axes indicated with P, Q and R relate the subunit shown to the other members of the tetrameric molecule. (Reproduced, with permission, from Biesecker *et al.*, 1977)

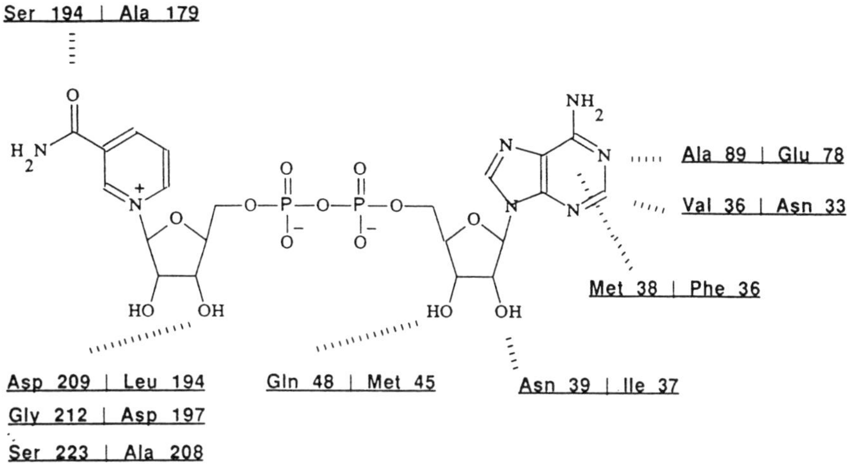

Figure 7.6. Indication of amino acids near NAD bound to GAPDH which are different between the *T. brucei* glycosomal enzyme and the human enzyme. The left residue in each box is the residue in gGAPDH, the right one the equivalent residue in human GAPDH

whereas now a virtually 'unlimited' supply of protein material is available. Hopefully, new crystals can be obtained using the recombinant material, which should lead to the design of selective inhibitors of glycosomal *T. brucei* GAPDH by the cyclic procedure depicted in Figure 7.1.

7.8 GLYCOSOMAL TRIOSEPHOSPHATE ISOMERASE

7.8.1 GENERAL FEATURES AND AMINO ACID SEQUENCE

Triosephosphate isomerase (TIM) is a well characterized glycolytic enzyme. Its function in glycolysis is indicated in Figure 7.4. The enzyme catalyses the reaction shown in Figure 7.7. TIM consists of a dimer with subunits of ~ 250 amino acid residues each, resulting in a total molecular weight of ~ 55 000. The enzyme is only fully active as a dimer (Waley, 1973), but there is no evidence for cooperativity. During catalysis no cofactors are required. A wealth of kinetic evidence has led to the conclusion that triosephosphate isomerase is a very efficient enzyme indeed (Knowles and Albery, 1977). Human beings missing the enzyme suffer from very poor health and die at an early age (Maquat *et al.*, 1985). The trypanosome, which is more dependent on glycolysis than humans, will depend even more on a proper functioning TIM. For this reason inhibitors of *T. brucei* TIM (gTIM) which leave the activity of human TIM largely unaffected are of great interest for the treatment of sleeping sickness.

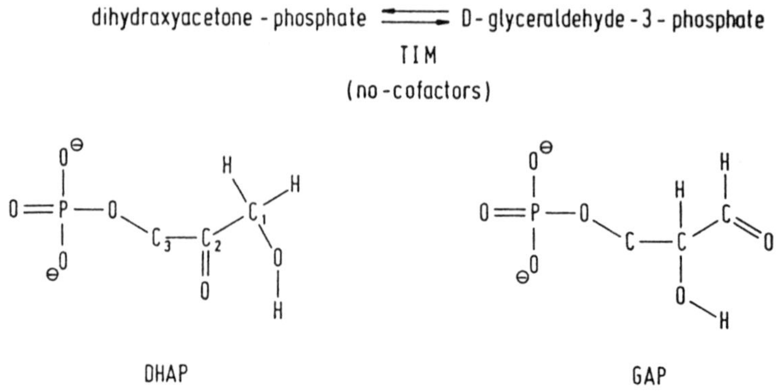

Figure 7.7. Overall reaction and structures of the substrates of triosephosphate isomerase (TIM)

The three-dimensional structure of chicken triosephosphate isomerase was determined in the mid-1970s by the Oxford group (Banner *et al.*, 1975; Alber *et al.*, 1981). The yeast enzyme has also been studied in detail (Lolis and Petsko, 1990; Lolis *et al.*, 1990). The structure exhibits pseudo eightfold symmetry with eight β strands forming a central parallel β barrel surrounded by eight α helices, linked together by a number of loops. A schematic view of a single subunit is depicted in Figure 7.8.

General features of triosephosphate isomerase (Banner *et al.*, 1975; Alber *et al.*, 1981; Wierenga *et al.*, 1987a, 1991; Joseph *et al.*, 1990; Lolis and Petsko, 1990; Lolis *et al.*, 1990; Noble *et al.*, 1991a,b) are indicated in this figure of the *T. brucei* enzyme. They include:

- three residues in the active site cavity which are directly interacting with the substrate: Lys 13, His 95 and Glu 167;
- the position of His 95 situated at the N-terminus of a short 'active site helix' comprising residues 95–105;
- the phosphate moiety of the substrate interacting with peptide units of the N-terminus of a 'phosphate binding helix' (Lolis *et al.*, 1990);
- at the end of β_6 residues 167–180 form a so-called 'flexible loop' which covers the active site during catalysis (Alber *et al.*, 1981; Joseph *et al.*, 1990);
- helices α_2 and α_3 plus the loop between β_1 and α_1 plus the loop between β_2 and α_2 are involved in contacts with the other subunit of the dimer, while residues 74–77, between β_3 and α_3, form the so-called 'interface loop' which protrudes into the active site of the other subunit (Wierenga *et al.*, 1987a).

When the structure of yeast TIM became available (Lolis and Petsko, 1990; Lolis *et al.*, 1990) it became clear that the conformations of the chicken and yeast enzymes were essentially the same, despite a 53% difference in the amino acid

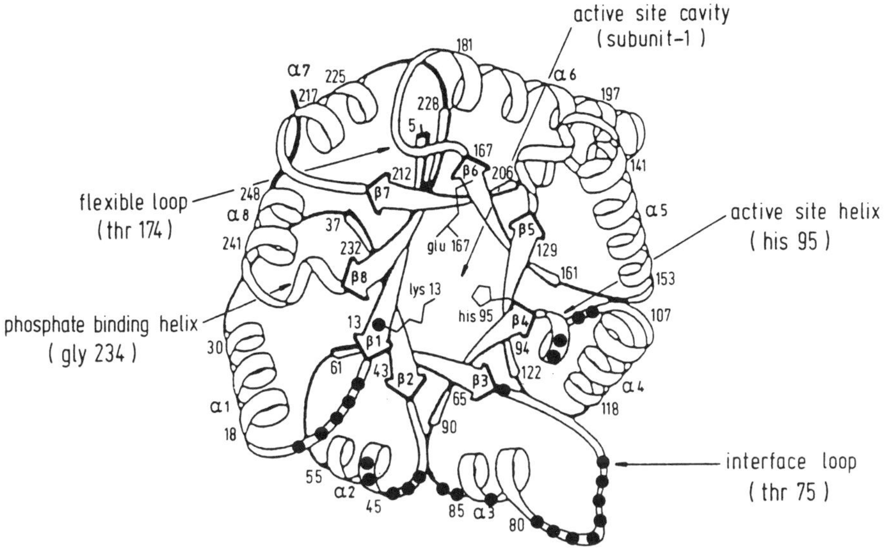

TRIOSEPHOSPHATE ISOMERASE
(SUBUNIT-1)

Figure 7.8. A schematic drawing of subunit 1 of the *T. brucei* TIM dimer. The approximate positions of the active site residues Lys 13, His 95 and Glu 167 are shown. The black dots, emphasizing the interface region, indicate the 27 residues with atoms which are closer than 3.5 Å from atoms of the other subunit (Alber *et al.*, 1981). The 'interface loop' (72–79) points towards the active site of subunit 2. The 'flexible loop' (167–180) closes over the active site after substrate binding (Alber *et al.*, 1981; Lolis and Petsko, 1990; Lolis *et al.*, 1990; Noble *et al.*, 1991b). The 'active site helix' is formed by residues 95–103. The 'phosphate binding helix' is formed by residues 234–239. Residues Ala 100–Tyr 101–Tyr 102 have the sequence His–Val–Phe in human TIM and constitute amino acid residue differences between human and parasite enzyme occurring relatively close to the active site. (Residues of subunit 1 are numbered 2–250, those of subunit 2 302–550)

sequences. Recently, the sequence of *T. brucei* TIM has been derived from the available gene sequence (Osinga *et al.*, 1985). Also the sequence of human TIM is known (Lu *et al.*, 1984; Maquat *et al.*, 1985). An alignment of all four sequences is given in Figure 7.9. The *T. brucei* sequence is approximately 50% identical with other known sequences, both eukaryotic and prokaryotic. The human TIM sequence is 88% identical with that of chicken TIM so that only minor alterations are required to obtain a quite accurate three-dimensional model of human TIM. For trypanosomal TIM this is clearly not the case and a structure determination has been carried out. Before describing aspects of the crystallographic work, comparison of the *T. brucei* and human TIM sequences

			B1		H1	
T. brucei TIM	(2)	S K P	Q P I A A A N W (K)	C N G S	Q Q S L S E L I D L F N S	(30)
Human TIM	(1)	A P S R	K F F V G G . . .	M . . R	K . . . G . . . G T L . A	(30)
Chicken TIM	(2)	A P R	K F F V G G . . .	M . . D	K K . . G . . . H T L . G	(30)
Yeast TIM	(2)	A R	T F F V G G . F .	L . . .	K . . I K . I V E R L . T	(29)
			B2		H2	
T. brucei TIM	(31)	T S I N H D	V Q C V V A S	T	F V H L A M T K E R L	(55)
Human TIM	(31)	A K V P A .	T E V . C A P	P	T A Y I D F A R Q K .	(55)
Chicken TIM	(31)	A K L S A .	T E V . C G A	P	S I Y . D F A R Q K .	(55)
Yeast TIM	(30)	A . . P E N	. E V . I C P	P	A T Y . D Y S V S L V	(54)
			B3		H3	
T. brucei TIM	(56)	S H P K F	V I A A Q	N A I A K S - G A F T G E V S	L P I L K D	(85)
Human TIM	(56)	D P - . I	A V . . .	. C Y K V T N . . . . . . I .	P G M I . .	(85)
Chicken TIM	(56)	D A - . I	G V . . .	. C Y K V P K . . . . . . I .	P A M I . .	(85)
Yeast TIM	(55)	K K . Q V	T V G . .	. . Y L . A S . . . . . . N .	V D Q I . .	(85)
			B4	H4f	H4	
T. brucei TIM	(86)	F G V N	W I V L G	(H) S E R R A Y Y G E T N	E I V A D K V A A A V A	(118)
Human TIM	(86)	C . A T	. V . . .	. . . . . H V F . . S D	. L I G Q . . . H . L .	(118)
Chicken TIM	(86)	I . A A	. V I . .	. . . . . H V F . . S D	. L I G Q . . . H . L .	(118)
Yeast TIM	(86)	V . A K	. V I . .	. . . . . S . F H . D D	K F I . . . T K F . L G	(118)
			B5	H5f	H5	
T. brucei TIM	(119)	S G F	M V I A C I G E	T L Q E R E S G R T A	V V V L T Q I A A I A K K	(153)
Human TIM	(119)	E . L	G	K . D . . . A . I . E	K . . F E . T K V . . D N	(153)
Chicken TIM	(119)	E . L	G	K . D . . . A . I . E	K . . F E . T K . . . D N	(153)
Yeast TIM	(119)	Q . V	G . . L	. . E . K K A . K . L	D . . E R . L N . V L E E	(153)
			B6		H6	
T. brucei TIM	(154)	L K K A D W A	K V V I A Y (E)	P V W A I G T G K V A T P	Q Q A Q E A H A L I R S W V S S K	(197)
Human TIM	(154)	V . - - - . S	. . . L . . .	. . . . T . . .	. . . . . V . E K L . G . L K . N	(195)
Chicken TIM	(154)	V . - - - . S	. . . L . . .	. . . . T . . .	. . . . . V . E K L . G . L K . H	(195)
Yeast TIM	(154)	V . - - - . T	N . . V . . .	. . . . L A . . .	E D . . D I . . S . . K F L A . .	(195)
		H7b	B7		H7	
T. brucei TIM	(198)	I G A D V R G E	L R I L Y G G	S V N G	K N A R T L Y Q Q	(225)
Human TIM	(196)	V S D A . A Q S	T . . I . . .	. . T .	A T C K E . A S .	(223)
Chicien TIM	(196)	V S D A . A Q S	T . . I . . .	. . T .	G . C K E . A S .	(223)
Yeast TIM	(196)	L . D K A A S .	. . . .	. A . .	S . . V . F K D K	(223)
			B8	H8f	H8	
T. brucei TIM	(226)	R D	V N G F L	V G G A S L K P	E F V D I I K A	(248)
Human TIM	(224)	P .	. D . . .	. . . .	. . . . . . N .	(246)
Chicken TIM	(224)	H .	. D . . .	. . . .	. . . . . . N .	(246)
Yeast TIM	(224)	A .	. D . . .	. . . .	. . . . . . N S	(246)
T. brucei TIM	(249)	T Q				
Human TIM	(247)	K .				
Chicken TIM	(247)	K H				
Yeast TIM	(247)	R N				

in the light of the chicken and yeast structures showed that (Wierenga *et al.*, 1987a):

(a) the active sites differ very little in sequence between the human and *T. brucei* enzyme: ~ 85% of all residues within 10 Å from His 95 are identical. Surface residues near to the active centre which differ between parasite and host enzyme are residues 100–102. These are Ala–Tyr–Tyr in *T. brucei* and His–Val–Phe in human. These residues are approximately 10 Å removed from the essential active site atoms;

(b) quite in contrast, the interface residues differ substantially. Only ~ 55% of these residues are identical in human and *T. brucei* TIM.

In view of these limited sequence differences near the active site it is understandable that few differences exist between the kinetic properties of *T. brucei* and other triosephosphate isomerases. The kinetic differences at low ionic strength and high pH (Lambeir *et al.*, 1987) may be due to differences in the subunit interface (Wierenga *et al.*, 1987a).

7.8.2 CRYSTAL STRUCTURE

In spite of the fact that TIM is the least abundant glycolytic enzyme in the glycosome (Misset *et al.*, 1986), it was the first to give good quality crystals and hence the first glycosomal enzyme for which a structure was determined. In fact it is the first enzyme from any tropical parasite with known three-dimensional structure (Wierenga *et al.*, 1987a).

Crystals were obtained from solutions with a high ammonium sulphate concentration and crystallized with a complete dimer in the asymmetric unit. Using five crystals grown from the first batch of 0.6 mg purified protein a complete dataset of ~ 12 000 unique reflections to a resolution of 2.4 Å could be obtained. Starting from the chicken TIM structure, rotation and translation functions (Rossmann, 1972) gave the orientation and position of the model structure which was subsequently refined by restrained least squares and computer graphics techniques (Wierenga *et al.*, 1987a). In order to obtain a very

Figure 7.9. (*opposite*) Amino acid sequence alignment of human (Lu *et al.*, 1984; Maquat *et al.*, 1985) and *T. brucei* (Osinga *et al.*, 1985) triosephosphate isomerase, along with the two enzymes with known three-dimensional structure, those of chicken and yeast (Alber and Kawasaki, 1982; Straus and Gilbert, 1985). Dots indicate that a residue is identical to the *T. brucei* residue. On each line a β strand, a loop and an α helix are shown, indicated by B1–B8 for the strands and H1–H8 for the helices. H4f, H5f and H8f are helices occurring in loop regions at the C-terminal ends of the β strands, near the active site. H7b is a small helix at the other side of the subunit (see also Figure 7.8). The important amino acid differences near the active site between human and *T. brucei* TIM (see text), AYY, residues 100–102, are in a box of dotted lines. The catalytically important residues Lys 13, His 95 and Glu 167 are encircled accurate structure for our molecular modelling studies, an additional 1.83 Å data set was collected on films using the intense X-ray beam of the Daresbury synchrotron. The 115 film packs collected from 11 crystals were processed, giving a total of ~ 230 000 measured intensities. This gave a set of ~ 39 000 unique reflections. The refinement of the high resolution structure of gTIM has now been completed with an R-factor of 18.3%, good geometry of the model, inclusion of 279 water molecules, two sulphate ions and one dithiothreitol molecule (R. K. Wierenga *et al.*, to be published).

This accurate structure has proved to be an exciting starting point to improve our knowledge of the properties of *T. brucei* TIM by crystallographic binding as well as by molecular modelling studies. These investigations are currently in full swing and some examples of the results obtained will be given in the next few sections.

7.8.3 CRYSTALLOGRAPHIC BINDING STUDIES

Protein crystals consist for ~ 50% of mother liquor. This allows small molecules to diffuse through solvent channels and reach their binding site in the crystal. This approach can fail for a number of reasons such as:

(a) the binding site is blocked by a neighbouring protein molecule;
(b) ions or molecules in the crystallization medium occupy the binding site;
(c) conformational changes, occurring upon binding of the small molecule, destroy the crystal lattice.

In the case of *T. brucei* TIM all three of these factors appear to play a role as we now know:

(a) the active site of 'subunit 1' in the asymmetric unit is in contact with a neighbouring molecule and hence inaccessible for medium-affinity inhibitors;
(b) the entrance of the active site of 'subunit 2' is occupied by a sulphate ion. Sulphate has a K_i of ~ 5 mM (Lambeir *et al.*, 1987). Since the crystallization medium contains 2.4 M ammonium sulphate, straightforward inhibitor binding studies by soaking experiments are made unfeasible;
(c) major conformational changes occur near the active site upon substrate and inhibitor binding as reported for chicken and yeast TIM (Banner *et al.*, 1975; Alber *et al.*, 1981; Joseph *et al.*, 1990; Lolis and Petsko, 1990; Lolis *et al.*, 1990) and are observed repeatedly in our own studies on *T. brucei* TIM (Noble *et al.*, 1991a,b; Wierenga *et al.*, 1991).

This accumulation of problems was solved—at least for medium affinity inhibitors—by transferring *T. brucei* TIM crystals to another mother liquor. Hanging drop equilibration experiments were used to establish potentially suitable concentrations of PEG solutions (Schreuder *et al.*, 1988) which could substitute for ammonium sulphate. Eventually a specific protocol was obtained which allowed transfer of the *T. brucei* TIM crystals from 2.4 M ammonium sulphate to 44% (w/w) PEG 6000 solutions without loss of their diffraction power (Schreuder *et al.*, 1988). This breakthrough has allowed us to study crystallographically the mode of binding of inhibitors with inhibition constants down to 500 μM. It turned out that use of an inhibitor with a K_i of 27 μM, i.e. 2-phosphoglycolate, led to a deterioration of crystal quality, probably due to conformational changes near the active site of 'subunit 1'.

This is clearly still not an ideal situation because we are interested in the study of inhibitors with K_is much lower than 27 μM. Fortunately, very recently, the gene of *T. brucei* TIM has been overexpressed in Brussels by Dr K. Pratt and Dr P. A. M. Michels. This allows large-scale co-crystallization experiments which was not really feasible with the small amounts of enzyme which so far had to be obtained from trypanosomes grown in rats. Such co-crystallizations have now been initiated.

The study of the binding mode of a variety of inhibitors by means of soaking into the 'transferred' crystals has, however, already given us precious insight as to the surprising adaptability of TIM upon binding of a number of related compounds.

7.8.3.1 'Sulphate Free' *T. brucei* TIM

Sometimes experiments fail to give an answer to the question addressed, but nevertheless provide useful information. This was the case in our attempts to unravel the binding mode of 1-aminonaphthalene-4,6,8-trisulphonic acid (ANTS). This compound was kindly provided by Dr Nickel in Bonn and the inhibition constant of *T. brucei* TIM was found to be 0.3 mM (A. M. Lambeir and F. R. Opperdoes, personal communication). *T. brucei* TIM crystals were transferred to a 44% (w/w) PEG 6000 solution, as described above, containing also 2 mM ANTS. Data out to 2.8 Å resolution were collected on the FAST area detector and the structure was determined starting from the well refined model of TIM in 2.4 M ammonium sulphate. Upon application of molecular dynamics refinement techniques, the 'ANTS model' eventually reached an R-factor of 13.9% with good geometry and 30 water molecules (Wierenga *et al.*, 1991). No inhibitor was visible in the electron density map, however. This is quite puzzling, in particular since we have elucidated the binding mode of several other inhibitors successfully. We refrain here from trying to explain why the inhibitor did not bind but, instead, report briefly the results which were obtained by this study.

It appeared that 'subunit 2' did not contain a sulphate ion any more. This resulted in a number of conformational changes the most important of which was a major repositioning of the 'flexible loop' near the active site. This loop, comprising residues 167–180, is closed in the presence of sulphate- and

Table 7.4. Conformational changes of the flexible loop observed in *T. brucei* TIM (Noble *et al.*, 1991a; Wierenga *et al.*, 1991)

Residue number in 'subunit 2'	Difference of C^{α} position between the 'open' and the 'closed' conformation			
	3PP	G3P	3PGA	Sulphate
Pro 468	1.0	0.8	1.1	0.8
Val 469	0.9	0.7	0.9	0.7
Trp 470	2.6	2.7	2.7	2.3
Ala 471	2.9	2.9	3.0	2.8
Ile 472	3.2	3.1	3.0	2.7
Gly 473	5.6	5.7	5.4	4.9
Thr 474	6.1	6.3	6.1	5.7
Gly 475	6.8	7.1	7.1	6.9
Lys 476	4.6	4.6	4.6	4.8
Val 477	4.8	4.7	4.6	4.9
Ala 478	1.7	1.9	1.7	1.5

phosphate-containing compounds (Noble *et al.*, 1991a,b), but turned out to adopt an open conformation in the absence of sulphate. This conformational change causes the C^{α} atom of residue 475 (i.e. residue 175 of the 'second' subunit) at the top of the loop to move 6.9 Å towards the open conformation. As can be seen in Table 7.4, the conformation of this loop in the presence of phosphate-containing substrate analogues is slightly different from the conformation in the presence of sulphate. The main point, however, is a major conformational change which occurs during the catalytic process.

The loop closure may have several functions. One of these is avoiding dephosphorylation as a side reaction during the catalytic process (Pompliano *et al.*, 1990). Another one might be to bring active site residues into the optimal position with respect to the substrate. This may especially be true for Glu 167. This residue can adopt several conformations and is located at the very beginning of the flexible loop. Leaving mechanistic implications aside, this study provided us with a picture of the active site of 'subunit 2' in the 'open' conformation (Joseph *et al.*, 1990) as occurs in the absence of substrates and inhibitors. Although it has so far not yet been done, this 'open' conformation provides an alternative starting point for modelling studies to the 'closed' conformation which has been investigated in detail as we see in the next sections.

7.8.3.2 The Difference between Sulphate and Monohydrogen Phosphate Binding by *T. brucei* TIM

The active site pocket of TIM is deeply buried in the central part of the molecule (Figure 7.8). As mentioned earlier, amino acid differences between trypanosomal and human TIM useful for the design of *selective* inhibitors are located quite far from the active site. For example, the tripeptide 100–102, which is AYY in *T. brucei* TIM and HVF in human TIM, is located more than 12 Å away from the phosphate binding site at the entrance of the active site pocket (Figure 7.8; Table 7.5). Hence we are interested in elucidating the binding mode of phosphate containing molecules which bind with their 'non-phosphate' moiety *outside* the active site pocket. Modelling studies suggested that this might be the case for ribose-5-phosphate (R5P). M. Callens and F. Opperdoes (personal communication) determined that the K_i of this compound was ~ 5 mM.

For our crystallographic studies, crystals were transferred to 44% (w/w) PEG 6000 (Schreuder *et al.*, 1988) containing 15 mM ribose-5-phosphate. Data were collected on the FAST area detector yielding 14 592 unique reflections which is 88% of the data out to 2.4 Å resolution. Refinement resulted in a molecular model with an R-factor of 15.0%, with good geometry, 156 water molecules, and one monohydrogen phosphate ion bound to the active site of 'subunit 2'.

No ribose moiety was visible in the electron density map, almost certainly due to hydrolysis of the inhibitor molecule. This would explain why a phosphate ion is observed at the entrance of the active site pocket of 'subunit 2'. This result gave us the opportunity to compare the binding mode of the divalent sulphate ion with the divalent monohydrogen phosphate ion (Verlinde *et al.*, submitted). This turned out to be most intriguing.

The main result is that the sulphate position in the 1.83 Å ammonium sulphate structure is about 1 Å further 'out' of the active site pocket than the phosphate position. At the same time, the flexible loop is ~ 0.5 Å less well closed in the 'sulphate complex' than in the 'monohydrogen phosphate complex'. Also, the water molecules within the active site pocket seem to be arranged in a distinctly different manner. And, finally, the catalytically essential Glu 467 adopts a different side chain conformation (Figure 7.10).

A survey of the Cambridge structural database reveals that the differences in size between SO_4^{2-} and HPO_4^{2-} are less than 0.1 Å. Charges on the oxygen atoms are quite similar for the two ions. Hence, the presence of the extra

Table 7.5. Distances (Å) between the phosphorus atom of glycerol-3-phosphate complexed to *T. brucei* TIM and specific atoms of the tripeptide 100–102 Ala–Tyr–Tyr (numbered 400–402 in the 'second' subunit)

Cα (Ala 400)	15.6
Cβ (Ala 400)	14.4
Cα (Tyr 401)	16.7
OH (Tyr 401)	12.1
Cα (Tyr 402)	19.0
OH (Tyr 402)	16.3

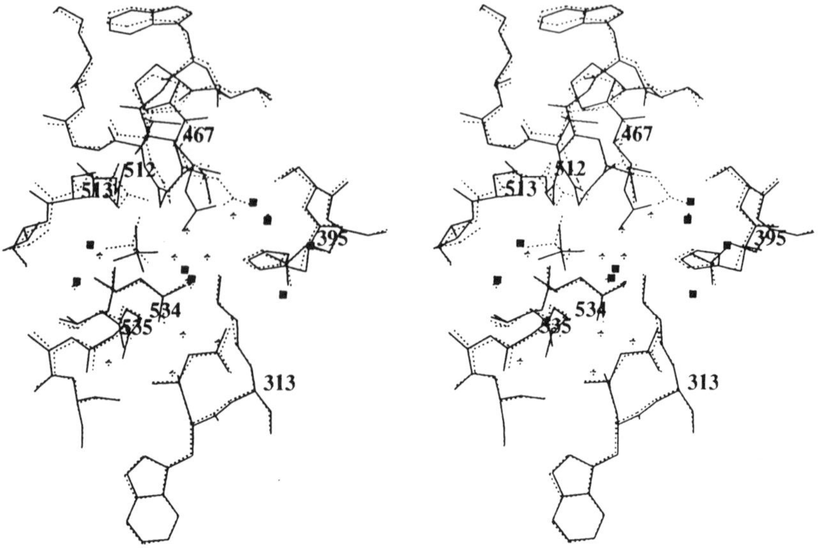

Figure 7.10. Stereo drawing of the active site of glycosomal TIM from *T. brucei* complexed with sulphate (solid lines and waters indicated with squares) and complexed with phosphate (as deduced from the ribose-5-phosphate soaking experiment in dashed lines)

hydrogen atom on one of the oxygens of HPO_4^{2-} is likely to be the cause of the differences observed in the active site. It remains probably quite a challenge for molecular modellers to arrive by theory alone at this difference of 1 Å in positions of ions which are so similar, and also have virtually identical inhibition constants.

One important message of this study is that subtle features have distinct effects on the binding mode of related molecules to the same enzyme. Another message seems to be that divalent ions can possibly be bound in a variety of positions ranging by 1 Å between the 'phosphorus' and the 'sulphur' sites. However, as we shall see in the next sections, studies on organic phosphates and phosphonates have so far not supported this hypothetical variability in phosphorus position.

7.8.3.3 Differences between a Phosphate and a Phosphonate

Starting from the proposed transition state (Rieder and Rose, 1959), several transition state analogues of TIM have been investigated (Wolfenden, 1969; Collins, 1974). Wolfenden and co-workers studied the inhibition of TIM activity

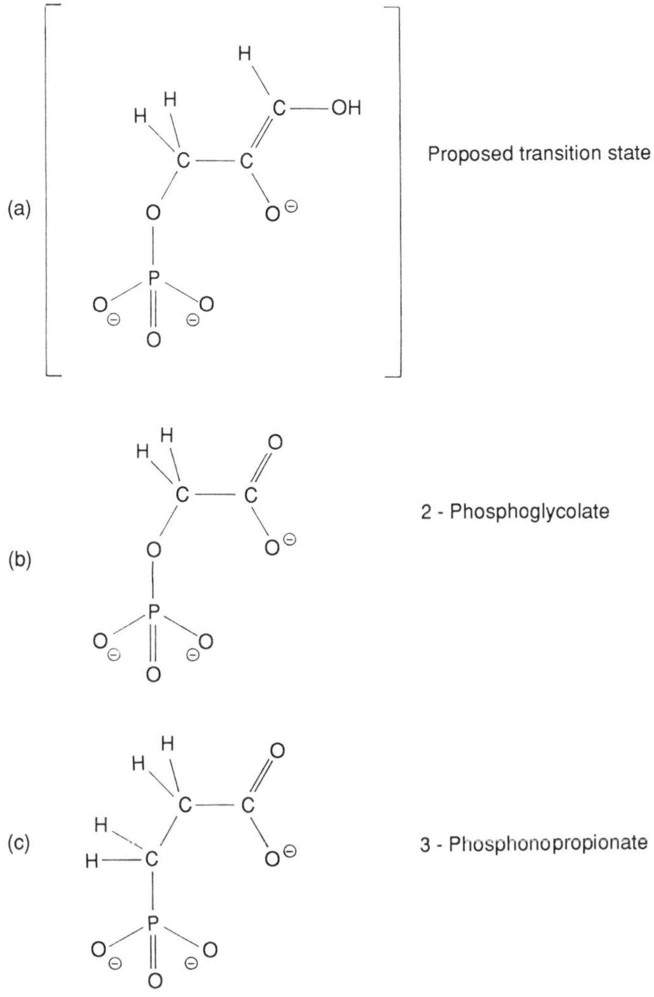

Figure 7.11. Molecular formulae of (a) the proposed transition state, (b) a transition state analogue 2-phosphoglycolate ($K_i = 0.027$ mM) and (c) an inhibitor, 3-phosphonopropionate ($K_i = 27$ mM), of triosephosphate isomerase. Only one atom is different in the latter two compounds, due to the replacement of an oxygen atom in 2-phosphoglycolate by a methylene group in 3-phosphonopropionic acid

by a transition state analogue, 2-phosphoglycolate (Wolfenden, 1969), (Figure 7.11). This compound turned out to be an excellent inhibitor indeed, having a K_i of 0.027 mM for trypanosomal TIM (Lambeir *et al.*, 1987). The structure of 2-phosphoglycolate led the late Dr Alan Horn to suggest 3-phosphonopropionate (Figure 7.10) as a potential inhibitor. This compound was bought from Aldrich and appeared to be a much less powerful inhibitor than 2-phosphoglycolate, the K_i being 27 mM (A.-M. Lambeir, unpublished observations).

In order to obtain insight into the reasons for this difference in affinity between a phosphate and a phosphonate of a factor of 10^3—the difference being the change of an oxygen atom into a methylene group—crystallographic binding studies were initiated. Crystal packing effects made it impossible to study 2-phosphoglycolate, as explained in a previous section. The structure of *T. brucei* TIM complexed with 3-phosphonopropionate (3PP) could, however, be determined.

The glycosomal TIM crystals were transferred, by the usual procedure (Schreuder *et al.*, 1988), to a PEG solution containing 100 mM 3PP. Crystals were of limited quality so only a data set of 2.8 Å resolution could be obtained. The 10 879 unique reflections represent 75.7 % of the data out to this resolution limit. The crystallographic R-factor of the 3PP–TIM complex was 12.5 % after refinement for a model containing nine ligand atoms, 38 water molecules and 3778 non-hydrogen protein atoms (Noble *et al.*, 1991a).

The 3PP molecule is bound to the active site pocket with each oxygen atom involved in a hydrogen bond, one of which is a hydrogen bond to the side chain of Glu 467. This residue is probably protonated, by analogy with the presumed protonation state of Glu 467 after binding of 2-phosphoglycolate to yeast TIM (Lolis and Petsko, 1990). The main point we like to focus on here is that the phosphate of 3PP is within 0.2 Å of the phosphate of monohydrogen phosphate discussed in the previous section. Also, the flexible loop is 'fully closed' as in the complex with monohydrogen phosphate—not 'almost closed' as in the complex with sulphate. Clearly, the low affinity of 3PP for *T. brucei* TIM is *neither* due to lack of closure of the flexible loop *nor* due to an unusual position of the phosphate moiety. It is also not due to the fact that the ether phosphate oxygen of 2-phosphoglycolate is interacting with a protein hydrogen bond donor—it is *not* according to the study of Lolis and Petsko (1990) TIM as well as to our own studies on inhibitors like glycerol-3-phosphate bound to *T. brucei* TIM (Noble *et al.*, 1991a). Replacing this ether phosphate oxygen by a CH_2 moiety does therefore not result in an unsatisfied protein hydrogen bond donor.

Hence, the reasons for the factor of 10^3 difference in K_i of 2-phosphoglycolate and 3-phosphonopropionate might stem from:

(a) less favourable torsion angles when 3PP is bound to TIM, compared with 2PGH;
(b) differences in interactions with water molecules of the two inhibitors;
(c) a difference in pK_a of phosphate and phosphonate groups (see e.g. also Belasco *et al.*, 1978).

In view of the absence of a structure of 2-phosphoglycolate with *T. brucei* TIM and the somewhat limited resolution of the complex with 3PP it is probably best not to carry on here with a long discussion of possible explanations. Rather we turn to other experimental results.

7.8.3.4 Binding Modes of Glycerol-3-phosphate and 3-Phosphoglycerate

In order to increase further our insight as to the properties of *T. brucei* TIM when binding low molecular weight compounds, we studied the inhibitors glycerol-3-phosphate (G3P) and 3-phosphoglycerate (3PGA) in complex with the enzyme. Both these inhibitors have one atom more than 3PP when going along the chain from the phosphorus to the terminal oxygen atom. In addition, 3PGA and G3P have one extra hydroxyl group at the 2-position compared with 3PP. 3-Phosphoglycerate and glycerol-3-phosphate differ merely by changing the carboxylate of 3PGA into a hydroxymethyl group in G3P. This removes a charge near to the catalytic Glu 467 from the complex with G3P compared to the complex with 3PGA.

The three-dimensional structures of *T. brucei* TIM with G3P and 3PGA have been determined in an analogous way to that described in the previous section for 3PP. The R-factor for the 3PGA complex after refinement is 14.0% for 16 813 unique reflections out to 2.2 Å with 73% completeness of the data. The R-factor for the G3P complex is 13.7% with essentially the same completeness out to the same resolution.

The structures of the protein in the complexes with 3PP, 3PGA and G3P are all very similar despite the differences between the three ligands. The phosphorus positions of 3PGA and G3P are of special interest in view of the difference of 1 Å described before between the sulphate position of SO_4^{2-} and the phosphate position of HPO_4^{2-} and 3PP. It turns out that thc phosphate positions of 3PGA, G3P, 3PP and HPO_4^{2-} all fall within 0.2 Å of each other. In other words, there seems to be a rather well defined binding site for phosphate and phosphonate moieties. This 'phosphorus site' is independent of the precise nature of the inhibitor moiety bound to the active site.

In the TIM–3PP complex the side chain of Glu 467 is engaged in a hydrogen bond with the carboxylate groups of 3PP as described above. This is the so-called 'swung-in' conformation of Glu 467. In spite of the presence of equivalent carboxylate groups in 3PP and 3PGA, in the TIM–3PGA complex Glu 467 is in a 'swung-out' conformation. This conformation differs from the 'swung-in' conformation by a rotation of 120° about the $C^{\alpha}—C^{\beta}$ bond, of 40° about $C^{\beta}—C^{\gamma}$ and of 60° about $C^{\gamma}—C^{\delta}$.

In view of the small differences between 3PGA and G3P one might have expected that Glu 467 would adopt similar conformations upon binding these two analogues. This is not the case: the inhibitor G3P binds with Glu 467 in the 'swung-out' conformation.

As seen before, subtle differences in the molecules bound do lead to distinct variations in the mode of binding.

7.8.4 MOLECULAR MODELLING OF ACTIVE SITE INHIBITORS

The exploration of the binding properties of the active site of gTIM described above are just the very first steps of our 'long' path towards the design of selective inhibitors of trypanosomal TIM. As detailed in Table 7.5, a distance of 12–19 Å has to be spanned from the phosphorus atom at the entrance of the active site towards a hydrophobic surface area where significant differences in amino acids between human and trypanosome TIM occur.

Upon careful analysis this 'pathway' appears to consist of a complex mixture of hydrogen-bond donors and acceptors, charged groups, and small hydrophobic areas. One of us (C. L. M. J. Verlinde), in collaboration with Dr K. Müller of Hoffmann-La Roche in Basle, has proposed that T-shaped molecules might be suitable for arriving at selective inhibitors. One such T-shaped molecule is shown in Figure 7.12. This molecule can be considered as consisting of four fragments each of which has a different function and is engaged in different interactions with the enzyme:

(a) the 'anchor' which would block entrance to the active site;
(b) the 'NN-pocket filler' which extends into a well defined pocket, made up by Asn 311 and Asn 315;
(c) the 'interface' fragment is located at the interface of the two subunits of

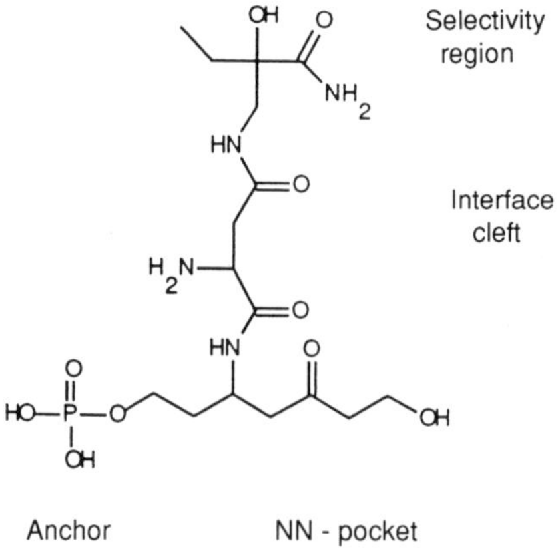

Figure 7.12. One of the possible 'T-shaped' molecules which might block the entrance to the active site pocket by its phosphate moiety and extend to the region near residues Ala 400 and Tyr 401 of *T. brucei* triosephosphate isomerase

TIM and links the base of the 'T' formed by the two previous fragments with the next one;

(d) the 'selectivity fragment' which fits into the hydrophobic area made up by Ala 400 and Tyr 401 and which should provide the inhibitor with the required selectivity versus the human enzyme.

We wish to emphasize that the T-shaped molecule depicted in Figure 7.12 is simply a 'guide molecule' outlining the general dimensions, shape and functional groups. Shortcomings of the modelling process, intervening water molecules, experimental errors and protein adaptability make it virtually impossible that the 'perfect' T-shaped molecule can be designed in one single step. Instead, a 'linked fragment approach' might be more realistic: different functional fragments could first be studied separately. Once several of these fragments, or analogues thereof, are crystallographically characterized when bound to the protein surface, the next step would be to link these fragments together into larger molecules. Such molecules are likely to bind better than the separate fragments, in particular if they could be made more rigid.

Clearly, there is still a long way ahead. But with large amounts of overexpressed *T. brucei* TIM now available, and data collection times being greatly reduced due to area detectors and powerful computers, we hope to increase our structural data base and to improve thereby the accuracy of our molecular modelling efforts.

7.8.5 INTERFACE LOOP BASED MODELLING STUDIES

As discussed previously, the triosephosphate isomerase active site contains residues from the 'interface loop' of the other subunit in the dimer. This loop is indicated schematically in Figure 7.8 and shown in stereo in Figure 7.13. The exact function of this loop is unknown but it may be responsible for the fact that TIM is only fully active as a dimer, and not as a monomer (Waley, 1973). This fact provides, in principle, another starting point for design of selective *T. brucei* TIM inhibitors: compounds disrupting the *T. brucei* dimer and not the human TIM dimer would decrease the efficiency of trypanosomal glycolysis, while leaving glycolysis of the host unaffected.

With the design of such compounds in mind, the interface loop in the 2.4 Å structure of *T. brucei* TIM (Wierenga *et al.*, 1987a) was used as a starting point for the design of cyclic peptide analogues by Dr Klaus Müller in Basle. During the modelling sessions careful attention was paid to one of the many differences in the *T. brucei* TIM interface compared with that of human TIM, namely residue Cys 14 which is a Met in the human enzyme. The interface loop contains a β turn which consists of Phe 74–Thr 75–Gly 76–Glu 77. One of the cyclopeptides modelled contained this tetrapeptide and consisted of: cyclo-[-Phe–Thr–Gly–Glu–Gly–D-Cys–], where the D-Cys might be able to form a

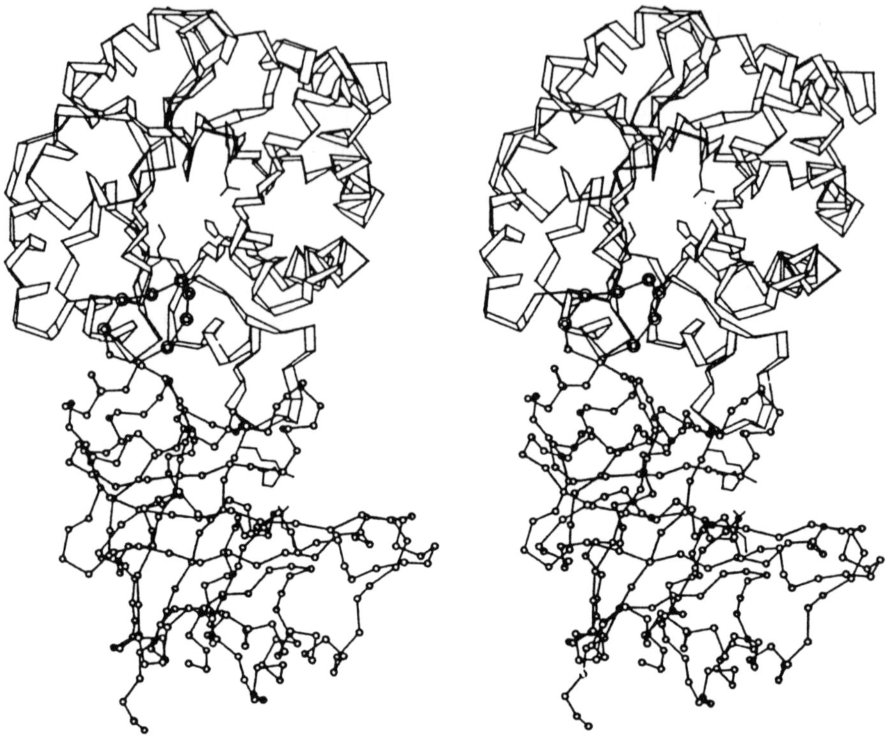

Figure 7.13. Stereo drawing of the complete gTIM dimer. The upper subunit (drawn as a continuous ribbon) is 'subunit 1'; the bottom subunit (drawn as a continuous line connecting the Cα atoms) 'subunit 2'. The active site residues Lys 13, His 95 and Glu 167 of both subunits are indicated. The 'interface loop' of subunit 2, indicated by two concentric circles, consists of Gly 72–Ala 73–Phe 74–Thr 75–Gly 76–Glu 77–Val 78.

disulphide bridge with Cys 14 of *T. brucei* TIM in 'subunit 1'. If this cyclopeptide would form a tight complex with single *T. brucei* TIM subunits, then proper interaction of the two subunits in the TIM dimer is prevented.

Inspired by these modelling studies, Dr R. Osowski in the group of Prof. Kessler in Munich, Germany, has synthesized a number of cyclopeptides containing many features of cyclo[-Phe-Thr-Gly-Glu-Gly-D-Cys-]. In enzymatic assays carried out by M. Callens and F. R. Opperdoes in Brussels, it appeared that such peptides had no effect on the enzymatic activity of *T. brucei* TIM. Fortunately, however, Dr Osowski also took with him a 'reference cyclopeptide' which was just taken from the shelf to serve as a control regarding possible general inhibitory effects of cyclopeptides. By sheer good luck, then, it turned out that this reference peptide—the upper one in Table 7.6—did have quite a significant inhibitory effect on the activity of *T. brucei* TIM. And in

Table 7.6. Inhibitory properties of 'serendipitously' discovered cyclopeptide inhibitors of *T. brucei* TIM

Cyclic peptide*	IC_{50}-value** (μM)
c[W-Y(S)-P-F-T(S)-K(Z)]	17
c[W-Y-P-F-A-K(Z)]	12
c[W-Y-P-F-T-K(Z)]	11
c[W-F-P-F-T-K(Z)]	6.5
c[W-F-P-F-F-K(Z)]	2.8

* (Z), benzene-oxycarbonyl blocking group; (S), sulphate group; P is D-proline.
**The IC_{50}-value is the concentration where 50% of the original activity is present.

addition it appeared to be a better inhibitor of *T. brucei* TIM than of rabbit TIM. The not entirely illogical supposition that one of the sulphate moieties of the 'reference cyclopeptide' were essential for its inhibitory effects by occupying the phosphate binding site turned out to be wrong. As can be seen in Table 7.6, cyclopeptides with even smaller K_is were obtained by removing the sulphate moieties. Eventually, it appeared that the quite hydrophobic cyclo[-Trp-Phe-D-Pro-Phe-Phe-Lys(Z)] had an IC_{50} of 2.8 μM for *T. brucei* TIM and an IC_{50} >125 μM for rabbit TIM.

It is clear that we will now try to exploit this stroke of good luck in the continuation of our collaborative project.

7.9 PROSPECTS

The above is a description of results obtained by mid-1990 in our project aimed at designing new trypanocidal compounds by focusing on unique features of the energy generating machinery of the sleeping sickness parasite *Trypanosoma brucei*. The first phase of the project has provided a wealth of physicochemical, kinetic, sequence and structural data. This forms a base for the second phase which will concentrate on unravelling details of the import mechanism, elucidating three-dimensional structures of enzymes complexed with a range of ligands, and investigating the effect of newly synthesized compounds on catalysis. At the same time, efforts are continued to obtain sequence information and three-dimensional structures of additional proteins. With such an expanded structural database, molecular modelling will become increasingly important to guide the synthetic efforts of the collaborating chemists.

We may conclude by expressing the hope that we will eventually not find one, but *several* therapeutically useful compounds, each of these interfering with different processes and proteins in the trypanosome, in order to be able to prepare *multi-compound drugs*. Such drugs would suffer much less than single-compound drugs from disastrous drug resistance mechanisms which have been operative in the treatment of other tropical protozoic diseases such as malaria. This is, admittedly, an ambitious goal, but the first steps towards reaching it have been sketched in this overview.

ACKNOWLEDGEMENTS

It is a pleasure to thank the many scientists who are jointly working on this rational drug design project: first of all our colleague crystallographers F. M. D. Vellieux, M. E. M. Noble, C. L. M. J. Verlinde, H. Groendijk, T. Pijning, J. Zeelen and A. M. W. H. Thunnissen. Special mention should be made of Mr K. H. Kalk who made a tremendous contribution regarding data collection and processing. In addition we are greatly indebted to Dr Fred Opperdoes, Dr P. Michels and their colleagues in Brussels and Prof. P. Borst and co-workers in Amsterdam for a most enjoyable collaboration. We also gratefully acknowledge the enthusiastic support from Dr Klaus Müller in the modelling efforts. We thank Dr Jos van Beeumen for gas phase sequenator expertise, Prof. Nickle from Bonn for providing ANTS, and Drs Kessler and Osowski, and Drs Périé and Wilson for joining with their organic synthetic know-how. We were deeply shocked by the sudden and tragic death of Professor Alan Horn on the second of January 1990. Professor Horn played an important role in supervising the design and synthesis of a variety of active site inhibitors of trypanosomal TIM in collaboration with Cees Witmans and Cor Grol, and it is therefore that we dedicated this paper to his memory.

The research described was financially supported by the WHO/UNDP/WHO Special Programme for Research and Training of Tropical Diseases, the Alberta Medical Heritage Foundation; a fellowship from Hoffmann-La Roche; the Programme for Developing Countries of the EC; and the Dutch Foundation for Chemical Research (SON) with financial aid from the Dutch Organisation for Scientific Research (NWO).

We are indebted to Mrs R. A. Hogenkamp for excellent assistance in preparing the manuscript.

Note: This manuscript has also appeared in the *Ramachandran Festschrift* published by the Indian Academy of Sciences (P. Balaram and S. Ramaseshan, editors).

REFERENCES

Alber, T. and Kawasaki, G. (1982). *J. Mol. Appl. Gen.*, **48**, 419–34.

Alber, T., Banner, D. W., Bloomer, A. C., Petsko, G. A., Phillips, D. C., Rivers, P. S. and Wilson, I. A. (1981). *Phil. Trans. R. Soc. Ser. B.*, **293**, 159–71.

Albert, A. (1985). *Selective Toxicity*, University Press, Cambridge.

Bacchi, C. J. (1981). *J. Protozool.*, **28**, 20–7.

Baltz, T., Giroud, C., Baltz, D. Roth, C., Raiband, A. and Eisen, H. (1986). *Science*, **319**, 602–6.

Banner, D. W. Bloomer, A. C., Petsko, G. A., Phillips, D. C., Pogson, E. I., Wilson, I. A., Corran, P.H., Furth, A. J., Milnzan, J. D., Offord, R. E., Priddle, J. D. and Waley, S. G. (1975). *Nature*, **255**, 609–14.

Belasco, J. G., Herilhy, J. M. and Knowles, J. R. (1978). *Biochemistry*, **17**, 2971–8.

Benne, R., Van den Burg, J. Brakenhoff, J. P. J., Sloof, P., Van Boom, J. H. and Tromp, M.C. (1986). *Cell*, **46**, 819–26.

Biesecker, G., Harris, J. I., Thierry, J. C., Walker, J. E. and Wonacott, A. J. (1977). *Nature*, **266**, 328–33.
Bitonti, A. J., Bacchi, C. J., McCann, P. P. and Sjoerdsma, A. (1985). *Biochem. Pharmacol.*, **34**, 1773–7.
Borst, P. (1986). *A. Rev. Biochem.*, **55**, 701–32.
Clarkson, A. B. and Brohn, F. H. (1976). *Science*, **194**, 204–6.
Clayton, C. E. (1985). *EMBO J.*, **4**, 2997–3003.
Clayton, C. E. (1987). *J. Cell Biol.*, **105**, 2649–54.
Collins, K. D. (1974). *J. Biol. Chem.*, **249**, 136–42.
Eilers, M. and Schatz, G. (1986). *Nature*, **322**, 228–32.
Esser, K. M. and Schoenbehler, M. J. (1985). *Science*, **229**, 190–3.
Fairlamb, A. H. and Opperdoes, F. R. (1986). In *Carbohydrate Metabolism in Cultured Cells* (Ed. M. J. Morgan), Plenum Publishing Corporation, New York, pp. 183–224.
Fairlamb, A. H., Opperdoes, F. R. and Borst, P. (1977). *Nature*, **265**, 270–1.
Fairlamb, A. H., Blackburn, P., Ulrich, P., Chait, B. T. and Cerami, A. (1985). *Science*, **227**, 1485–7.
Feagin, J. E., Abraham, J. M. and Stuart, K. S. (1988). *Cell*, **53**, 413–22.
Freymann, D. M., Metcalf, P., Turner, M. and Wiley, D. C. (1984). *Nature*, **311**, 167–9.
Gutteridge, W. E. (1985). *Br. Med. Bull.*, **41**, 162–8.
Hart, D. T., Baudhuin, P., Opperdoes, F. R. and de Duve, C. (1987). *EMBO J.*, **6**, 1403–11.
Hol, W. G. J. (1985). *Prog. Biophys. Mol. Biol.*, **45**, 149–95.
Hol, W. G. J. (1986). *Angew. Chem.* (English edition), **25**, 767–78.
Joseph, D., Petsko, G. A. and Karplus, M. (1990). *Science*, **249**, 1425–8.
Kendall, G., Wilderspin, A. W., Ashall, F., Miles, M. A. and Kelly, J. M. (1990). *EMBO J.*, **9**, 2751–8.
Knowles, J. R. and Albery, W. J. (1977). *Acc. Chem. Res.*, **10**, 105–11.
Krauth-Siegel, R. L., Enders, B., Henderson, G. B., Fairlamb, A. H. and Schirmer, R. H. (1987). *Eur. J. Biochem.*, **164**, 123–8.
Lambeir, A.-M., Opperdoes, F. R. and Wierenga, R. K. (1987). *Eur. J. Biochem.*, **168**, 69–74.
Lambeir, A.-M., Loiseau, A. M., Kuntz, D. A. and Opperdoes, F. R. (1990). *Eur. J. Biochem.*, submitted.
Leslie, A. G. W. and Wonacott, A. J. (1984). *J. Mol. Biol.*, **178**, 743–72.
Lolis, E. and Petsko, G. A. (1990). *Biochemistry*, **29**, 6619–25.
Lolis, E., Alber, T., Davenport, R. C., Rose, D., Hartman, F. C. and Petsko, G. A. (1990). *Biochemistry*, **29**, 6609–18.
Lu, H. S., Miau, P. and Gracy, R. W. (1984). *J. Biol. Chem.*, **259**, 11958–68.
Maquat, L. E., Chilcote, R. and Ryan, P. M. (1985). *J. Biol. Chem.*, **260**, 3748–53.
Marchand, M., Poliszczak, A., Gibson, W. C., Wierenga, R. K., Opperdoes, F. R. and Michels, P. A. M. (1988). *Mol. Biochem. Parasitol.*, **29**, 65–76.
Marchand, M., Kooijstra, U, Wierenga, R. K., Lambeir, A.-M., Van Beeumen, J., Opperdoes, F. R. and Michels, P. A. M. (1989). *Eur. J. Biochem.*, **184**, 455–64.
Mercer, W. D., Winn, S. I. and Watson, H. C. (1976). *J. Mol. Biol.*, **104**, 277–83.
Metcalf, P., Blum, M., Freymann, D. Turner, M. and Wiley, D. C. (1987). *Nature*, **325**, 84–6.
Michels, P. A. M. (1988). *Biology of the Cell*, **64**, 157–64.
Michels, P. A. M., Poliszczak, A., Osinga, K. A., Misset, O., van Beeumen, J., Wierenga, R. K., Borst, P. and Opperdoes, F. R. (1986). *EMBO J.*, **5**, 1049–56.
Misset, O. and Opperdoes, F. R. (1987). *Eur. J. Biochem.*, **162**, 493–500.
Misset, O., Bos, O. J. M. and Opperdoes, F. R. (1986). *Eur. J. Biochem.*, **157**, 441–53.

Molyneux, D. H. and Ashford, R. W. (1983). *The Biology of Trypanosoma and Leishmania*, Taylor and Francis, London.
Moras, D., Olsen, K. W., Sabesan, M. N., Buehner, M., Ford, G. C. and Rossmann, M. G. (1975). *J. Biol. Chem.*, **250**, 9137–62.
Noble, M. E. M., Wierenga, R. K., Lambeir, A.-M., Opperdoes, F. R., Thunnissen, A.-M. W. H., Kalk, K. H., Groendijk, H. and Hol, W. G. J. (1991a). *Proteins*, **10**, 50–69.
Noble, M. E. M., Verlinde, C. L. M. J., Groendijk, H., Kalk, K. H., Wierenga, R. K. and Hol, W. G. J., (1991b). *J. Med. Chem.*, **34**, 2709–18.
Opperdoes, F. R. (1983). In *Mechanism of Drug Action*, Academic Press, New York, pp. 121–31.
Opperdoes, F. R. (1985). *Br. Med. Bull.*, **41**, 130–6.
Opperdoes, F. R. (1987). *A. Rev. Microbiol.*, **41**, 127–51.
Opperdoes, F. R. and Borst, P. (1977). *FEBS Lett.*, **80**, 360–4.
Opperdoes, F. R., Baudhuin, P., Coppens, I., de Roe, C., Edwards, S. W., Weijers, P. J. and Misset, O. (1984). *J. Cell. Biol.*, **98**, 1178–84.
Opperdoes, F. R., Wierenga, R. K., Noble, M. E. M., Hol, W. G. J., Wilson, M., Kuntz, D. A., Callens, M. and Perié, J. (1990). In *Parasites: Molecular Biology, Drug and Vaccine Design* (Eds T. Cerami and N. Agabian), Wiley-Liss, New York, pp. 233–46.
Osinga, K. A., Swinkels, B. W., Gibson, W. C., Borst, P., Veeneman, G. H. van Boom, J. H., Michels, P. A. M. and Opperdoes, F. R. (1985). *EMBO J.*, **4**, 3811–17.
Pays, E. and Steinert, M. (1988). *A. Rev. Genet.*, **22**, 107–26.
Perutz, M. F. (1981). In *Atlas of Molecular Structures in Biology* (Eds D. C. Phillips and F. M. Richards), Clarendon Press, Oxford, pp. 71–104.
Pompliano, D. L., Peyman, A. and Knowles, J. R. (1990). *Biochemistry*, **29**, 3186–94.
Read, R. J., Wierenga, R. K., Groendijk, H., Lambeir, A.-M., Opperdoes, F. R. and Hol, W. G. J. (1987). *J. Mol. Biol.*, **194**, 573–5.
Reddy, E. P., Reynolds, R. K., Santos, E. and Barbacid, M. (1982). *Nature*, **300**, 149–52.
Rieder, S. V. and Rose, I. A. (1959). *J. Biol. Chem.*, **234**, 1007–10.
Rossmann, M. G. (1972). *The Molecular Replacement Method*, Gordon and Breach, New York.
Schreuder, H. A., Groendijk, H., van der Laan, J. M. and Wierenga, R. K. (1988). *J. Appl. Crystallogr.*, **21**, 426–9.
Shames, S. L., Fairlamb, A. H., Cerami, A. and Walsh, C. T. (1986). *Biochemistry*, **25**, 3519–26.
Shaw, J. M., Feagin, J. E. Stuart, K. S. and Simpson, L. (1988). *Cell*, **53**, 401–11.
Simpson, L. and Shaw, J. (1989). *Cell*, **57**, 355–66.
Sjoerdsma, A. and Schechter, P. J. (1984). *Clin. Pharmacol. Ther.*, **35**, 287–300.
Skarzynski, T., Moody, P. C. E. and Wonacott, A. J. (1987). *J. Mol. Biol.*, **193**, 171–87.
Straus, D. and Gilbert, W. (1985). *Proc. Natl Acad. Sci. USA*, **82**, 2014–18.
Swinkels, B. W., Gibson, W. C., Osinga, K. A., Kramer, R., Veeneman, G. H. van Boom, J. H. and Borst, P. (1986). *EMBO J.*, **5**, 1291–8.
Swinkels, B. W., Evers, R. and Borst, P. (1988). *EMBO J.*, **7**, 1159–65.
Tabin, C. J., Bradley, S. M., Borgmann, C. I., Weinberg, R. A., Papageorge, A. G., Scolnick, E. M., Dahr, R., Lowy, D. R. and Chang, E. H. (1982). *Nature*, **300**, 143–9.
Vickerman, K. (1978). *Nature*, **273**, 613–17.
Waley, S. G. (1973). *Biochem. J.*, **135**, 165–72.
Wang, C. C. (1984). *J. Med. Chem.*, **27**, 1–9.
Weiner, A. M. and Maizels, N. (1990). *Cell*, **61**, 917–20.
WHO. (1985). *Tropical Disease Research*, TDR Seventh Program Report, WHO, Geneva.
WHO. (1987). *Tropical Disease Research*, TDR Eighth Program Report, WHO, Geneva.

Wierenga, R. K., Kalk, K. H. and Hol, W. G. J. (1987a). *J. Mol. Biol.*, **198**, 109–21.
Wierenga, R. K., Swinkels, B. W., Michels, P. A. M., Osinga, K. A., Misset, O., van Beeumen, J., Gibson, W. C., Postma, J. P. M., Borst, P., Opperdoes, F. R. and Hol, W. G. J. (1987b). *EMBO J.*, **6**, 215–21.
Wierenga, R. K., Noble, M. E. M., Postma, J. P. M., Groendijk, H., Kalk, K. H., Hol, W. G. J. and Opperdoes, F. R. (1991). *Proteins*, **10**, 33–49.
Wolfenden, R. (1969). *Nature*, **223**, 704–5.

8 Drug Discovery and Invention: Some Approaches Compared

ALWYN S. GILBERT AND RICHARD M. HYDE
The Wellcome Research Laboratories, Beckenham, Kent, UK

8.1 INTRODUCTION

The main purpose of this book is to cover in some detail those situations where detailed knowledge of the three-dimensional (3-D) structure of the macromolecular active/receptor site exists and advantage can be taken of this to design ligands that will bind in an effective manner. The objective of this chapter is to indicate how other techniques within medicinal chemistry can impinge upon these situations, and to show how the process of modelling ligands to (3-D) macromolecular sites can be enhanced by combination with and consideration of QSAR (quantitative structure activity relationships).

Drugs can be described as either primarily pharmacodynamic or anti-invasive; the former act by correcting or modifying bodily states while the latter attack organisms that are foreign to the host. A broad generalization is that many pharmacodynamic agents interact with receptors and many anti-invasive agents with enzymes but this division is far too exclusive to describe all drugs. Some in fact interfere directly with genetic components, others with transport systems and a few cause an effect by modifying the overall properties of tissues.

Chlorine and ozone gases are well known agents which are routinely used to destroy most known bacteria or viruses but they non-specifically attack many

The Design of Drugs to Macromolecular Targets. Edited by C. R. Beddell
© 1992 John Wiley & Sons Ltd

organic materials and thus cannot be used in the body. What is required of a useful anti-invasive drug is that it be a selective poison and it must therefore act by virtue of some specific difference between the host and the infective organism. This requirement for selectivity also extends to most pharmacodynamic agents, for instance anti-stomach ulcer drugs that operate by reducing gastric acidity should antagonize only those receptors responsible for inducing the production of acid secretions and not others elsewhere. For a comprehensive description of how drug substances act upon living systems and the relevant biology the reader is referred to one of the many textbooks on pharmacology, for example the recent edition of Bowman and Rand (1980).

The molecular basis of drug action was established by the pioneering work of P. Ehrlich and separately J. N. Langley at the turn of the last century. A good description of the development of the concepts and ideas involved has been given by Dean (1987). Over the intervening years a vast corpus of data has been gathered that has allowed a steady development of theory. In certain cases highly detailed views have been obtained at the molecular level, for example by X-ray crystallography.

What has been revealed is that drugs exert their effects by binding in some way to their target and it is evident that this is achieved by a subtle and complex interplay of intermolecular forces. A few drugs react chemically with the target, thus causing an irreversible change in molecular constitution, but the majority yield only non-covalent and therefore reversible interactions such as the formation of H-bonds. Thus the overwhelming factors in determining the specificity and selectivity of drug binding are their physicochemical properties and it is by the study of how these are correlated with molecular structure and biological activity that a rational process of design can lead to the discovery and invention of new drugs.

8.2 THE BASICS OF DRUG ACTION

The binding of a drug to the active site of an enzyme or receptor is only the last of a long line of interactions that it must make with the body following administration. Some of these cause little hindrance to the drug on its way to the target but others may severely prejudice its therapeutic utility. Metabolism by the vast range of enzymes within the host may effectively destroy the drug as well as possibly generating toxic by-products, but perhaps even greater impediment comes from the various physical barriers that a drug must surmount during its journey, such as the membranes of bacteria and the tissues separating the stomach fluids from the blood (if the drug is taken orally). As the factors that determine the ease of passage over these hurdles are usually physicochemical, however, it is evident that if drugs are designed with regard solely to the properties of the target, they may very well be useless as medicines.

The physicochemical properties of molecules are many and varied. Those that have been found to be relevant to how and whether a particular chemical will bind to a particular target include the H-bonding ability of individual groups within the molecule, its overall dipole moment and properties which might be thought of as purely physical such as size and shape. Some properties, e.g. solubility and melting point, have no direct relevance to binding but can be crucial to transmission and the mode of formulation. Others play a role in all circumstances, a notable example being hydrophobicity; this is not, as its name would suggest, the direct inverse of solubility though it is strongly correlated with it.

The process of drug design can only sensibly start once an assay of the biological activity of the target has been established. If this involves a bacterial or viral enzyme, then a system is needed whereby the rate of turnover of the natural substrate or analogue can be measured. The business of a drug in such a case may be to compete with the natural substrate in binding to the active site and thereby disrupt some vital function of the organism or to provide an alternative substrate. Receptors, on the other hand, do not chemically modify their substrates (except to control their effects) which essentially act as stimuli to some bodily process elsewhere and therefore assays of their activity tend to be rather more complicated though tests of their binding can sometimes be rapid. Here the role of the drug may be, like the case of an enzyme, to compete with (antagonize) the natural substrate, so as to calm some overexcited body state or alternatively to supplement the natural substrate in order to 'stimulate' an over-sluggish system.

The ideal primary assay should give a reliable quantitative measure of a compound's effect on the key biochemical target, albeit enzyme or receptor, and that alone. Having found a chemical structure or template that is active with respect to this paramount criterion, optimization of this and the other component processes associated with drug action can begin. How, then, does the medicinal chemist get into the required area of activity and, once there, how does he exploit it to proceed from a lead to a potential drug? This question will be addressed in the following section.

8.3 THE COMPOUND

In an ideal world the designer would possess 3-D knowledge of the molecular attributes of the target site but unfortunately many targets are not known and even when they are, the macromolecules involved do not easily yield up such information. How then is it possible to get 'started'?

There may, of course, be a natural substrate or substrates on which to base ideas and if two or more such molecules are known this offers the possibility of identifying the essential pharmacophore, i.e. the relevant parts of the chemical structure. But if there is nothing to go on then screening is the only option. This need not necessarily be random, however, and QSAR methodology may be used to assist this process.

Having obtained, by whatever means, a 'lead' compound, it is usually necessary to modify its structure to optimize the biological activity and efficacy. Without knowledge of the molecular organization of the target site some type of structure activity methodology must be employed whereby knowledge of the target is gained by probing it with different chemicals rather than by looking at it directly. Such indirect techniques may in fact provide more useful information than a single 3-D picture of the macromolecule, obtained by, say, X-ray diffraction, which may only be strictly relevant to the particular ligand that is bound, i.e. different ligands may induce different conformations of the target site.

At this point a distinction must be drawn between structure activity relationships and QSAR. The former, which are epitomized by the method of Free and Wilson (1964), relates structural differences to changes in biological activity in a purely empirical way. Thus, for instance, a methyl substitutent is observed to change biological activity to a different extent than a nitro substituent but no attempt is made to find a connection between the two that allows the effect of a third and untested substituent to be predicted. By contrast, QSAR treats a substituent as contributing towards some overall physicochemical property of the molecule (e.g. hydrophobicity) that is relevant to and varies *smoothly* with the biological activity and thus, by seeking correlations between properties and activity, meaningful extrapolations may be made.

QSAR can be used to optimize any biological property for which there is a quantitative assay, whether it is related to the effect of a compound on its target or the processes by which it gets there.

Knowledge of the criteria for optimal activity is obtained by deduction. QSAR does not require any *direct* information about the structure or properties of the target. Thus the design process consists of gathering together a set of chemicals, measuring their 'activity' in the biological assay and then looking for physicochemical properties that correlate with such activity. The work proceeds stepwise, each new compound hopefully adding to the strength of connection between properties and activity and thereby enhancing the power of prediction. Eventually a successful outcome is reached when a particular chemical structure is predicted to yield the desired activity (and to possess other physicochemical properties appropriate to its efficacy as a medicine), followed by synthesis and confirmation by the biological test.

The archetypal QSAR equation describes a relationship between a biological potency ($y = \log 1/c$, where c = concentration) and the logarithm of a partition coefficient ($\log P$). Perhaps the simplest type is where a set of compounds are presumed to act by a non-specific mechanism, so that potency is dictated by the molarity of compound which can be achieved in a non-aqueous compartment. Anaesthetic effects based on non-specific membrane perturbation fall into this category. The work of Meyer and Hemmi (1935) on the narcosis of tadpoles is echoed in much of today's environmental QSAR. Relationships of this type are based on the assumption that partition into biological membranes can be modelled by partition into appropriate organic solvents. The hypothesis that the latter may predict the former is tested by regression analysis.

A linear equation of the form $y = ax + k$ can often be fitted to simple data sets using the least squares procedure. This fitting process gives a quantitative means of predicting y (the biological property) from x (the physicochemical property) and also a means of assessing the level of confidence than can be placed upon the relationship which the equation suggests. Confidence is judged on the basis of r, F and t statistics.

Where a linear function is demonstrably unsuitable, a parabola (McFarland, 1970; Hansch and Clayton, 1973) can be fitted to good effect, allowing the identification of an 'optimum' value for log P. Alternative model-based approaches have been proposed (Hyde, 1975; Martin and Hackbarth, 1976; Kubinyi, 1977).

Desirable though it is to estimate a biological property from readily measurable chemical properties such as log P or pK_a, there is clearly much more to be gained if the physicochemical properties can be predicted. The σ constant (Hammett, 1940) is familiar to chemists as an expression of substituent contribution to electronic effects.

Following the work of Hansch and Fujita (1964) in which the π constant was introduced as an expression of substituent contribution to log P, the octanol/water partition system has become a standard in molecular design.

The 'Hansch' equation considers contributions to potency from electronic (σ), hydrophobic (π) and steric (E_s) terms. The E_s substituent constant is from the work of Taft (1956). There have been numerous variations on the theme, many involving attempts to improve on the physicochemical descriptors. Two examples will be quoted. First, following the work of Swain and Lupton (1968), the substituent constants F and R have been widely used to describe the field/inductive and resonance components (respectively) of the σ constant. Secondly, the molar refraction, MR, has proved a most useful expression of substituent bulk, to the extent that π, F, R and MR are widely seen as a 'front-line' set of constants essential to a QSAR substituent constant database. Developments to this basic set have, however, occurred, for example for steric properties approaches have been described for going beyond the use of a single descriptor such as MR. Verloop's (1976) Sterimol system allows a description of shape with respect to predefined axes, and is a real advance especially in QSAR studies focused on specific ligand target interactions.

Historically there was a tacit assumption that QSAR relationships derive from control on the processes by which compounds reach their target. However, for numerous equations in the literature, epecially those involving electronic and steric terms, the presumption of non-specific action at the target is clearly not credible. Many QSARs must, it would seem, derive at least in part from the stage at which compound interacts with target.

QSAR studies on inhibitors of dihydrofolate reductase (DHFR) (Blaney *et al.*, 1979; Hyde and Roth, 1982) provide examples where substituent constants have been used to interpret inhibitor binding to enzyme and will be described in more detail in the next section.

Recently a close collaboration has been growing between computational chemists engaged in small molecule modelling and QSAR analysts which has set the scene for further development. Potential ligands are now routinely 'built' by computational chemists and subjected to conformational analysis. Given well chosen conformations and a common orientation within a group of compounds, a wide range of computer calculable descriptors, both steric and electronic, have become available (e.g. Glen and Rose, 1987; Hyde and Livingstone, 1988). However, utilization of large numbers of descriptors introduces the danger of spurious apparent relationships with biological data though the adoption of a range of multivariate methods to meet the opportunities offered by the abundance of calculable properties has helped to prevent over-ready acceptance of conclusions that could be misleading.

8.4 THE RELEVANCE OF QSAR TO THE 3-D MODELLING (DOCKING) APPROACH

At first sight it may not be obvious how the various parameters used in QSAR relate to the interactions between ligand and macromolecule that are familiar to the 3-D modeller. This is understable because without fore-knowledge of the spatial characteristics of the binding site QSAR must utilize parameters that are in general non-directional and often associated with the ligand as a whole. In addition, there is a problem with duality in that, for example, in the case of electronic parameters, a correlation involving σ may derive either from the existence of dipole–dipole/H-bond interactions or from the existence of two possible ionization states only one of which is associated with ligand binding. Involvement of, for instance, log P, may signal either the influence of membrane permeability or alternatively a contribution from desolvation to the binding energy. Nevertheless it is often easy to reconcile the various viewpoints and a QSAR study can provide valuable assistance to the 3-D modeller where a series of compounds have been modelled, synthesized and biologically tested as well as giving valuable additional insight into the overall binding process.

The concept of hydrophobicity is probably the most useful notion in QSAR, but it is not easy when docking a putative ligand into a binding site to estimate the likely degree of desolvation, though this could be attempted by calculating the reduction in protein and ligand surface accessible to water. The coefficient of π from a multiple linear regression on binding data, however, can give an indication of the extent to which groups of the ligand are removed from solvent when bound. If the value is < 1 then only partial desolvation has probably occurred; values ranging from 1 upwards to over 2 may indicate, as suggested by Fersht (1985), compete group desolvation and removal of water from macromolecular surface. Such information can be taken further, e.g. poor desolvation of a hydrophobic moiety not only results in loss of potential binding free energy from this process but also indicates that beneficial van der Waals interactions may well be being prevented by solvent barriers. But the coefficients must not be taken too literally for, as pointed out by Taylor (1990), π refers to the partition coefficient between water and *octanol*, not water and protein interior.

Molar refraction MR of substituent groups has been found to correlate with binding in numerous QSAR studies. It is normally thought of as, and is in fact closely related to, molecular bulk and as such its most obvious connection would appear to be with steric factors. But MR is often found to provide a positive contribution to binding, implying that it also models van der Waals attractive forces. To model both binding and steric repulsion there must therefore be some non-linear relationship involving MR in a QSAR regression study.

As with hydrophobic effects it is not obvious when docking a putative ligand as to the amount of binding to be expected from van der Waals dispersion forces. In any particular case one or another mechanism may perhaps predominate and this is neatly illustrated for binding of the drug trimethoprim (TMP) (**1**) and

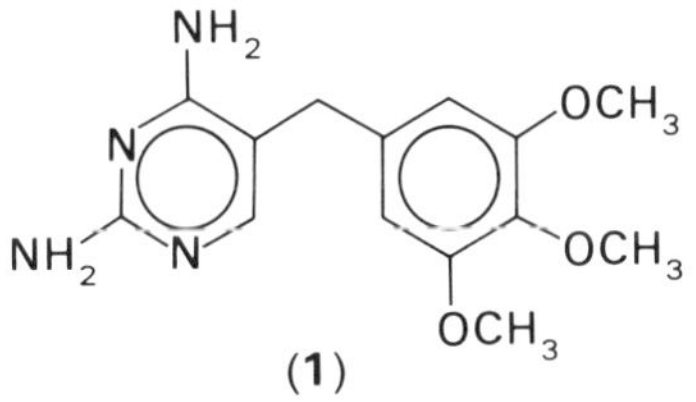

(**1**)

analogues to the enzyme DHFR. It is well known from X-ray crystallographic studies that the structures of bacterial and vertebrate DHFRs are different and that TMP binds to them in a different manner. Hyde and Roth (1982), have shown that for mono-meta-substituted analogues (i.e. 3′ position substitution, on benzyl) the hydrophobicity (π) of the substituent shows a better correlation with inhibition of the vertebrate enzyme (rat liver) while MR is better correlated with inhibition of the bacterial enzyme (*E. coli*).

Evidently the 3′ substituent is less accessible to solvent and may well be effectively buried when in the vertebrate enzyme. Desolvation does not seem to be particularly important for this substituent in the bacterial enzyme but as pointed out by Hyde and Roth the correlation with MR is actually little better than with π, suggesting some other facet to the situation. In fact they were able to obtain a very much better correlation for MR by considering only the first atom of the substituent, thus demonstrating the power of QSAR analysis to probe a very specific and localized interaction between the ligand and active site.

A further ramification occurs when a substituent is added at the 4′ position on benzyl in this series of compounds. When the additional group is methoxy, the inhibition of *E. coli* enzyme is enhanced for a number of symmetrical substituents at the 3′ position by up to 20-fold; however, for the cases of asymmetrical moieties such as 3′-alkoxy the improvement is rather larger, being better than 200-fold for ethoxy. Evidently, the proximity of the 3′ and 4′ substituents is responsible for influencing the orientation of the non-symmetrical groups and causing additional binding interactions with the enzyme and Hyde and Roth have been able to provide an elegant rationalization of this in terms of the 'Sterimol' shape QSAR parameter of Verloop mentioned in the previous section. This is thus another illustration of how a QSAR analysis can elucidate effects that would be difficult or impossible to gauge from consideration only of the spatial coordinates of enzyme and modelled ligand.

The role of the electronic parameters has been discussed by Blaney and Hansch (1990). A study of the inhibition of carbonic anhydrase by some simple sulphonamides (**2**) yielded a positive coefficient on σ for substituents X below, implying that it is the anionic form that effectively binds to the enzyme's Zn ion. This conclusion has also been reached by spectroscopic studies (e.g. Carey, 1982). But for the case of the inhibition of alcohol dehydrogenase by 4-substituted pyrazoles (**3**) a negative coefficient on σ for X must be connected with an 'ionic' binding of one of the nitrogens to a nearby positively charged Zn ion as the sign of the coefficient means that electron releasing substituents promote binding.

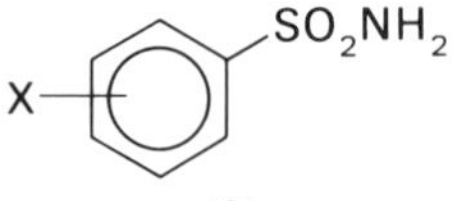

(**2**)

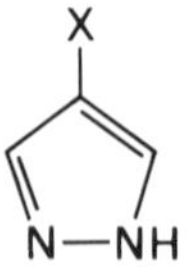

(**3**)

8.5 MAPPING THE BINDING SITE

Unfortunately many proteins may not be amenable to crystallization and hence structuring by X-ray diffraction, though other techniques such as NMR spectroscopy may eventually be able to obtain spatial data from most enzymes of interest. However, not one relevant receptor has as yet been structured by any technique to a resolution suitable for detailed drug design and indeed many receptors have never been isolated and are identified only by their function.

Given that a drug is essentially complementary to its binding site, it might be thought that by considering a number of different ligands, all known to act at the same site, a 'picture' of the latter can be obtained by a process of mapping. QSAR alone would not be of much help here partly because of the non-directional nature of most of its parameters but also because some parameters can model different interactions (e.g. electronic parameters can influence both dipoles and hydrogen bonds) and the methodology does not and has no need to make assumptions about the precise nature of the interactions involved.

Instead the usual approach is to attempt to find the closest superimposition of the various ligands in 3-D space and thereby define the common or basic structural unit(s) necessary for binding and relevant biological activity (agonism or antagonism of a receptor). This basic structural entity is usually termed the pharmacophore. Deductions are then made about the location of binding site groups such as 'an H-bond donor is situated about so many angstroms from this part of the pharmacophore, a hydrophobic surface surrounds this part' and so on. The locations cannot, of course, be defined precisely because of the nature of the interactions that are presumed to be involved, e.g. a carbonyl group (on the pharmacophore) is potentially capable of accepting an H-bond over a wide solid angle and might even accept more than one from donors on different parts of the binding site.

Thus at best only a fuzzy and patchwork view of the binding site/receptor can be obtained and unfortunately it is generally not possible to know the degree of reliance that can be placed on the picture because of the assumptions inherent in the methodology. For instance ligands may well possess many degrees of conformational freedom though methods such as the distance geometry approach (e.g. see Ghose and Crippen, 1990) attempt to deal with such uncertainties. A more important objection is that all pharmacophore analysis necessarily assumes that all the ligands bind to a particular site in the same mode which limits at best the study to ligands derived from one chemical series and therefore restricts the mapping. Alternatively superimposition of rather dissimilar molecules relies on bold and imaginative assumptions. It should be noted that QSAR is not bound by any of these limitations because it is strictly phenomenological and should not therefore be utilized in this context.

Nevertheless, exercises in pharmacophore analysis must not be dismissed as they can provide very useful information about the ligands themselves. Lloyd and Andrews (1986) have defined a common structural basis for central nervous system (CNS) drugs based on molecular superimposition of a number of semi-rigid ligands and have shown that, for a large number of entries picked randomly from the Merck index, those possessing the defined pharmacophore have a high probability of exhibiting CNS action. It is interesting to note, however, that the ligands/drugs they utilized are known to interact with at least two of the major superfamilies of receptors (see Lester, 1988), the LGIC group and the G-protein group which are known to have very different overall architectures, the receptor of the former being located on the surface of the cell membranes and the latter within (i.e. transmembranal). Thus a common pharmacophore, which Lloyd and Andrews (other workers have proposed similar models) define as an aromatic ring joined to a basic nitrogen atom a few angstroms away, binds to active sites which are almost certainly very different in detail.

How can the limitation of one chemical series be avoided? It may be possible by the approach of rational screening whereby a set of structurally varied compounds are chosen for experimental testing as putative ligands on the basis that the members of the set should span the multidimensional space of what are considered to be relevant parameters, for instance overall hydrophobicity (as defined by log P). In general usage the approach would be aimed at finding a lead compound but a more specific use is to continue the process after a lead has been obtained and to choose a new set with physicochemical parameters that are biased towards those of the lead. In practice this approach would, of course, be far more demanding of synthetic chemical effort than mere exemplification of a series but can be considered when there is access to a large collection of structurally diverse molecules.

Before this section is closed it should be mentioned that the alternative of using such data as is known about the macromolecule itself for picturing the binding site is probably just as valid an approach as pharmacophore analysis. Where a receptor can be isolated and cloned or an enzyme purified, then determination of the amino acid sequence is straightforward and there is now growing experience on how to translate this information into secondary and (less reliably) tertiary structure. It is well outside the capability of any algorithm so far devised to achieve this *de novo* (i.e. from knowledge only of the amino acid sequence) but if a homologous protein of known structure exists then rules have been derived to 'perturb' this structure on the basis of observations that certain levels of amino acid substitution can be tolerated within an overall configuration. Even where no 'homologous structure' is known it seems possible to model at least short sections of protein backbone and Cockcroft *et al.* (1990), among others in this field, have presented a general model of a binding loop for the LGIC receptor family. Whether these models are sufficiently accurate for use in drug design remains to be seen but at least they will be a source of ideas.

8.6 DISCUSSION

It is hoped that the foregoing has demonstrated that the 3-D molecular modelling and QSAR approaches complement one another in the drug design process. But what are their relative strengths and weaknesses and where and how are they best employed?

It is helpful to set them in the context of the overall binding energies that arise on interaction between the ligand and the active site of the macromolecule. In the case of enzyme inhibition a binding constant in the region of 10^9 molar is typically that needed and therefore bearing in mind that a bound ligand must lose some 14 kcal of translational and overall rotational entropy plus a few extra kcal if it has potentially flexible main or side chains, then a total of around 30 kcal of binding free energy is required (see Chapter 2). This will usually be obtained from H-bonds and other electrostatic interactions along with contributions from van der Waals (dispersion forces) and hydrophobic desolvation energies.

A typical ligand may have three or four good H-bonding groups making, say, 3 kcal of binding energy each while van der Waals contributions from any benzene ring will give perhaps 4 kcal. These figures translated to binding constant terms would give improvements of 140-fold and 720-fold respectively, but if the ligand did not fit properly such groups might easily generate similar *decreases* in binding constant deriving from perhaps forced desolvation in the case of hydrophilic H-bonding groups on the one hand and steric repulsion from bulky groups on the other. Experience shows that it is all too easy to design badly fitting ligands given the inaccuracies, etc. of 3-D data and thus it is evident that the 3-D approach can be somewhat imprecise.

Classical QSAR, in contrast, deals with smaller energy and binding constant changes. Usually, substituent effects in a chemical series are examined and it is possible to detect much finer and more subtle effects than can be appreciated by consideration of 3-D data derived from X-ray diffraction experiments. The π values for typically encountered substituents range over about 4 log units which coupled with a coefficient for binding of, say, 0.5 means that the observed binding varies at most by a factor of 100-fold and can be 'adjusted' in very much smaller steps.

Most assays of binding (or activity) have a lower limit in the region of 10^4 molar or so. This means that the lead compound in a drug design project will not possess activity any less than this and thus the process of optimization requires an increase in binding constant by at most 10^5 which in energy terms represent only about 25% of the total binding free energy needed and is in the region of about 7–8 kcal, being achievable by a couple of good H-bonds or well placed additional benzene rings, etc. Adding extra groups to the ligand that produce a badly fitted molecule, however, can easily push the activity well below what an assay can reliably measure with therefore no forthcoming data on just how bad the fit has become. Thus lead optimization is better carried out by a QSAR approach where relatively fine adjustments can be carried out on the ligand and correlations found that allow extrapolation to the best molecule.

Where then does the 3-D approach fit in? In the authors' opinion it is not yet particularly well suited to refinement of the binding powers of ligands although a potential source of ideas, but instead is pre-eminently applicable to finding the lead compound itself. There are many situtations where analogues of the natural substrate are inappropriate for use as medicines and in these cases possession of 3-D information of the target gives an opportunity to design ligands of *novel* structure or to make radical modifications to existing leads. A recent publication by Appelt *et al.* (1991) neatly illustrates the results of following both these avenues. Clearly the approach will have progressively greater influence as the range of defined macromolecular targets increases. Whilst currently primarily offering a source of chemical inspiration, rather that a quantitative aid to ligand design, there is some hope that this situation will change as increasing amounts and accuracy of experimental data allow the steady development of prediction methods.

ACKNOWLEDGEMENT

The authors are very grateful to Dr C. R. Beddell for many helpful discussions.

REFERENCES

Appelt, K., Bacquet, R. J., Bartlett, C. A., Booth, C. L. J., Freere, S. T., Fuhry, M. A. M., Gehring, M. R., Hermann, S. M., Howland, E. F. Janson, C. A., Jones, T. R., Kan, C-C., Kathardekar, V., Lewis, K. K., Marzoni, G. P., Matthews, D. A., Mohr, C., Moomaw, E. W., Morse, C. A., Oatley, S. J., Ogden, R. C., Reddy, M. R., Reich, S. H., Schoettlin, W. S., Smith, W. W., Varney, M. D., Villafranca, J. E., Ward, R. W., Webber, S., Webber, S. E., Welsh, K. M. and White, J. (1991). *J. Med. Chem.*, **34**, 1925–34.

Blaney, J. M. and Hansch, C. (1990). In *Comprehensive Medicinal Chemistry* (Ed. C. Hansch), Vol. 4, Pergamon Press, Oxford, p. 459–96.

Blaney, J. M., Dietrich, S., Reynolds, M. A. and Hansch, C. (1979). *J. Med. Chem.*, **22**, 614–7.

Bowman, W. C., and Rand, M. J. (1980). *Textbook of Pharmacology*, 2nd edn, Blackwell Scientific, Oxford.

Carey, P. R. (1982). *Biochemical Applications of Raman and Resonance Raman Spectroscopies*, Academic Press, New York, p. 155.

Cockcroft, V. B., Osguthorpe, D. J., Barnard, E. A. and Lunt, G. G. (1990). *Prot. Struct. Funct. Genet.*, **8**, 386–97.

Dean, P. M. (1987). *Molecular Foundations of Drug–Receptor Interaction*, Cambridge University Press, Cambridge.

Fersht, A. (1985). *Enzyme Structure and Mechanism*, 2nd edn, Freeman, New York, p. 305.

Free, S. M. and Wilson, J. W. (1964). *J. Med. Chem.*, **7**, 395–99.

Ghose, A. K. and Crippen, G. M. (1990). In *Comprehensive Medicinal Chemistry*, Vol. 4, Pergamon Press, Oxford, p. 715–33.

Glen, R. C. and Rose, V. S. (1987). *J. Mol. Graph.*, **5**, 79–86.

Hammett, L. P. (1940). *Physical Organic Chemistry*, McGraw Hill, New York.

Hansch, C. and Clayton, J. M. (1973). *J. Pharm. Sci.*, **62**, 1–21.
Hansch, C. and Fujita, T. (1964). *J. Am. Chem. Soc.*, **86**, 1616–26.
Hyde, R. M. (1975). *J. Med. Chem.*, **18**, 231–3.
Hyde, R. M. and Livingstone D. J. (1988). *J. Comput.-Aided Mol. Design*, **2**, 145–55.
Hyde, R. M. and Roth, B. (1982). In *Strategy in Drug Research*, (Ed. J. A. Keverling), Elsevier, Amsterdam, p. 385–409.
Kubinyi, H. (1977). *J. Med. Chem.*, **20**, 625–9.
Lester, H. A. (1988). *Science*, **241**, 1057–63.
Lloyd, E. J. and Andrews, P. R. (1986). *J. Med. Chem.*, **29**, 453–62.
Martin, Y. C. and Hackbarth, J. J. (1976). *J. Med. Chem.*, **19**, 1033–39.
McFarland, J. M. (1970). *J. Med. Chem.*, **13**, 1192–6.
Meyer, K. H. and Hemmi, H. (1935). *Biochem. Z.*, **277**, 39.
Swain, C. G. and Lupton, E. C. Jr, (1968). *J. Am. Chem. Soc.*, **90**, 4328–37.
Taft, R. W. (1956). In *Steric Effects in Organic Chemistry*, (Ed. M. S. Newman), Wiley, Chichester, p. 556.
Taylor, P. J. (1990). In *Comprehensive Medicinal Chemistry*, (Ed. C. Hansch), Vol. 4, Pergamon Press, Oxford, p. 241–94.
Verloop, A., Hoogenstraaten, W. and Tipker J. (1976). In *Drug Design* (Ed. E. J. Ariens), Vol. 7, Academic Press, New York, p. 165–207.

Author Index

Note: only specific mention of a name in the text is listed; the reference listing at the end of each chapter is indicated by *italic* page numbers

Subject Index